Studies in Logic
Logic and Cognitive Systems
Volume 48

Trends in Belief Revision and Argumentation Dynamics

Volume 37
Handbook of Mathematical Fuzzy Logic. Volume 1
Petr Cintula, Petr Hájek and Carles Noguera, eds.

Volume 38
Handbook of Mathematical Fuzzy Logic. Volume 2
Petr Cintula, Petr Hájek and Carles Noguera, eds.

Volume 39
Non-contradiction
Lawrence H. Powers, with a Foreword by Hans V. Hansen

Volume 40
The Lambda Calculus. Its Syntax and Semantics
Henk P. Barendregt

Volume 41
Symbolic Logic from Leibniz to Husserl
Abel Lassalle Casanave, ed.

Volume 42
Meta-argumentation. An Approach to Logic and Argumentation Theory
Maurice A. Finocchiaro

Volume 43
Logic, Truth and Inquiry
Mark Weinstein

Volume 44
Meta-logical Investigations in Argumentation Networks
Dov M. Gabbay

Volume 45
Errors of Reasoning. Naturalizing the Logic of Inference
John Woods

Volume 46
Questions, Inferences, and Scenarios
Andrzej Wiśniewski

Volume 47
Logic Across the University: Foundations and Applications. Proceedings of the Tsinghua Logic Conference, Beijing, 2013
Johan van Benthem and Fenrong Liu, eds.

Volume 48
Trends in Belief Revision and Argumentation Dynamics
Eduardo L. Fermé, Dov M. Gabbay, and Guillermo R. Simari

Studies in Logic Series Editor
Dov Gabbay   dov.gabbay@kcl.ac.uk

Trends in Belief Revision and Argumentation Dynamics

Edited by
Eduardo L. Fermé,
Dov M. Gabbay
and
Guillermo R. Simari

© Individual author and College Publications 2013.
All rights reserved.

ISBN 978-1-84890-065-3

College Publications
Scientific Director: Dov Gabbay
Managing Director: Jane Spurr

http://www.collegepublications.co.uk

Original cover design by Orchid Creative   www.orchidcreative.co.uk
Printed by Lightning Source, Milton Keynes, UK

All rights reserved. No part of this publication may be reproduced, stored in a retrieval system or transmitted in any form, or by any means, electronic, mechanical, photocopying, recording or otherwise without prior permission, in writing, from the publisher.

Preface

Research in Belief Revision has been teeming with activity for the past three decades. Although investigations in this area had begun earlier, the foundational paper by Alchourrón, Gärdenfords, and Makinson, known today as AGM, was the proposal that captured the attention of the community. This community was comprised by a variety of interests and, among them, Philosophy, Formal Logic, Economics, and Artificial Intelligence were the ones that became the most involved. Since then, that community has produced a wealth of results on the dynamics of knowledge, with many proposals being put forward and refined to mature approaches.

The Argumentation tradition has ancient roots, but a sharp increase in attention from the researchers of these same areas was almost coincidental with the interest in Belief Revision. The work by Toulmin published in the mid-fifties can perhaps be recognized as a point of departure, but full attention to his proposal was delayed until the beginning of the eighties. Also, the work of Jon Doyle on Truth Maintenance Systems (TMS) could be considered as a starting point for this research, as it attracted attention to the problem of how something comes to be believed. Interestingly enough, that work can also be regarded as a first effort in focussing on the problems considered in the overlapping area where Belief Revision and Argumentation meet. Again, the researchers working on Argumentation during the past few decades has led to many developments that include both theoretical and practical contributions.

These two areas grew along mostly separate lines of inquiry until recently; indeed, in the last decade there has been a renewed interest in the interplay between Belief Revision and general forms of reasoning. Argumentation represents a reasoning mechanism that is particularly apt for obtaining the consequences of a repository of potentially inconsistent and/or incomplete beliefs, offering the possibility of a greater capability for explaining those conclusions. The area where the two research efforts have become confluent has produced a significant number of interesting results that show a promising opportunity to further our comprehension not only in each individual area, but also advance our understanding of the mechanisms involved in general patterns of reasoning. Clearly, the problems involved in this enterprise call for the participation of both communities.

This book is the result of the work done during the 2012 Workshop on Belief Revision and Argumentation in Funchal, Madeira. The presentations of research results and the subsequent lively discussions held in that motivating settings provided an excellent environment for the advancement of the area. The dynamic exchange of ideas contributed to refining the approaches and to discovering new insights that undoubtedly will positively affect the field by advancing the individual research lines. The authors of the chapters in this book have made an effort to reflect these discussions, successfully addressing many of the important points raised during the workshop.

The present collection comprises work by contributing authors that come from two different areas of research, but both are concerned with the epistemic state of an agent: Belief Revision and Argumentation. As such, the book reflects dissimilar perspectives that have contributed to advance the understanding of such a puzzling and thrilling problem. At the same time, with this collection we hope to reach a wider audience and stimulate research on this fertile topic, as we strongly believe that the interaction between the two areas will lead to a better understanding of each of them.

We would like to acknowledge and extend our deeply felt appreciation to everyone who has contributed to the completion of this book; in particular, to the authors and reviewers who have contributed greatly to the quality of the final result.

Eduardo L. Fermé

Centre of Exact Sciences and Engineering
University of Madeira, Portugal

Centre for Artificial Intelligence (CENTRIA)
New University of Lisbon, Portugal
Email: ferme@uma.pt

Dov M. Gabbay

Department of Computer Science
Bar-Ilan University, Ramat- Gan, Israel

Department of Informatics
King's College London
Strand, London, UK

University of Luxembourg
Faculté des Sciences, de la Technologie et de la Communication
6, rue Richard Coudenhove-Kalergi, L-1359 Luxembourg
Email: dov.gabbay@kcl.ac.uk

Guillermo R. Simari

Dept. of Computer Science and Engineering
Universidad Nacional del Sur
Bahia Blanca, Argentina
Email: grs@cs.uns.edu.ar

Contents

Belief revision and argumentation: a reasoning process view 1
Pietro Baroni, Massimiliano Giacomin, and Guillermo Ricardo Simari
    1   Introduction 1
    2   Defining belief revision and argumentation 3
    3   A brief technical background 7
    4   Belief revision and argumentation as processes 11
    5   The process objects 15
    6   Process drivers and goals 19
    7   Postulates and principles 22
    8   Conclusions 25
    References 25

Evolutionary Belief Change: Priorities, Conditionals and Argumentation 29
Alexander Bochman
    1   Introduction 29
    2   The Problem of Belief Change 30
    3   Evolutionary Belief Change 41
    4   Belief Change and Argumentation 48
    5   Conclusions 54
    References 54

Cognitive realism in belief revision 57
Sven Ove Hansson
    1   Introduction 57
    2   AGM, the standard approach 58
    3   Requirements on alternatives to AGM 61
    4   Partial meet contraction with a finite language 61
    5   Base-generated partial meet contraction 62
    6   Specified meet contraction 63
    7   Perimaximal contraction 65
    8   Blockage contraction 68

9	Bootstrap contraction	69
10	Conclusion	70
References		71

Characterizing change in abstract argumentation systems 75
Pierre Bisquert, Claudette Cayrol, Florence Dupin de Saint-Cyr, and Marie-Christine Lagasquie-Schiex

1	Introduction	75
2	Background	77
3	Properties of change operations	79
4	Characterizing change operations: Preliminary results	83
5	Characterizing argument addition: direct results	85
6	Characterizing argument removal: direct results	87
7	Duality	88
8	Characterizing argument addition thanks to duality	91
9	Characterizing removal thanks to duality	92
10	Discussion	94
References		96

Argumentation and Belief Updating in Social Networks: A Bayesian Model 103
George Masterton and Erik J. Olsson

1	Introduction	103
2	An informal introduction to Laputa	106
3	Derivation of the Update Functions	108
4	Debates in Laputa	115
5	Applications and outlook	118
References		121

A Propositional Typicality Logic for Extending Rational Consequence ... 123
Richard Booth, Thomas Meyer, and Ivan Varzinczak

1	Introduction	123
2	Defeasible Consequence Relations	126
3	A Logic to Talk about Typicality	129
4	Unnesting the Birds	132
5	Belief Revision and Typicality	135
6	Rational Consequence on $\mathscr{L}^\bullet$	138
7	Entailment for PTL	141
8	Discussion and Related Work	147
9	Concluding Remarks	149
References		150

Contents

Credibility-based Selective Revision by Deductive Argumentation 155
Luciano H. Tamargo, Matthias Thimm, Patrick Krümpelmann, Alejandro
J. García, Marcelo A. Falappa, Guillermo R. Simari, and Gabriele
Kern-Isberner
  1 Introduction 156
  2 Background 159
  3 Deductive Argumentation 164
  4 Selective Revision by Deductive Argumentation 167
  5 Argumentative Credibility-based Revision 170
  6 Related Work 178
  7 Conclusions 180
  References 180

A Glance at Preemption Operators 183
Philippe Besnard, Éric Grégoire, and Sébastien Ramon
  1 Introduction 183
  2 Formal Preliminaries 184
  3 AGM Postulates 185
  4 Postulates for Preemption 186
  5 Discussion, Examples and Elementary Properties 187
  6 Characterization 190
  7 Related Work: Hansson's Replacement Operator 191
  8 Conclusion 192
  References 193

Representation of basic improvement operators 195
Mattia Medina Grespan and Ramón Pino Pérez
  1 Introduction 195
  2 Preliminaries 198
  3 Brief recall about improvement operators and their representation . 199
  4 Basic improvement operators 204
  5 The proof 207
  6 Implications on related works 225
  7 Concluding remarks 226
  References 227

Argument Revision as a means of supporting dishonesty 229
Mark Snaith and Chris Reed
  1 Introduction 229
  2 Background 230
  3 Lying with structured argumentation 234
  4 Argument revision and lying 237
  5 Example 241
  6 Conclusions 244
  References 245

Layered Argumentation Frameworks with subargument relation 247
Beishui Liao
    1   Introduction 247
    2   Motivating examples 249
    3   Dung-style argumentation framework 251
    4   Argumentation framework with subargument relation (AFwS) 253
    5   Layered AFwS 256
    6   Semantics of a layered AFwS 258
    7   Updating a layered AFwS 263
    8   Concluding Remarks 265
        References 267

Extending belief bases change to logic programs with ASP 269
Julien Hué, Odile Papini, and Éric Würbel
    1   Introduction 269
    2   Preliminaries 272
    3   Changing logic programs 279
    4   Implementation 285
    5   Related works 294
    6   Conclusion 296
        References 297

Belief revision and argumentation: a reasoning process view

Pietro Baroni, Massimiliano Giacomin, and Guillermo Ricardo Simari

Abstract This chapter aims at comparing and relating belief revision and argumentation as approaches to model reasoning processes. Referring to some prominent literature references in both fields, we will discuss their (implicit or explicit) assumptions on the modeled processes and hence commonalities and differences in the forms of reasoning they are suitable to deal with. The intended contribution is on one hand assessing the (not fully explored yet) relationships between two lively research fields in the broad area of defeasible reasoning and on the other hand pointing out open issues and potential directions for future research.

1 Introduction

Belief Revision and *Argumentation* represent two exceedingly rich and fertile areas in Artificial Intelligence and Philosophy; both areas are concerned with the problem of handling beliefs and reasoning. In this chapter we will study the common ground and the differences that can be found between these two research fields focusing our efforts in finding stimuli for future research in their confluence.

In Belief Revision we study the process of changing the content of a belief repository when considering the arrival of a new piece of information. This problem has been the focus of research in the Computer Science areas of Artificial Intelligence and Databases, although the inquiry was started as a philosophical question that later got the attention of logicians. In Argumentation our main interest is centred on

---

Pietro Baroni · Massimiliano Giacomin
Dept. of Information Engineering University of Brescia, Italy,
e-mail: {baroni,giacomin}@ing.unibs.it

Guillermo Ricardo Simari
Dept. of Computer Science and Engineering, Universidad Nacional del Sur Bahia Blanca, Argentina
e-mail: grs@cs.uns.edu.ar

studying the acceptance of claims backed by arguments which represent a form of justification for holding those claims; these justifications are obtained by reasoning from accepted premises. The acceptance process is carried through in the context of disagreement where the support and interference of other arguments is considered. In settings where autonomous agents are involved, Belief Revision describes the way in which an agent is supposed to change its beliefs when new information arrives or when changes in the world are observed restoring the consistency[1] of the knowledge base; Argumentation is concerned with establishing the agent's current beliefs from a potentially incomplete and/or inconsistent knowledge base ensuring that these beliefs satisfy some desirable properties.

Doyle [Doyle, 1979], has studied the relation between belief revision and argumentation, elaborating his findings in [Doyle, 1992] and producing, to our knowledge, the first comparison between the areas.

In this work, Doyle contrasted the foundations approach and the coherence approach. In his own words, taken from the introductory section in [Doyle, 1992]: *"Recent years have seen considerable work on two approaches to belief revision: the so-called foundations and coherence approaches. The foundations approach supposes that a rational agent derives its beliefs from justifications or reasons for these beliefs: in particular, that the agent holds some belief if and only if it possesses a satisfactory reason for that belief. According to the foundations approach, beliefs change as the agent adopts or abandons reasons. The coherence approach, in contrast, maintains that pedigrees do not matter for rational beliefs, but that the agent instead holds some belief just as long as it logically coheres with the agent's other beliefs. More specifically, the coherence approach supposes that revisions conform to minimal change principles and conserve as many beliefs as possible as specific beliefs are added or removed."*

Although the reference to a foundations approach directs to his 1979 paper which describes a *Truth Maintenance System* or TMS, it is clear that that system was a forerunner of the frameworks that later were called argumentation systems. The coherence approach is represented by the AGM theory [Alchourrón et al., 1985]. It is interesting to observe that Doyle was persuaded that there existed similarities in both approaches that could produce significant advances in the endeavor to provide an autonomous agent with powerful knowledge representation and reasoning abilities.

In what follows we will explore the commonalities and differences in both approaches with the intention of introducing research questions that could bring more understanding of the problem of belief maintenance and reasoning, much in the spirit of the work in [Gärdenfors, 1990] that studied the same problem.

---

[1] It has to be acknowledged that notions of belief revision not requiring consistency restoring have also been considered in the literature. For instance in [Rott, 2001] the *horizontal* perspective on belief revision, focused on keeping a knowledge base consistent in face of changes, is compared with the *vertical* perspective, where inconsistencies in the knowledge base need not to be eliminated as far as a suitable inconsistency-tolerant inference mechanism is adopted.

2 Defining belief revision and argumentation

From a very general perspective, both belief revision and argumentation can be seen as processes carried out by a reasoner having the goal (or the need) to modify its state in front of some kind of change.

As a preliminary step in our analysis, we review some informal definitions of the two processes available in the literature with the goal of identifying commonalities, specificities, and, possibly, significant differences between and inside both areas.

Starting with belief revision, it may be interesting to observe first that the probably most cited and most influential paper in the area, namely the AGM paper [Alchourrón et al., 1985], does not contain the word *belief* at all: as stated in its title, it deals with the logic of *theory change*. It simply states " This paper extends earlier work by its authors on formal aspects of the processes of contracting a theory to eliminate a proposition and revising a theory to introduce a proposition." and the only (very quickly mentioned) application concerns the modification of *norms* rather than beliefs.

Actually, the use of the term belief revision predates the AGM paper. An AI oriented survey article of 1980 [Doyle and London, 1980] provides a selected bibliography of about 250 papers related to this area. In the introduction it states that "Belief revision concentrates on the issue of revising systems of beliefs to reflect perceived changes in the environment or acquisition of new information. In addition, belief revision research includes the study of methods for representing models of environments as collections of beliefs and the development of formal theories of belief."

This definition adopts the perspective of an intelligent agent dealing with an external environment and suggests a loosely specified distinction between different forms of change ("perceived changes in the environment or acquisition of new information").

A quite different perspective and kind of change is considered in another paper of the early eighties [Fagin et al., 1983] which has played a significant role in the subsequent development of the belief revision area. Here the problem is that of updating a database endowed with integrity constraints and rules for deriving explicit information. The reasons of the update are left unspecified nor is there any notion of environment, attention is rather focused on the requirements for the update operation: "we consider the problem of updating arbitrary theories by inserting into them or deleting from them arbitrary sentences. The solution involves two key ideas: when replacing an old theory by a new one we wish to minimize the change in the theory, and when there are several theories that involve minimal changes, we look for a new theory that reflects that ambiguity."

Thus, since the beginning of the field, we witness the coexistence of two views of belief revision, an agent oriented one, where beliefs are held by an agent and referred to an external world, hence some attention is paid to the features of the agent-environment interaction, and a theory oriented one, where the term belief is used in a rather generic sense (a currently true proposition in some information repository) and attention is paid to the change operations to be carried out and their properties.

Gärdenfors's foundational book [Gärdenfors, 1988] considers the two visions together, as stated in the preface: "This is a book about how to change your mind. More precisely, I present a theory of rational changes of belief. The epistemic changes that are at focus are revisions that occur when the agent receives new information that is inconsistent with the present epistemic state." While adopting explicitly an agent-oriented perspective, attention is focused since the beginning on the formal property of consistency and its preservation.

Sampling the literature we can identify examples of definitions which can be assimilated to each of the three perspectives above. Notably, alternative terms like *belief change* or *belief revision* have been sometimes used.

- Agent-oriented
  - Any intelligent agent has to account for a changing environment and the fact that its own beliefs might be inaccurate. For this reason belief revision is a task central for any kind of intelligent behavior [Nebel, 1989].
  - The study of belief change has been an active area in philosophy and in artificial intelligence, and, more recently, in game theory. The focus of this research is to understand how an agent should revise his beliefs as a result of getting new information [Friedman and Halpern, 1994].

- Theory-oriented
  - Belief Revision is the process of incorporating new information into a knowledge base while preserving consistency [Nebel, 1992].
  - In the logic of belief revision, a belief state (or database) is represented by a set of sentences. The major operations of change are those consisting in the introduction or removal of a belief-representing sentence [Hansson, 2011].

- Combined
  - The capability of gathering information about the world and revising its beliefs based on the new information is crucial for an intelligent agent. Belief revision therefore is a central topic in Artificial Intelligence. Technically, belief revision is the process of changing the beliefs of an agent to accommodate new, more precise, or more reliable evidence that is possibly inconsistent with the existing beliefs [Jin and Thielscher, 2007].
  - Belief revision is the area of knowledge representation that is concerned with how an agent may incorporate new information about a domain into its set of beliefs. It is assumed that the agent has some corpus of beliefs K which are accepted as being true, or holding in the domain of application. A new formula α is given, which the agent is to incorporate into its set of beliefs. Since consistency is to be maintained wherever possible, if α conflicts with K some beliefs will have to be dropped from K before α can be added [Delgrande, 2012].

All the above definitions have in common the reference to the process of managing changes to an information base in front of some input. As to the underlying differences, some detailed comments will be given later, at a general level it is possible to point out that the theory oriented view puts more emphasis on the form of the process, with no specific attention to the nature of the input and of the object of the process itself (*e.g.*, the terms belief base, knowledge base, logical theory, and database are regarded as somehow interchangeable). On the other hand, the agent oriented view has, at first reading, a narrower focus and some consideration for different kinds of input, while leaving formal properties of the process in the background.

In computational argumentation too it is possible to identify a probably most cited and most influential paper, namely Dung's paper on abstract argumentation frameworks [Dung, 1995] (where, by the way, the words *arguments* and *argumentation* are definitely present) but interest in using computational models of arguments in AI predates Dungs's paper by at least fifteen years. The earliest works [Birnbaum *et al.*, 1980; Cohen, 1981; Birnbaum, 1982] arose from the area of computational linguistics, where argumentation is regarded as a dialogical process ("a specific kind of conversation", quoting [Cohen, 1981]) between two opponents which hold different opinions and try to defeat each other through utterances of attacks and counter-attacks. While being very preliminary (and compact), these works introduce several fundamental notions and issues, in particular argument defeasibility, the relation of attack among them, and the problem of identifying the prevailing arguments which are key elements of all subsequent works in the field.

While building (more or less explicitly) on these key elements, other definitions describe argumentation as a kind of reasoning process, where non-conclusive reasons are used to derive conclusions. For instance, Pollock [Pollock, 1992] states that "Defeasible reasoning is, *a fortiori*, reasoning. Reasoning proceeds by constructing arguments, where reasons provide the atomic links in arguments. Conclusive reasons logically entail their conclusions. Defeasibility arises from the fact that not all reasons are conclusive. Those that are not are prima facie reasons. Prima facie reasons create a presumption in favor of their conclusion, but it is defeasible."

Thus, also in informal descriptions of computational argumentation, we can identify (at least) two non-incompatible views: a dialogue-oriented one, where arguments are moves in a game concerning the acceptance of some claim and a reasoning-oriented one, where arguments are the results of some inference process, conflicts among them may arise due to the limits of the information and/or knowledge used in the process and the reasoner needs to take a stance in front of the conflicts themselves. Again, definitions combining the two views can also be found, as evidenced by the following literature sampling.

- Dialogue-oriented
    - Argument is a social and verbal means of trying to resolve, or at least to contend with, a conflict or difference that has arisen or exists between two (or more) parties. An argument necessarily involves a claim that is advanced by at least one of the parties [Walton, 1990].

- The study of argumentation may, informally, be considered as concerned with how assertions are proposed, discussed, and resolved in the context of issues upon which several diverging opinions may be held [Bench-Capon and Dunne, 2007].
- To the extent that agents are autonomous, no one agent can impose its will on another. To the extent that agents are intelligent, they will need to persuade one another to adopt particular beliefs or courses of action, or negotiate with one another to divide scarce resources between them. Such activities are examples of argument, which we might define as rational, or reason-based, interaction between autonomous and intelligent agents to achieve particular goals [Rahwan and McBurney, 2007].

- Reasoning-oriented

  - Arguments are *prima facie* proofs that may make use of assertions that one sentence is (defeasible) reason for another. They indicate support for a proposition, but do not establish warrant once and for all; it matters what other counterarguments there may be [Simari and Loui, 1992]
  - This chapter surveys logics for a particular group of patterns of inference, namely those where arguments for and against a certain claim are produced and evaluated, to test the tenability of the claim. Such reasoning processes are usually analysed under the common term 'defeasible argumentation' [Prakken and Vreeswijk, 2001].

- Combined

  - In multi-agents systems, conflicts of interest are inevitable. To address these problems, agents can use argumentation, a process based on the exchange and valuation of arguments for and against opinions, proposals, claims and decisions. Argumentation, in its essence, can be seen as a particularly useful and intuitive paradigm for doing nonmonotonic reasoning [Caminada and Amgoud, 2007].
  - The word argument may recall several intuitive meanings, like the ones of line of reasoning leading from some premise to a conclusion or of utterance in a dispute [Baroni and Giacomin, 2009a]
  - In a sense, monological argumentation is a static form of argumentation. It captures the net result of collating and analyzing some conflicting information. In contrast, dialogical argumentation is a dynamic form of argumentation that captures the intermediate stages of exchanges in the dialogue(s) between the agents and/or entities involved [Besnard and Hunter, 2008].

All the above definitions have in common the reference to the process of managing some kind of conflict. The dialogue-oriented view encompasses the presence of different actors and emphasises the dynamic aspects of the process, while the reasoning-oriented view is not concerned with process actors and is more focused on defeasibility and the evaluation of argument acceptability.

Partly due to the actual development of the relevant scientific literature along the last 20 years, another important distinction concerns the abstraction level of theoretical models of argumentation. Rooted in the seminal paper by Dung [Dung, 1995], *abstract argumentation* models adopt the notions of argument and attack as primitive concepts, the underlying nature and structures of arguments and attacks being totally abstracted away. Abstract argumentation models are therefore completely focused on the issues of conflict management and argument justification. As such they cover only a part (though a very important one) of the whole argumentation process. Models where argument construction and attack definition are explicitly taken into account can be referred to generically as *instantiated argumentation* models, though it may well be the case that these models are abstract in some respect and far from real instantiations (*e.g.*, they may leave the logical language for argument construction unspecified). It is not uncommon that instantiated models are bridged with abstract argumentation as far as the issues of conflict management and argument justification are concerned. Clearly the distinction between abstract and instantiated models is orthogonal to the separation of dialogical and reasoning-oriented argumentation and the literature is rich in works covering all four possible combinations.

Armed with this (clearly non-exhaustive) overview of basic concepts and intuitions in belief revision and argumentation we are now ready to undertake a more detailed analysis and comparison from a process modeling perspective. As to belief revision we will use as main reference the AGM approach. As to argumentation we will refer to the structure of a generic argumentation system as conceptualized in [Prakken and Vreeswijk, 2001], and use as a formal counterpart Dung's theory of abstract argumentation frameworks.

Before diving into the matter, to make the chapter self-contained we very quickly recall the necessary basic technical background and terminology in the following section.

3 A brief technical background

We will succinctly introduce the necessary technical details regarding belief revision and argumentation systems.

3.1 The AGM approach to belief revision

In the AGM approach [Alchourrón *et al.*, 1985; Gärdenfors, 1988], beliefs are represented as sets of sentences in some language $\mathscr{L}$ including the standard connectives $\neg, \wedge, \vee, \rightarrow, \leftrightarrow$ and constants $\top, \bot$. The language is governed by a logic expressed by a consequence relation $Cn \subset 2^{\mathscr{L}} \times 2^{\mathscr{L}}$ from sets of sentences to sets of sentences. Given a set of sentences A, $Cn(A)$ is the set of sentences which are logical consequences of A. For a sentence x, $x \in Cn(A)$ is also written as $A \vdash x$ and $x \notin Cn(A)$ as $A \nvdash x$. A set A of sentences is consistent if $A \nvdash \bot$. It is assumed that, for any sets of sentences A and B, Cn satisfies the following three conditions: $A \subseteq Cn(A)$, $Cn(A) = Cn(Cn(A))$, if $A \subseteq B$ then $Cn(A) \subseteq Cn(B)$.

A *belief set* is a set of sentences K closed under the consequence relation namely $K = Cn(K)$. Given a consistent belief set K a sentence x is accepted with respect to K if $x \in K$, is rejected with respect to K if $\neg x \in K$, is indetermined if neither of the above cases holds.

Three basic operations of change on a belief set K are considered:

- *expansion* (denoted as $K^+(x)$) consists in adding to K a new sentence x which is consistent with K;
- *contraction* (denoted as $K^-(x)$) consists in removing from K a sentence which was previously in K;
- *revision* (denoted as $K^*(x)$) consists in adding to K a new sentence x which is not consistent with K.

The AGM approach provides a set of postulates for the operations of contraction and revision. In fact these operations can not be univocally defined (while the expansion operation is) and these postulates (which will be introduced and commented later where useful) constrain the space of possible choices.

Another standpoint of the AGM approach is the *Levi identity*, which states that the revision operation can be expressed in terms of contraction and expansion: revising a belief set K by adding an inconsistent sentence x is equivalent to contract K with respect to $\neg x$ and then expanding the result with x, formally $K^*(x) = K^-(\neg x)^+(x)$. In a dual manner, it is possible to express contraction in terms of revision through the *Harper identity*, which states that contracting of a belief set K with respect to a sentence x is equivalent to intersecting K and the revision of K with respect to $\neg x$, formally $K^-(x) = K \cap K^*(\neg x)$. Levi and Harper identities and the postulates on contraction and revision operators are nicely related in a coherent framework (in particular see [Gärdenfors, 1988]). In virtue of this, one can focus on the characterization of just one of the two operators as the other can be simply defined as a derived concept.

Further details and comments on these operations and their properties and relationships will be given throughout the chapter, we only remark that the term *revision* is overloaded as it is used to denote both a specific operation and the general activity of managing any change operation. This justifies the use in several papers of the term *belief change* (or similar ones) for the latter meaning.

3.2 Argumentation Systems

There is consensus in the argumentation community over what are the required steps to define an argumentation system [Prakken and Vreeswijk, 2001]; namely, five main conceptual components are necessary:

a) an underlying "logical" language
b) the definition of what an argument is

Belief revision and argumentation: a reasoning process view

c) the definition of a conflict relation between arguments
d) the definition of a defeat relation between arguments
e) the definition of the justification status of arguments

Some quick comments on these components are in order. First, the construction of arguments is typically assumed to be monotonic: new arguments do not suppress previous arguments but rather may be in conflict with them. The presence of conflicts does not give rise to pathological situations like the production of any possible argument but is regarded as a basic feature of the system. Several types of argument conflicts can be conceived, the basic idea being, in any case, that conflicting arguments can not be accepted together. Not all conflicts are effective, the ones which are give rise to the defeat relation. To give a simple example, if two arguments α and β have incompatible conclusions, *e. g.*, α concludes that tomorrow will be sunny and β concludes that tomorrow will be rainy, they are in mutual conflict. However, the reasons underlying α might be stronger than the ones underlying β (*e. g.*, a reliable local forecast against the weather report of a national newspaper) so α defeats β but not vice versa. Finally, given the defeat relation, the justification status of the arguments has to be evaluated (e.g. an undefeated argument will probably be accepted, while an argument which is defeated by an undefeated argument will be rejected). Many variants, extensions, and subtle analyses of these basic concepts are available in the literature. The reader is referred to [Chesñevar *et al.*, 2000; Prakken and Vreeswijk, 2001; Bench-Capon and Dunne, 2007; Besnard and Hunter, 2008; Rahwan and Simari, 2009; Baroni *et al.*, 2011] for introductory and/or review material on the field. We will now introduce two important formal systems in this context. First, we will present a fundamental approach that has allowed the deep study of the interplay of arguments when nothing but their existence and the relation of attack (defeat in the terminology introduced above) is taken into account. Second, we will discuss all the five elements that are considered when the full construction of a system is required to build the arguments and their interplay.

3.2.1 Dung's abstract argumentation framework

An *argumentation framework* [Dung, 1995], is simply a pair $\langle \mathscr{A}, \mathscr{R} \rangle$ consisting of a set $\mathscr{A}$ whose elements are called *arguments* and of a binary relation $\mathscr{R} \subset \mathscr{A} \times \mathscr{A}$ called *attack relation*. Given two arguments α, β if $\langle \alpha, \beta \rangle \in \mathscr{R}$ it is said that α attacks β. An argumentation framework has an obvious representation as a directed graph where nodes are arguments and edges are drawn from attacking to attacked arguments.

While the word *argument* may recall several intuitive meanings abstract argumentation frameworks are not (even implicitly or indirectly) bound to any of them: an abstract argument is not assumed to have any specific structure but, roughly speaking, an argument is anything that may attack or be attacked by another argument, where, again, no specific meaning is ascribed to the notion of attack. As such, in argumentation frameworks the first three components of an argumentation system (language, argument definition, conflict definition) are totally abstracted away: the (binary) attack relation corresponds to the defeat relation of [Prakken and Vreeswijk, 2001] and the main focus is on evaluation of the justification status of arguments.

An *argumentation semantics* is the formal definition of a method (either declarative or procedural) ruling the argument evaluation process. A rich variety of alternative proposals of argumentation semantics is available in the literature (see [Baroni et al., 2011; Baroni and Giacomin, 2009a]). Two main styles of argumentation semantics definition can be identified in the literature: *extension-based* and *labelling-based*.

In the extension-based approach a semantics definition specifies how to derive from an argumentation framework a set of *extensions*, where an extension E of an argumentation framework $\langle \mathscr{A}, \mathscr{R} \rangle$ is simply a subset of $\mathscr{A}$, intuitively representing a set of arguments which can "survive together" the conflict or are "collectively acceptable". Putting things in more formal terms, given an extension-based semantics $\mathscr{S}$ and an argumentation framework $AF = \langle \mathscr{A}, \mathscr{R} \rangle$ we will denote the set of extensions prescribed by $\mathscr{S}$ for AF as $\mathscr{E}_{\mathscr{S}}(AF) \subseteq 2^{\mathscr{A}}$. The justification state of an argument $\alpha \in \mathscr{A}$ according to an extension-based semantics $\mathscr{S}$ is then a derived concept, defined in terms of the membership of α to the elements of $\mathscr{E}_{\mathscr{S}}(AF)$.

In the labelling-based approach a semantics definition specifies how to derive from an argumentation framework a set of *labellings*, where a labelling L is the assignment to each argument in $\mathscr{A}$ of a label taken from a predefined set $\mathscr{L}$, which corresponds to the possible alternative states of an argument *in the context of a single labelling*. Again, an argumentation semantics prescribes a set of labellings and the justification state of an argument α turns out to be a derived concept, defined in terms of the labels assigned to α in the various labellings. A typical choice for the set of labels is $\mathscr{L} = \{\text{in}, \text{out}, \text{undec}\}$: under this choice it is possible to translate an extension-based semantics into its labelling-based equivalent formulation and vice versa (see [Baroni et al., 2011]).

3.2.2 Arguments with Structure

As we have just seen, in an abstract framework the inner details of the arguments are not considered, *i.e.,* arguments are considered as atomic entities. Indeed, when the issue of actual argument construction becomes pertinent for the formalism the five items introduced at the beginning of this section become relevant.

It is clear that there are many possibilities to choose from when it is necessary to construct arguments and the research community has produced a number of systems where the structure of the argument has been considered [Pollock, 1987; Loui, 1987; Simari and Loui, 1992; Prakken and Sartor, 1997; Bondarenko et al., 1997; Besnard and Hunter, 2001; García and Simari, 2004; Prakken, 2010]. In all of these proposals it was necessary to make decisions regarding the language, the actual construction of arguments using the underlying language, the relation of conflict between these arguments, the determination of how the conflict is decided, and the final decision on the justification status of arguments.

All the five elements mentioned at the beginning of Subsection 3.2 have an impact in the characterizations of revision mechanisms over structured argumentation systems. Every one of them becomes more or less relevant in the different approaches to the definition of systems that combine belief revision and argumentation. In the forthcoming discussion we will make appropriate comments on these issues.

4 Belief revision and argumentation as processes

We are now ready to start our analysis of belief revision and argumentation as processes. By process we mean a well-defined series of operations which, given the initial state of the object of processing activity and some input, define the transition of the object to a new state, where some further input can be processed in turn. The object of processing activity, its state, the process input and the transition steps have therefore to be characterized.

In the AGM approach this is, at a first glance, relatively easy:

- the object of processing activity is a belief set
- state information is implicit in the belief set itself (the members of the belief sets are accepted)
- the input is a sentence along with the specification of the operation to be carried out: expansion, contraction, or revision (note that a preliminary consistency check is needed to distinguish between expansion and revision)
- there is only one transition step, namely the application of a suitable operator to the belief set to produce a new belief set

In argumentation the situation is more articulated. First, within the whole argumentation process (at least) two distinct sub-processes can be identified: the production of arguments (and specifically of counterarguments in dialogical argumentation) and the evaluation of their justification status. The grain of interleaving of the two sub-processes depends on the type of argumentation. In dialogical argumentation at each dialogue step some arguments (typically one) can be produced. The new argument is bound to be related (typically to be a defeater) of a previous one. According to the rules of the dialogue protocol, the newly added argument entails the modification of the acceptance status of some previously produced arguments and the dialogue can proceed with the next step. In reasoning-oriented argumentation typically no specific constraints are given on how far argument production can go before the relevant conflicts are identified and argument justification status is evaluated.

Leaving apart interleaving, let us now focus on the two sub-processes.

Argument production operates on some knowledge base and produces a set of arguments. In this sense, differently from belief revision, it does not modify the initial object (the knowledge base) but rather produces a new object (a set of arguments). There are no requirements of consistency or closure on the knowledge base. Indeed the presence of inconsistencies is one of the main reasons for using argumentation, where inference is explicit in the construction of arguments rather than being implicit in the assumption of closure.

The acquisition of new information corresponds, in this context, to modifying the knowledge base. At least conceptually, all arguments are then rebuilt from scratch from the new knowledge base. Drawing a parallel with belief revision operators we note that, in the addition of a new element to the knowledge base used to build arguments, there is no need to distinguish between expansion and revision, as inconsistency does not need any special treatment. As to retraction we note that argument production is a monotonic process: no previous argument needs to be canceled in order to change your mind, since this is the task of the subsequent evaluation phase.

The subprocess of argument evaluation operates on a set of arguments and their defeat relation and attaches a justification status to each of them. There is no direct information acquisition in this process, since it operates on already built arguments. At an abstract level one can however, consider modifications of the set of arguments and/or of the defeat relation as triggers of the evaluation process, without considering explicitly the causes of these modifications in the argument construction phase. In any case, after the modifications, at least conceptually, argument evaluation is carried out from scratch. As belief revision operators refer to sentences while evaluation operates on arguments, drawing a direct parallel with this subprocess is not immediate. It can be however noted that, like belief revision operators, argument evaluations concerns justification states and their dynamics. Hence, some parallel could be drawn between the various justification changes in the two areas, with the specific difference that belief revision operators require as parameter a sentence p, while this is not the case in the dynamics of argument evaluation. These kinds of issues are left for future analysis.

4.1 A "hidden" state in belief revision

While the description above offers a basic and self-contained view of belief revision and argumentation as processes, it has to be remarked that any non-trivial belief revision process requires a further state component, which is not explicitly addressed by its basic definitions and postulates.

In fact, the AGM postulates offer a wide range of choices for the definition of a contraction (or equivalently revision, as one can be defined in terms of the other) operator, some of the possible choices being definitely unsatisfactory (*e. g.*, in [Alchourrón *et al.*, 1985] it is stated that *maxichoice revision* has *disconcerting properties*). To constrain the choice to a set of more reasonable alternatives, in [Alchourrón *et al.*, 1985] the notion of partial meet contraction is introduced. Leaving apart formal details, given a belief set K and a sentence x to be retracted, partial meet contraction is based on a *selection function* γ picking some of the maximal subsets of

K which do not imply x. The result of contraction is simply the intersection of the selected maximal subsets. With this addition, it emerges that the starting point of belief revision is not just the initial belief set: it also includes the selection function γ adopted by the reasoner. Intuitively, γ has to select the "most important" [Alchourrón et al., 1985] or, using the more precise terminology of [Gärdenfors, 1988], the "most epistemically entrenched" maximal subsets of K that fail to imply x. It has to be evidenced that the notion of epistemic entrenchment does not concern the credibility of sentences, the idea is rather that "some propositions are more useful than others in inquiry and decision making" and then a reasoner is more inclined to give up those propositions which are less useful.

In principle, the notion of epistemic entrenchment can be implicit in the definition of the selection function, it is however more natural to express it explicitly. This can be done through an ordering relation over sets of sentences [Alchourrón et al., 1985] or over sentences [Gärdenfors, 1988] so that the selection function is then defined on this basis. Another way to express epistemic entrenchment is by a set of *conditional beliefs* of the form $A > B$ with the meaning "If A were the case then B would be true".

Independently of technical details, the important point is that in this context the initial state of the belief revision process does not consist in the belief set only but includes also the reasoner's attitude towards the selection of what sentences should survive contraction, possibly expressed by an entrenchment ordering. This extra-logical information is necessary to determine a reasonable revision strategy.

Altogether, the reasoner's belief set and this additional "strategical" information are called *epistemic state* in [Darwiche and Pearl, 1997; Boutilier, 1998] and *belief state* in [Jin and Thielscher, 2007] (it has however to be noted that these terms have no uniform usage in the literature: for instance epistemic state is used with a different meaning in [Gärdenfors, 1988] and belief state is used with a different meaning in [Boutilier, 1998]). We will adopt the term epistemic state here.

4.2 Iteration in argumentation and belief revision

Having laid out this basis, an important comment is now in order concerning the iteration of these processes. Clearly dialogical argumentation is iterative by nature. Similarly, in reasoning-oriented argumentation the processes of argument production and argument evaluation can be iterated at will using the result of the previous iteration as a starting point of the subsequent one without any specific conceptual problem. In fact, studies on argumentation dynamics in Dung's framework [Cayrol et al., 2010; Liao et al., 2011] are especially focused on computational rather than basic conceptual/modeling aspects, the problem being the one of minimizing recomputation when only a part of the framework changes by characterizing the effects and scope of different kinds of change. This can also be useful in a multi-agent context to identify efficiently which argument to use in a dialogue in order to obtain a desired effect. At a less abstract level, dynamics can be explicitly modeled in the underlying logic for argument construction, with consequent effects on the set of arguments and attacks which have to be considered at a given step of the process evolution[2] [Rotstein et al., 2010]. The notion of dynamics in argumentation has also been considered with richer and more articulated meanings, involving the change of the dialogue protocol. For instance in the context of dialogical argumentation an approach has been proposed to model the fact that "at any point in the argumentation, participants can start a meta level debate, that is, the rules of order can be made the topic of discussion" [Brewka, 2001].

The situation is very different for the basic AGM approach, which does not lend itself to iteration: in fact *iterated belief revision* is, in a sense, a research subfield on its own (for instance, see [Darwiche and Pearl, 1997; Boutilier, 1998; Jin and Thielscher, 2007; Hunter and Delgrande, 2011]).

In fact, given that the starting point of the process of belief revision is actually a full epistemic state, not just a belief set, it follows that when iterating the process for a further step a new full epistemic state is needed. The belief set component of this new state is produced by the revision operator, but the "strategic" component (as introduced in the previous subsection) is problematic. In fact, it can not stay the same, since also the revision strategy needs to take into account the new information acquired, but the AGM theory does not gives any account on how to revise the revision strategy. So it stops after the first step.

One possible solution to this aporia would be to encode the strategical information within the belief set in form of conditional beliefs. Gardenfors's triviality theorem [Gärdenfors, 1988] shows however that including conditional beliefs within the belief set leads to a trivial revision model unless one gives up some basic requirements on the revision operator or on the notion of conditional belief. To circumvent this problem, one needs to define a distinct revision mechanism for the strategical component of the epistemic state and redefine the whole belief revision process so that it operates on the whole epistemic state and produces a new epistemic state. This gives a concrete evidence that the question on what is the object of the revision (and, as we will see, also of the argumentation) process, is more open than what the traditional models suggest. This will be the subject of the next section.

4.3 Summing up ...

Synthetic statements

- The process of belief revision involves three kinds of change operations in front of new information: expansion, contraction, and revision. Expansion not being problematic, attention is focused on contraction and revision. Under some assumptions, contraction can be defined in terms of revision and vice versa. In

---

[2] This approach opens in fact an interesting perspective on a potential interplay between belief change and argument change, which is however out of the scope of the present chapter

argumentation different kinds of change operations are not identified: new information gives rise to the generation of new arguments and conflicts and then the justification status of all arguments is re-evaluated.
- In belief revision the justification status of sentences is defined implicitly by the membership to the belief set, in argumentation the justification status is the result of an explicit evaluation. In both theories, the result, though subject to some general constraints, is not predetermined since there are alternative choices for the process mechanisms. In belief revision, the choice concerns the actual operator to apply. In argumentation the choice concerns basically the semantics for argument evaluation. Given a revision operator the result is a univocal belief set. Given an argumentation semantics the result is a set of extensions or labellings.
- Iteration is problematic in classical belief revision since the revision process requires some extra-logical information driving the revision strategy but does not specify how to revise it, so this information is lost after the first step. To account for iteration an extended model is needed. Iteration is explicit in dialogical argumentation, implicit in reasoning-oriented argumentation, but does not pose specific conceptual problems in this context.

Questions

- Can argumentation borrow distinct kinds of change operations from belief revision for a more detailed modeling of information acquisition? Would this be appropriate?
- Is there any significant relation between the existence of different operators in belief revision and of different semantics in argumentation?
- Can contraction be accommodated within argumentation? If not, is anything missing in argumentation?

5 The process objects

As mentioned in the previous section, while in the AGM approach the object of the revision process is a belief set, this modelling choice turns out to be unsatisfactory when iteration comes into play. In this section we further elaborate on this point, examining other issues regarding the definition of the process object in belief revision and argumentation.

As already mentioned, classical belief revision theory encompasses the existence of information concerning the revision strategy in addition to the belief set. So the reasoner is endowed with an epistemic state which, at least for iteration purposes, should be the object of the change process rather than the belief set itself. In a sense, the belief set is "too small" as it does not cover the whole epistemic assets of the reasoner.

From another perspective, the belief set has been questioned as being "too large". In fact, the assumption that it is closed under the consequence relation corresponds to an unrealistic property of logical omniscience of the reasoner. In the classical belief revision theory this has been defended by stating that the belief set does not correspond to what the agent actually believes at a given instant but, rather, to what the agent is committed to believe at that instant. Since the belief set is infinite, except in very simple cases, it has to be acknowledged however that actual applications and reasoners can only work on a finite representation. In fact, one can typically identify a finite set of *basic beliefs*, which the reasoner holds *a priori* independently of any inference activity and a (usually infinite) set of *merely derived* beliefs, which the reasoner holds as a result of the inference activity. Merely derived beliefs can in turn be distinguished into *explicit beliefs*, which the reasoner has derived up to the current instant and *implicit beliefs* which the reasoner could produce with further inference activity.

Some approaches to belief revision consider therefore the set of *basic beliefs*, also called *belief base*, as the object of the change process [Fuhrmann, 1991; Nebel, 1989; 1992; Hansson, 1993]. This alternative choice has several implications which go beyond the mere issue of practical feasibility.

First, it has to be observed that a belief base induces a unique derived belief set, while the same belief set can be induced by more distinct belief bases. In fact, revising different belief bases may give rise to different results even if the bases correspond to the same belief set. This technical difference evidences that *base revision* and classical belief revision have not to be necessarily regarded as two sides of the same coin. In fact, in base revision the actual form of the belief base may make a difference, in spite of logical equivalence. In other words, the form of the belief base carries some meaning with respect to what the reasoner "has in mind" (and is revising). To put it in other words, most beliefs are based on reasons and these reasons may, intuitively, play a role in the revision process. However, as stated by Gärdenfors himself [Gärdenfors, 1988]: "belief sets cannot be used to express that some beliefs may be reasons for other beliefs". Base revision appears to be more suitable to support some form of *reason maintenance* though not encompassing an explicit representation of the reasons underlying the beliefs. In particular, the contents of a belief base can be separated in different classes with different roles in the revision process: *e.g.*, in the database field one can distinguish facts from integrity rules, where, clearly, integrity rules are "less revisable" than facts [Fagin *et al.*, 1983]. On the basis of this kind of considerations, Nebel [Nebel, 1989; 1992] develops an analysis on *symbol level* vs *knowledge level* belief revision. It is suggested that base revision corresponds to *symbol level* revision, as it depends on the actual representation of the belief base, while classical belief revision is at the *knowledge level* since the revision of a belief set is not affected by the choice of a specific representation as a belief base. Nebel proposes an approach to reconcile these different views, by showing that base revision and belief revision produce the same result under a suitable choice of the selection function underlying belief revision (and provided that one gives up one of the classical revision postulates). Using

Nebel's words, in the context of this approach reason maintenance "appears as a side effect".

A more explicit distinction of the objects of revision and, consequently, of the revision processes, has been suggested in [Dubois *et al.*, 1996]. The authors distinguish two main kinds of information: *factual evidence* and *generic knowledge*. Factual evidence consists of information gathered on the case at hand or the description of the actual world in a given situation. Generic knowledge pertains to a class of situations considered as a whole but does not refer to a particular case. Then the authors distinguish three processes of information acquisition:

- *focusing* which consists in conditioning the generic knowledge by the factual evidence, namely changing the reference class for an individual (*e. g.*, from bird to penguin);
- *F-revision*, which consists in enforcing a new piece of factual evidence;
- *G-revision*, which consists in enforcing a new piece of generic knowledge.

The authors emphasize that this kind of difference "is most clear in settings where generic knowledge is represented and processed as distinct from factual evidence. Such a distinction is not relevant in propositional logic, for instance, since every piece of information takes the form of a propositional sentence." In this perspective, the distinction on the objects of revision becomes also a difference in the actual revision process, an issue we will deal with in the next section (see also [Baroni *et al.*, 2007]).

Turning to argumentation, the object of the change process is a set of arguments with their justification status. Argument construction and structure are abstracted away in Dung's framework while they are part of the game in more concrete formalisms like the ones mentioned in Subsection 3.2. As already remarked, the notion of closure under consequence is typically absent in argumentation contexts (though it may be present in the underlying argument construction formalism if it is logic-based). The distinction between belief set and belief base in belief revision may be put in parallel with the distinction between warrant and justification by Pollock [Pollock, 1992]. Pollock conceives reasoning as an activity which "can involve numerous false starts, wherein a belief is adopted, retracted, reinstated, retracted again, and so forth. At each stage of reasoning, if the reasoning is correct then a belief held on the basis of that reasoning is justified, even if subsequent reasoning will mandate its retraction. Epistemic justification, in this sense, is a procedural notion consisting of the correct rules for belief updating having been followed by the system up to the present time in connection with the belief being evaluated." On the other hand warrant corresponds to an ideal situation where a "reasoner unconstrained by time or resource limitations would ultimately be led to believe the proposition. Warranted propositions are those that would be justified "in the long run" if the system were able to do all possible relevant reasoning."

The argumentation process deals with justification rather than warrant, and so is closer to base revision than to classical belief revision. In particular, by operating on arguments rather than on sentences, argumentation incorporates, by construction, a form of reason maintenance. In non-abstract argumentation, the distinction between primary and derived beliefs is explicit as it is the reason why an argument is rejected.

In argumentation the possibility to combine different semantics offers an interesting opportunity of non-monolithic modeling of reasoning activity. In fact, in the context of the same reasoning process (or of the same dialogue) one can identify different sub-contexts or sub-topics where different flavors of reasoning (*e.g.*, more or less skeptical) are appropriate. This has been considered for instance in [Prakken, 2006] where a proof procedure combining skeptical and credulous reasoning is proposed, based on the standpoint that, in certain contexts, epistemical reasoning (*i.e.*, reasoning about beliefs) is skeptical while practical reasoning (*i.e.*, reasoning about actions) is credulous and that the two forms of reasoning are interleaved in real situations.

Looking at flexible modeling from another side, we observe that the ASPIC+ formalism is able to accommodate the distinction between factual knowledge (that can be represented by the knowledge base) and generic knowledge (that can be represented by the set of rules). Finer distinctions are then available for both: the knowledge base is partitioned into axioms, premises, assumptions, and issues, while rules can be strict or defeasible. While this provides a basis to support useful distinctions between revision objects and processes of the kind advocated in [Dubois *et al.*, 1996], a full exploitation of this potential appears to be an open research question. Applying to argumentation an analysis of the kind carried out by Nebel on symbol-level vs knowledge-level representation is also an open issue.

5.1 Summing up ...

Synthetic statements

- Classical belief revision refers to an idealized, closed and possibly infinite, belief set as the object of the change process, while in concrete contexts the object of the change is a finite belief base. In argumentation an analogous idealization corresponds to the notion of warrant, but argumentation models typically refer to the more concrete notion of (revisable) justification.
- Reason maintenance and the distinction between primary and derived beliefs are not encompassed in classical belief revision, while they are dealt with implicitly in base revision. These notions are indeed foundational and explicit in argumentation. Further distinctions on the process objects have been considered in non-abstract argumentation.
- In belief revision, the idealized setting has been assimilated to the knowledge-level, while base revision to the symbol-level. A similar analysis is lacking in argumentation.

Belief revision and argumentation: a reasoning process view

Questions

- What is the counterpart in argumentation of the idealized knowledge-level process of belief revision? If it can not be identified, is it worth exploring?
- Which is (or which should be) the counterpart in argumentation of the extra-logical information driving the revision strategy?
- Can reason maintenance be encompassed explicitly in belief revision?
- Can the distinctions on the process objects be exploited to make belief revision (and argumentation itself) a more faithful representation of actual belief change processes?
- Can non-monolithic models of reasoning be adopted in belief revision?
- Can there be a cross-fertilization between the application of multiple argumentation semantics in the same reasoning process and the use of different epistemic states in iterated belief revision?

6 Process drivers and goals

In this section we consider the question of what are the causes of the change process and what goals it is aimed at.

In belief revision these issues are highly abstracted. As to the causes, the acquisition of new information (or the necessity to retract something) is taken as the starting point without additional considerations on why and from where the new information arises. The goal is essentially to preserve consistency (by suppressing some information which is incompatible with the new one).

It has been observed however that different causes for the new information require different processes. In fact, classical belief revision implicitly assumes that the starting point is a new acquisition of information about a static world which has not changed. It has been observed [Katsuno and Mendelzon, 1991] that the revision process, which is appropriate for this kind of context, is not suitable for the case where the new information acquired is an effect of a change in a dynamic world. In this case the process is called *belief updating* and needs to satisfy different requirements.

As a matter of fact, while belief revision and belief updating are conceptually distinct processes, it is quite evident that in most real contexts an agent has to manage them "simultaneously" so that an approach able to encompass both is highly desirable and has been considered in some literature works (see *e.g.*, the notion of belief change in [Friedman and Halpern, 1994], of generalized update in [Boutilier, 1998], of belief change with actions and observations in [Hunter and Delgrande, 2011]).

Orthogonal to the distinction between revision and updating is the fact that the change involves factual rather than generic knowledge, as mentioned in the previous section. As to our knowledge, this has received limited attention in the literature.

A common feature of all the processes considered above is that the new information has absolute priority with respect to the previous beliefs: this corresponds to a *prioritized* revision process. While prioritized revision is the only choice in some contexts (*e.g.*, when including the latest amendments in a normative system) it has been observed that in other contexts (*e.g.*, acquisition of information from possibly unreliable sensors) the new information needs not to prevail over the previous one. In this case, non-prioritized revision is more appropriate, which in turn may appear in a variety of forms (see [Hansson, 1999]). For instance in *screened revision* [Makinson, 1997] a decision is made at the beginning of the process whether to accept or not the input on the basis of its consistency with some previous *core beliefs*, while *selective revision* [Fermé and Hansson, 1999] encompasses the case of partial input acceptance.

Intuitive as it may be, non-prioritized revision is not the only way to go to take the possible unreliability of input into account. In fact, one can replace the "uncertain" fact that the room temperature is 28 degrees, with the "certain" fact that the sensor reading is 28 degrees and then consider a more complex reasoning scheme where the fact that the room temperature is 28 degrees is derived (and could not be accepted) from the fact (which is accepted with certainty) that the sensor reading is 28 degrees. In this perspective, non-prioritized belief revision is unnecessary provided that one adopts a "proper" modeling. This poses however two problems: first, the "proper" model may turn out excessively and unnaturally complicated, second, it opens a serious problem of what can be considered definitely certain, given that one may doubt about anything (in the example one can in turn be uncertain about the process of reading from the sensor, about its memory of what has been read, and so on). As a matter of fact, these problems do not arise if the input is rather regarded as an assumption of the reasoner in a frame of hypothetical reasoning. Clearly the reasoner is free to ascribe arbitrary strength to his assumptions (even the most absurd ones) but the question then arises on whether this casting to hypothetical reasoning is always sensible.

All the above distinctions did not find room for development in abstract argumentation, at least up to now. Here the input is represented by the addition of new arguments and/or the relevant attack relations and the process goal is simply to recompute the justification status of the whole set of arguments. Different criteria for argument evaluation are represented by different semantics but the cause of the modification (for instance the distinction between revision and updating) has not been related with the choice of the semantics. Moreover abstract argumentation is non-prioritized by nature, since all arguments (either new or old) are subject to the same evaluation method. Recovering consistency is not an issue in argumentation (given that the presence of conflicts lies at its very heart) but the weaker notion of conflict-freeness of extensions (or labellings) captures the same kind of requirement.

Dialogical argumentation is more interesting from this viewpoint, since different kind of dialogues with different goals can be considered. Probably the main reference in this field [Walton and Krabbe, 1995] identifies six main types of of dialogue: persuasion, negotiation, inquiry, deliberation, information-seeking, and eristic. They are distinguished by the initial situation, the overall dialogue goal, and participants' specific goals. Different dialogues obey different rules/protocols according to their goal-oriented nature. This is a significant departure from other forms of argumentation and belief revision which are essentially goal-neutral. It has to be observed however that dialogue types appear to be conceived for a static world where only the dialogue participants play an active role: encompassing a notion analogous to belief updating in this context would require further analysis. Moreover, in argumentation dialogues the last move is typically regarded as prevailing over previous ones and can be possibly defeated only by the next move. These two basic similarities suggest that dialogical argumentation can be closer to belief revision than it appears at first glance.

Argument schemes [Walton, 1996] provide another potential starting point for future investigations on different kinds of processes across the two fields. An argument scheme is a semi-formal predefined pattern representing a protoypical way to build a presumptive support for a given conclusion in a given context. Examples of these schemes include argument from Expert Opinion, from Position to Know, from Witness Testimony, and so on. For each scheme a set of *critical questions* are defined. In a nutshell, each critical question represents a challenge for an argument built according to the scheme: whether the presumption supported by an argument stands or falls depends on the answers to the relevant critical questions. So, in a sense, critical questions can be regarded as context-specific guidelines to revise the justification status of arguments on the basis of various kinds of input affecting different aspects of the argument structure. While this is hardly reconcilable with the general context-independent formulation of classical belief revision, one may wonder whether, especially in the context of base revision, introducing different kinds of revision schemes may be useful.

6.1 Summing up ...

Synthetic statements

- Classical belief revision is suitable to model the acquisition of new information which has the highest priority and concerns a static world. Removing the former assumption gives rise to a nonprioritized revision process, while considering a dynamic world corresponds to the different process of belief updating. Argumentation is, *per se*, a non-prioritized process where the distinction between revision and updating has not be considered up to now. Dialogical argumentation has some basic similarities with classical belief revision.
- Classical belief revision has the implicit goal of ensuring consistency, abstract argumentation aims at evaluating argument justification status, so they can be regarded as goal-neutral processes, while argument-based dialogues are goal-oriented processes.
- Argument schemes provide a semiformal approach to describe, in a context-dependent way, different patterns of presumptive justification and different patterns of revision through the notion of critical question.

Questions

- Is the distinction between revision and updating relevant to argumentation? Do abstract or dialogical argumentation need to be extended to accommodate it?
- Can the notion of goal-oriented process be suitably and profitably captured in abstract settings like belief revision and argumentation frameworks?
- Is there room for paradigmatic schemes in belief/base revision?

7 Postulates and principles

Both belief revision and argumentation can be seen as general processes which admit several different realizations. For instance belief revision can be carried out using different operators, while in argumentation alternative semantics for argument evaluation are available. General requirements of these general processes are formulated in terms of postulates or principles which can be used to discriminate or evaluate distinct realizations. In this respect, the historical evolution has been quite different in the two fields. In belief revision, AGM postulates [Alchourrón et al., 1985] have played a seminal role, so that postulate (re-)definition has been regarded as a central (and starting) point in all subsequent developments like belief updating, base revision, and so on.

At a more fundamental level, belief revision is built on two main principles: consistency, as already remarked, and *informational economy* (also called *minimal change* or *preservation criterion*). While the former does not require further comments, the latter corresponds to the idea that a reasoner should give up as little information as possible to recover consistency, or, to put it in other words, should give up a previous belief only when this is strictly necessary.

In abstract argumentation the identification and analysis of general principles for semantics evaluation and comparison [Baroni and Giacomin, 2007] has been undertaken more than a decade after Dung's seminal paper and checking the properties of a new semantics proposal with respect to these principles is not (yet) a standard practice though is receiving an increasing acceptance. For rule-based argumentation systems a set of so-called *rationality postulates* [Caminada and Amgoud, 2007] has been proposed in the same year.

Reviewing and comparing in detail postulates in the two fields is far beyond the scope of the present work, we therefore limit ourselves to some general observations.

First, we note that we can distinguish:

- informal principles, like informational economy;
- formal principles not involved with the underlying reasoning formalism, like semantics evaluation principles of [Baroni and Giacomin, 2007];
- formal principles concerning the underlying reasoning formalism, like consistency, the AGM postulates, and the rationality postulates of [Caminada and Amgoud, 2007].

Principles at different levels can not be directly compared, though in some cases they can be related each other.

Starting with informational economy, it has no direct counterpart in argumentation also because of basic differences in the two settings. In belief revision sentences are cancelled from the belief set, while in argumentation arguments change their justification status according to the adopted semantics, but, roughly speaking, nothing gets lost in the process. It can be noted however that, given the same argumentation framework, different semantics ascribe different justification status to the same arguments: some semantics tend to restrict the set of justified arguments while others are more "liberal" in this respect. This intuitive notion has a formal counterpart in the definition of *skepticism relations* between semantics [Baroni and Giacomin, 2009b]. While these relations have been introduced and analysed without reference to dynamic aspects of argumentation, relating them with informational economy is an interesting research issue.

Also, principles for abstract argumentation semantics [Baroni and Giacomin, 2007] have been conceived for a static setting and are not bound to any underlying logical representation. Some of them however bear some resemblance with notions in belief revision. For instance, the *language independence* principle (semantics evaluation at the abstract level is not influenced by differences in the representation at the language level) can somehow be put in relation with the AGM *extensionality* postulate (logically equivalent sentences give the same revision/contraction result). The *I-maximality* principle (no extension is a strict subset of another one) plays a role in the behavior of semantics in a setting of skeptical argument evaluation and can therefore be related with *informational economy*. *Reinstatement* (if the attackers of an argument are in turn attacked, the argument can be accepted) recalls the AGM postulate of *recovery* (if one removes and then adds a sentence, nothing of the initial belief set gets lost). To conclude this quick and really partial list of potential links, we already mentioned that the *conflict-free* property of extensions can be regarded as a weakening of the consistency requirement. These relationships, besides pointing forward to deeper investigations, suggest the opportunity that belief revision postulates are reconsidered at a more abstract logic-independent level, while argumentation principles are extended or reformulated to account for reasoning dynamics.

So called *rationality postulates* for rule-based argumentation have some stricter similarity with the belief revision setting [Caminada and Amgoud, 2007]. The postulate of *closure under strict rules* can be seen as a weakening of the assumption of closure of belief sets. So, on one hand it carries within argumentation the kind of "omniscience" problems we already mentioned, on the other hand, it points out again that distinguishing between facts and (different types of) rules may be very important. Similarly, the postulates of *direct* and *indirect consistency* provide a refinement of the consistency property. Direct consistency means that conclusions explicitly justified by the results of the argumentation process are consistent, while indirect consistency means that the closure under strict rules of the explicitly justified conclusions are consistent. Some rule-based argumentation formalisms fail to satisfy closure under strict rules and indirect consistency, but it can be (and it is) debated whether this is "a bug or a feature", in the sense that "blind" application of closure under strict rules can be regarded as basically in contrast with the nature of argumentation as an open process, where not only arguments but also conflicts have to be built incrementally and explicitly. Leaving apart conceptual issues, in [Caminada and Amgoud, 2007] it is shown that postulate satisfaction can be recovered through a proper (re-)definition of the notion of rebut (conflict) between arguments. This suggests that analogously, the notion of consistency could be reconsidered, so to open the way to some form of inconsistency-tolerant belief revision (for instance, see [Chopra and Parikh, 1999]).

In fact, looking deeper at the notion itself of consistency reveals another significant issue which is worth analysing. In belief revision, a sentence can be in one of three justification states depending on the membership of the sentence (and of its negation) to the belief set: accepted, rejected, or undecided. Clearly, a belief set can not include both a sentence and its negation and this is basically the only constraint for consistency. In extension-based argumentation the relationship between an argument and an extension (which is basically a conflict-free set of arguments) gives rise to the same justification states, while, equivalently, labelling-based argumentation assigns one out of three possible labels to each argument in a labelling. Most argumentation semantics prescribe however multiple alternative extensions (or labellings). If one adopts a skeptical attitude, they can be reduced to the case of a single extension/labelling by taking their intersection (this is done for instance in [Caminada and Amgoud, 2007]). The existence of multiple extensions allows however one to take a less drastic attitude and to consider a set of seven justification states (see [Baroni *et al.*, 2004; Baroni and Giacomin, 2009b; Wu and Caminada, 2010]), which basically correspond to the possible combinations of memberships/labels of an argument in different extensions/labellings (*i.e.*, to all non-empty subsets of {in, out, undec}).

In this more variegate model of justification, also the notion of consistency can assume a less "traditional" form, allowing various nuances. Whether this vision can be extended with beneficial effects to belief revision is a further stimulating question.

7.1 Summing up ...

Synthetic statements

- Postulates and principles are a fundamental element of classical belief revision theory (and of all its variations). They played a lesser role in the development of argumentation theory, some relatively recent proposals are receiving anyway significant (and increasing) attention.
- Postulates and principles formulated at different levels of abstraction and with reference to static vs. dynamic reasoning contexts are not directly comparable. It is possible however to identify some analogies and conceptual relationships

which (along with non-symmetrical gaps) suggest interesting opportunities of cross-fertilization between the two fields.

Questions

- Can a sufficiently abstract set of postulates for both static and dynamic aspects of inconsistency management processes be identified, so that current belief revision and argumentation theories are seen as special cases of a general framework?
- Is the notion of justification state a mere by-product of the available reasoning process models or can it be taken as a starting point to define requirements or postulates for the reasoning processes themselves?
- Can a spectrum of consistency notions and, consequently, of consistency recovery processes be identified?

8 Conclusions

Modeling reasoning processes is a formidable and fascinating enterprise. Belief revision and argumentation capture partially distinct but surely non-disjoint aspects of this research challenge, with many cross-fertilization opportunities and open questions this chapter has aimed, at least partially, to shed light on from an original perspective. As a final indication for future research, we suggest to both communities the importance of defining a set of reference reasoning cases, let's say of "reasoning benchmarks". In fact, using common benchmarks to compare the way different formal approaches deal with the same situation would be a very useful step in order to complement conceptual comparisons with concrete case-based counterparts.

Acknowledgments

The authors thank the anonymous referee for his/her helpful review.

References

[Alchourrón et al., 1985] C. E. Alchourrón, P. Gärdenfors, and D. Makinson. On the logic of theory change: Partial meet contraction and revision functions. *J. Symb. Log.*, 50(2):510–530, 1985.

[Baroni and Giacomin, 2007] P. Baroni and M. Giacomin. On principle-based evaluation of extension-based argumentation semantics. *Artif. Intell.*, 171(10-15):675–700, 2007.

[Baroni and Giacomin, 2009a] P. Baroni and M. Giacomin. Semantics of abstract argument systems. In Guillermo Simari and Iyad Rahwan, editors, *Argumentation in Artificial Intelligence*, pages 25–44. Springer, 2009.

[Baroni and Giacomin, 2009b] P. Baroni and M. Giacomin. Skepticism relations for comparing argumentation semantics. *Int. J. Approx. Reasoning*, 50(6):854–866, 2009.

[Baroni et al., 2004] P. Baroni, M. Giacomin, and G. Guida. Towards a formalization of skepticism in extension-based argumentation semantics. In *Proc. 4th Workshop on Computational Models of Natural Argument (CMNA 2004)*, pages 47–52, 2004.

[Baroni et al., 2007] P. Baroni, G. Guida, and M. Giacomin. A- and V-uncertainty: an exploration about uncertainty modeling from a knowledge engineering perspective. *International Journal on Artificial Intelligence Tools*, 16(2):161–194, 2007.

[Baroni et al., 2011] P. Baroni, M. Caminada, and M. Giacomin. An introduction to argumentation semantics. *Knowledge Eng. Review*, 26(4):365–410, 2011.

[Bench-Capon and Dunne, 2007] T. J. M. Bench-Capon and P. E. Dunne. Argumentation in artificial intelligence. *Artif. Intell.*, 171(10-15):619–641, 2007.

[Besnard and Hunter, 2001] P. Besnard and A. Hunter. A Logic-Based Theory of Deductive Arguments. *Artif. Intell.*, 128(1-2):203–235, 2001.

[Besnard and Hunter, 2008] P. Besnard and A. Hunter. *Elements of argumentation*. MIT Press, 2008.

[Birnbaum et al., 1980] L. Birnbaum, M. Flowers, and R. McGuire. Towards an ai model of argumentation. In *Proc. 1st National Conf. on Artificial Intelligence (AAAI)*, pages 313–315, 1980.

[Birnbaum, 1982] L. Birnbaum. Argument molecules: A functional representation of argument structure. In *Proc. 2nd National Conf. on Artificial Intelligence (AAAI)*, pages 63–65, 1982.

[Bondarenko et al., 1997] A. Bondarenko, P.M. Dung, R.A. Kowalski, and F. Toni. An abstract, argumentation-theoretic approach to default reasoning. *Artif. Intell.*, 93(1–2):63 – 101, 1997.

[Boutilier, 1998] C. Boutilier. A unified model of qualitative belief change: a dynamical systems perspective. *Artif. Intell.*, 98(1-2):281–316, 1998.

[Brewka, 2001] G. Brewka. Dynamic argument systems: A formal model of argumentation processes based on situation calculus. *J. Log. Comput.*, 11(2):257–282, 2001.

[Caminada and Amgoud, 2007] M. Caminada and L. Amgoud. On the evaluation of argumentation formalisms. *Artif. Intell.*, 171(5-6):286–310, 2007.

[Cayrol et al., 2010] C. Cayrol, F. Dupin de Saint-Cyr, and M.-C. Lagasquie-Schiex. Change in abstract argumentation frameworks: Adding an argument. *J. Artif. Intell. Res. (JAIR)*, 38:49–84, 2010.

[Chesñevar et al., 2000] C. I. Chesñevar, A. G. Maguitman, and R. P. Loui. Logical models of argument. *ACM Computing Surveys*, 32(4):337–383, 2000.

[Chopra and Parikh, 1999] S. Chopra and R. Parikh. An inconsistency tolerant model for belief representation and belief revision. In *Proc. 16th Int. Joint Conf. on Artificial Intelligence (IJCAI 99)*, pages 192–199, 1999.

[Cohen, 1981] R. Cohen. Investigation of processing strategies for the structural analysis of arguments. In *Proc. 19th Annual Meeting of the Association for Computational Linguistics*, pages 71–75, 1981.

[Darwiche and Pearl, 1997] A. Darwiche and J. Pearl. On the logic of iterated belief revision. *Artif. Intell.*, 89(1-2):1–29, 1997.

[Delgrande, 2012] J. P. Delgrande. Revising beliefs on the basis of evidence. *Int. J. Approx. Reasoning*, 53(3):396–412, 2012.

[Doyle and London, 1980] J. Doyle and P. London. A selected descriptor-indexed bibliography to the literature on belief revision. *SIGART Bull.*, (71):7–22, 1980.

[Doyle, 1979] J. Doyle. A truth maintenance system. *Artif. Intell.*, 12:231–272, 1979.

[Doyle, 1992] J. Doyle. Reason Maintenance and Belief Revision: Foundations versus Coherence Theories. In P. Gärdenfors, editor, *Belief Revision*, pages 29–51. Cambridge University Press, 1992.

[Dubois et al., 1996] D. Dubois, H. Prade, and P. Smets. Representing partial ignorance. *IEEE Trans. on Systems, Man, and Cybernetics, Part A*, 26(3):361–377, 1996.

[Dung, 1995] P. M. Dung. On the acceptability of arguments and its fundamental role in non-monotonic reasoning, logic programming, and n-person games. *Artif. Intell.*, 77(2):321–357, 1995.

[Fagin et al., 1983] R. Fagin, J. D. Ullman, and M. Y. Vardi. On the semantics of updates in databases. In *Proc. 2nd ACM SIGACT-SIGMOD Symp. on Principles of Database Systems*, pages 352–365, 1983.

[Fermé and Hansson, 1999] E. L. Fermé and S. O. Hansson. Selective revision. *Studia Logica*, 63(3):331–342, 1999.

[Friedman and Halpern, 1994] N. Friedman and J. Y. Halpern. A knowledge-based framework for belief change - part i: Foundations. In *Proc. 5th Conf. on Theoretical Aspects of Reasoning about Knowledge*, pages 44–64. Morgan Kaufmann, 1994.

[Fuhrmann, 1991] A. Fuhrmann. Theory contraction through base contraction. *J. Phil. Log.*, 20(2):175–203, 1991.

[García and Simari, 2004] A. J. García and G. R. Simari. Defeasible Logic Programming: An Argumentative Approach. *Theory and Practice of Logic Programming*, 4(1):95–138, 2004.

[Gärdenfors, 1988] P. Gärdenfors. *Knowledge in Flux: Modeling the Dynamics of Epistemic States*. MIT Press, 1988.

[Gärdenfors, 1990] P. Gärdenfors. The Dynamics of Belief Systems: Foundations vs.Coherence Theories. *Revue Internationale of Philosophie*, 44:24–46, 1990.

[Hansson, 1993] S. O. Hansson. Theory contraction and base contraction unified. *J. Symb. Log.*, 58(2):602–625, 1993.

[Hansson, 1999] S. O. Hansson. A survey of non-prioritized belief revision. *Erkenntnis*, 50:413–427, 1999.

[Hansson, 2011] S. O. Hansson. Logic of belief revision. In Edward N. Zalta, editor, *The Stanford Encyclopedia of Philosophy*. 2011.

[Hunter and Delgrande, 2011] A. Hunter and J. P. Delgrande. Iterated belief change due to actions and observations. *J. Artif. Intell. Res. (JAIR)*, 40:269–304, 2011.

[Jin and Thielscher, 2007] Y. Jin and M. Thielscher. Iterated belief revision, revised. *Artif. Intell.*, 171(1):1–18, 2007.

[Katsuno and Mendelzon, 1991] H. Katsuno and A. O. Mendelzon. On the difference between updating a knowledge base and revising it. In *Proc. 2nd Int. Conf. on Principles of Knowledge Representation and Reasoning (KR'91)*, pages 387–394, 1991.

[Liao et al., 2011] B. S. Liao, L. Jin, and R. C. Koons. Dynamics of argumentation systems: A division-based method. *Artif. Intell.*, 175(11):1790–1814, 2011.

[Loui, 1987] R. P. Loui. Defeat among arguments: a system of defeasible inference. *Computational Intelligence*, 3:100–106, 1987.

[Makinson, 1997] D. Makinson. Screened revision. *Theoria*, 63:14–23, 1997.

[Nebel, 1989] B. Nebel. A knowledge level analysis of belief revision. In *Proc. 1st Int. Conf. on Principles of Knowledge Representation and Reasoning (KR'89)*, pages 301–311, 1989.

[Nebel, 1992] B. Nebel. Syntax-based approaches to belief revision. In Peter Gärdenfors, editor, *Belief Revision*, pages 52–88. Cambridge University Press, 1992.

[Pollock, 1987] J. L. Pollock. Defeasible reasoning. *Cognitive Science*, 11(4):481–518, 1987.

[Pollock, 1992] J. L. Pollock. How to reason defeasibly. *Artif. Intell.*, 57(1):1–42, 1992.

[Prakken and Sartor, 1997] H. Prakken and G. Sartor. Argument-based extended logic programming with defeasible priorities. *J. Applied Non-Classical Logics*, 7(1), 1997.

[Prakken and Vreeswijk, 2001] H. Prakken and G. A. W. Vreeswijk. Logics for defeasible argumentation. In Dov M. Gabbay and F. Guenthner, editors, *Handbook of Philosophical Logic, Second Edition*. Kluwer Academic Publishers, Dordrecht, 2001.

[Prakken, 2006] H. Prakken. Combining sceptical epistemic reasoning with credulous practical reasoning. In *Proc. 1st Int. Conf. on Computational Models of Argument (COMMA 2006)*, pages 311–322, 2006.

[Prakken, 2010] H. Prakken. An abstract framework for argumentation with structured arguments. *Argument & Computation*, 1(2):93–124, 2010.

[Rahwan and McBurney, 2007] I. Rahwan and P. McBurney. Guest editors' introduction: Argumentation technology. *IEEE Intelligent Systems*, 22(6):21–23, 2007.

[Rahwan and Simari, 2009] I. Rahwan and G. R. Simari, editors. *Argumentation in Artificial Intelligence*. Springer, 2009.

[Rotstein et al., 2010] N. D. Rotstein, M. O. Moguillansky, A. J. García, and G. R. Simari. A dynamic argumentation framework. In *Proc. 3rd Int. Conf. on Computational Models of Argument (COMMA 2010)*, pages 427–438, 2010.

[Rott, 2001] H. Rott. *Change, choice and inference.* Oxford University Press, 2001.

[Simari and Loui, 1992] G. R. Simari and R. P. Loui. A mathematical treatment of defeasible reasoning and its implementation. *Artif. Intell.*, 53(2–3):125–157, 1992.

[Walton and Krabbe, 1995] D. N. Walton and E. Krabbe. *Commitment in Dialogue: Basic concept of interpersonal reasoning.* State University of New York Press, 1995.

[Walton, 1990] D. Walton. What is reasoning? what is an argument? *J. Phil. Log.*, 87:399–419, 1990.

[Walton, 1996] D. N. Walton. *Argumentation Schemes for Presumptive Reasoning.* Lawrence Erlbaum Associates, 1996.

[Wu and Caminada, 2010] Y. Wu and M. W. A. Caminada. A labelling-based justification status of arguments. *Studies in Logic*, 3(4):12–29, 2010.

Evolutionary Belief Change: Priorities, Conditionals and Argumentation

Alexander Bochman

Abstract We explore a representation framework for belief change that is based on the notion of a conditional (rule) as its main ingredient. We will argue that conditionals constitute the core of our epistemic states, while the latter underlie, in turn, what we perceive as belief changes. Moreover, epistemic states and their constitutive rules are relatively stable entities, so they are normally not changed by facts. As a result, normal (evolutionary) belief dynamics in this setting is representable using prioritized systems of rules and their expansions. We describe, in particular, two principal instantiations of such rule systems, preferential and argumentative (explanatory) ones, and discuss their properties, virtues and shortcomings.

1 Introduction

Belief change is an integral part of general human reasoning. Moreover, we will argue in this study that it is based, ultimately, on such principal forms of reasoning as inference and argumentation. Accordingly, the main objective of this study consists in exploring these reasoning foundations of the belief change process, as well as its connections with closely related areas of nonmonotonic reasoning and argumentation. This broader perspective will also naturally lead us to questions about proper foundations and necessary requirements for an adequate representation of belief change operations.

From its very beginning, it has been acknowledged in the belief revision literature that deductively closed belief sets are insufficient, taken by themselves, for determining 'rational' belief change operations - something else should be added to our beliefs, such as preferences, entrenchments, belief bases, etc., in order to guide us in choosing the right outcome. These auxiliary informational structures

Alexander Bochman
Computer Science Department, Holon Institute of Technology, Israel,
e-mail: bochmana@hit.ac.il

have been coined a generic name 'epistemic states'. According to this understanding, it is an epistemic state that determines the associated belief changes and, furthermore, provides the latter with a genuinely rational justification. Accordingly, it has been argued in [Bochman, 2001] that epistemic states and operations on them should be viewed as one of the main subjects of study in belief revision. Just as a more adequate understanding of chemical processes in Chemistry is achieved only by analysing them on the level of atoms, so a proper understanding of changes in beliefs can be obtained by viewing traditional belief change operations as only a derived output, or a by-product, of more complex operations on underlying epistemic states.

In this study, we will use a theory of epistemic change developed in [Bochman, 2001] in a systematic attempt to answer the following two basic questions:

- What are the main ingredients (informational units) of epistemic states that would provide an adequate basis for a *constructive* theory of epistemic change?
- What are the main operations on epistemic states that are sufficient for a comprehensive description of epistemic dynamics?

In an answer to the first question, we will argue that rules, or conditionals, should constitute an invariant core of epistemic states. More precisely, we will introduce the notion of a conditional epistemic state which is generated by a prioritized system of conditionals, and show that such epistemic states provide a proper basis for representing the usual belief change functions.

For the answer to the second question above, we will argue in this study that (conditional) epistemic states are relatively stable entities, so they are normally not changed by changes in beliefs or facts. More precisely, we will contend that normal, *evolutionary* dynamics of epistemic states is representable by cumulative expansions of prioritized systems of rules.

2 The Problem of Belief Change

The problem how to revise our beliefs in response to new information has long been an important self-contained area of research in philosophical and logical literature. It has received a more practical motivation, however, with the realization that this aspect of our intellectual activity has immediate relevance for artificial intelligence, and is tightly connected with nonmonotonic reasoning. As a result, belief revision is commonly considered now an integral part of the artificial intelligence research.

2.1 AGM and Base Change

Two main approaches have been suggested at the beginning stage of a formal belief change theory. The first is the so-called AGM theory of belief change, a starting point in the formal study of the problem. It has suggested a systematic approach to the problem both in terms of general rationality postulates that a belief change should satisfy, and in developing semantic representations that justify these postulates.

The AGM theory was intended to give a representation for a process of revising *belief states* considered simply as deductively closed sets of propositions, and it suggested a two-step procedure for accomplishing this task. First, the source belief state should be minimally contracted in order to make it compatible with a new belief, and then the contracted belief set should be expanded by simply adding the latter. It has turned out, however, that a straightforward, naive realization of this procedure immediately runs into difficulties. A reduction of a beliefs state in order to restore consistency with the new data usually does not give a unique result; in most cases we obtain a large number of maximal subsets that are compatible with the proposition being added. Furthermore, neither each such maximal subset taken alone (maxichoice contraction), nor their common part (full meet contraction) can serve as reasonable solutions; the first solution is inappropriate because it contains in general too many propositions, whereas the second one contains far less than what is expected. As a way out of this problem, the AGM theory suggested to use a *selection mechanism* allowing to choose among maximal subtheories of the belief set. In [Alchourrón et al., 1985] this was achieved using a suitable selection function on the maximal subsets of the source belief set; the contracted belief set was identified then with the intersection of all selected maximal theories (partial meet contraction). The authors have focused also on an important special case in which the selection function is relational, that is, based on a preference order among the subtheories. In addition, Gärdenfors has proposed the use of epistemic entrenchment relations as an alternative preference mechanism. As was noted in [Gärdenfors, 1990], the belief set coupled with the associated relation of epistemic entrenchment constitutes the second level of representing epistemic states in the AGM theory.

A major alternative approach to the problem of belief revision was based on the supposition that our corpus of beliefs is usually generated by some set of *basic* propositions (see [Hansson, 1999]). This base approach embodied an important aspect of our beliefs that did not find its proper place in the AGM theory, namely that some our beliefs are purely derivative and arise as logical consequences of other beliefs we have[1]. On this view, it is natural to require that such derived beliefs should be withdrawn when we remove beliefs that served as their justification. Accordingly, contractions and expansions of belief sets were determined on this approach by contraction and expansion of their underlying bases. This drastically reduced the set of alternatives and avoided, in particular, trivialization of both maxichoice and full meet revision. Still, the resulting belief set on most of these accounts was also determined by imposing preference relations on such alternatives; the latter were purported to resolve residual choice problems that were not decided by bases alone.

---

[1] Cf. [Doyle, 1992] for a general distinction between foundations and coherence approaches to belief formation.

2.2 Iterated Change and Categorial Matching

Both the AGM and the base approach have agreed that the set of beliefs alone is insufficient for grounding a reasonable belief revision process; some more structure needs to be discerned from our epistemic states in order to guide decisions about what to retain and what to retract in the process. This understanding has created, however, problems of its own.

Belief change is an iterative process; we constantly obtain new data that require, from time to time, new revisions of our beliefs. In order to perform such changes, we should have at each point not only the current belief set, but also the current choice mechanism that will determine the next revised belief set. Unfortunately, the source AGM theory does not account for this situation. Though the revised belief set is determined by the initial belief set and its associated selection function, the corresponding constructions do not (and have no means to) determine the preference structure associated with the resulting belief set. This is the source of the well-known shortcoming of the AGM theory, namely its inability to deal with iterated belief change.

A most natural solution to this problem consists in raising belief change operations to functions that take a new belief and an entire epistemic state as arguments and produce a new, revised epistemic state. The resulting framework will satisfy then the *principle of categorial matching*, stated in [Gärdenfors and Rott, 1995], according to which the representation of the epistemic state after a change should be of the same kind as that before the change. Thus, the AGM framework can be modified to provide a representation for changes that revise not only belief sets, but also their associated entrenchment orderings. The idea that revision should be performed on epistemic entrenchment relations rather than on belief sets was expressed first in [Rott, 1991]. Actually, a great number of different 'entrenchment revision' models of this kind have been proposed since then.

The 'pure' base approach (without preferences) conforms to the principle of categorial matching, since belief change operations are defined as operations on belief bases, rather than on belief sets. However, if we impose a preference (or selection) mechanism on the bases, the situation will become similar to the AGM approach, namely we have to determine also how belief change operations influence the associated preferences.

Unfortunately, the decision to replace only the belief sets with epistemic states as arguments of belief change operations still does not resolve all the problems. Indeed, an epistemic state is a structured entity that assigns, in particular, both beliefs and non-beliefs particular places in its preferential structure. Accordingly, the intended belief change function should not only assign an input proposition a status of belief, but also, and most importantly, it should assign it an appropriate place in the resulting preferential structure. Now, the problem here is that usually there is a number of possibilities for such a positioning, each of them compatible with the acceptance of an input proposition as a new belief. The differences between these options will be revealed, however, in subsequent changes that could be made to the revised epistemic state. And the difficulty is that the input proposition, taken alone, does not provide us with sufficient information for choosing among these options.

As has been argued in [Bochman, 2001], the above considerations actually challenge even the traditional understanding of belief *expansion* as a straightforward incorporation of a new belief. Speaking more generally, these considerations strongly suggest that the original AGM idea of defining belief change operations directly on propositional beliefs is problematic, and it should be replaced with a view that such operations are only a derived by-product of more fundamental operations on epistemic states. The necessity of such a conceptual shift stems also from principles of informational economy and minimality of change discussed below.

2.3 Informational economy and minimality of change

Two related principles, informational economy and minimality of change, lie at the heart of practically all approaches to belief change. The principle of informational economy says that information is valuable, and hence removal of information in revisions should be as small as possible (given the aim of restoring consistency). A more general principle of minimality of change requires that any kind of change made to belief or epistemic state should be a minimal change that would achieve its purposes.

An interesting puzzle arises with the above principles in the AGM framework. A straightforward application of informational economy points out to (preferred) maximal subtheories of the belief set as the most appropriate options for choice. The suggested partial meet construction takes, however, intersection of such theories as the final solution. As was rightly noted in [Levi, 1991], such a solution violates the very principle of informational economy it was based on, since this intersection is already not maximal by itself, and hence does not preserve as much information as possible.

There is more to the above argument than an intellectual puzzle. True, facing a number of acceptable alternatives, it is reasonable to believe only what is common to all of them. However, if this is all we will remember after the change, we will put ourselves into risk of losing important information *for subsequent changes*. Thus, it may well happen that subsequent data will show that some of these alternatives were wrong, and hence they should not be taken into account. In such a case keeping in mind the initial options becomes paramount for producing a more adequate and informed solution.

The above problem acquires an especially vivid form in the base approach. Suppose we have to contract $A \wedge B$ from the belief set generated by the base $\{A,B\}$. At the first step, we retreat to the two sub-bases $\{A\}$ and $\{B\}$ of the original base that do not imply $A \wedge B$. But if both these sub-bases are equally preferred, we have to form their intersection which happens to be empty! In other words, we have lost all the information contained in the initial base, so all subsequent changes should start from a scratch.

It seems reasonable to expect that the contracted belief set in the above situation should contain $A \vee B$, since each of the alternatives support this belief. In fact, this result is naturally sanctioned by the AGM theory. However, it seems also reasonable to require that if subsequent evidence rules out A, for example, we should believe that B. In other words, contracting first $A \wedge B$ and then A from the initial belief state should make us believe in B. Note that the AGM theory cannot produce this result, because the first contraction gives the closure of $A \vee B$ as the contracted belief set, and hence the subsequent contraction of A will not have any effect on the corresponding belief state. Thus, reduction of options to the intersection of alternatives may lead to a loss of information for subsequent changes.

The main conclusion that could be made from the above discussion is that belief change operations cannot be restricted to manipulations on belief sets, or even plain belief bases; they have to use a more elaborate structure of our epistemic states.

2.4 Preferential Epistemic States

A particular formalization of the notion of an epistemic state was suggested in [Bochman, 2001] as a general response to the problems with belief change sketched in the preceding sections. Among other things, it has allowed to unify the coherentist AGM theory and foundationalist base-oriented approach. Moreover, the same notion of an epistemic state was employed in the book as a uniform semantic framework for various kinds of nonmonotonic inference, confirming the general idea of [Makinson and Gärdenfors, 1991], according to which belief change and nonmonotonic reasoning are "two sides of the same coin".

Definition 1. A *general epistemic state* is a triple $(\mathscr{S}, l, \prec)$, where $\mathscr{S}$ is a set of *admissible belief states*, $\prec$ is a preference relation on $\mathscr{S}$, while l is a labeling function assigning a deductively closed theory (called *an admissible belief set*) to every belief state from $\mathscr{S}$.

The AGM approach can be modeled in the above framework by viewing the belief set and all its deductively closed subsets as admissible belief states, with labeling being a trivial identity function. On the other hand, in the base approach admissible belief states are all the subsets of the source belief base, while the labeling function assigns each such subset its deductive closure. Speaking more generally, admissible belief states represent what is considered by the agent as serious possibilities for choice (using the terminology of [Levi, 1991]), while sets of propositions supported by such admissible belief states will represent admissible belief sets that are actually envisaged by the agent. Notice, however, that we may have different admissible belief states that generate identical belief sets. Moreover, in many situations we may have several maximally preferred admissible belief states, forming a basis for different plausible sets of beliefs we can hold in such situations. This possibility is excluded, however, both by the AGM and base approaches, which is one of the main reasons why neither AGM-type states, nor base-generated epistemic states are sufficient for hosting a general theory of belief change, at least if it is required to satisfy all the desiderata mentioned earlier. Moreover, this broader understanding also naturally connects epistemic states with common systems of nonmonotonic reasoning that admit multiple models, such as an abductive system of [Poole, 1988], and even Reiter's default logic [Reiter, 1980].

An important aspect of the above notion of an epistemic state concerns the very distinction between admissible belief states and their associated belief sets. Taken in its full generality, this distinction sanctions a broader view of epistemic states advocated, for instance, by Isaac Levi in a number of works. According to this view, epistemic states are essentially non-linguistic entities, and they can be described, in general, only partly by available language means. In other words, current descriptions of epistemic states are only 'approximations' to the full information contained in them. This broader view creates an important degree of freedom in our understanding of epistemic states, and it can be explored by extending our expressive capabilities. Moreover, the starting point of this exploration is also an acknowledgment that general epistemic states as defined above should be seen as only a derived output of more basic informational units we actually have at our disposal. In other words, though quite useful as a theoretical construct, general epistemic states are not suitable as a primary reasoning framework for representing and resolving actual problems of belief change. At this point, the connection between belief change and nonmonotonic reasoning will turn out to be useful in singling out the basic informational units that would provide a more modular description of general epistemic states.

2.5 *Nonmonotonic Reasoning: The Second Side of the Coin*

A direct connection between belief change and nonmonotonic reasoning is provided by the famous *Ramsey test* for conditionals. More specifically, the test establishes a natural two-way correspondence between belief revision operations and *nonmonotonic conditionals*, one of the key notions in nonmonotonic reasoning:

> **Ramsey test**. *A conditional A/B holds in an epistemic state $\mathcal{E}$ if and only if B is believed in the epistemic state $\mathcal{E}*A$ obtained by revising $\mathcal{E}$ with A.*

Under the above correspondence, the AGM postulates for revision correspond precisely to the postulates of so-called rational inference relations (see [Kraus *et al.*, 1990]). As a matter of fact, similar conditional counterparts can be constructed for the other two basic belief change operations of AGM, namely expansions and contractions. Thus, expansions naturally correspond in this sense to classical (material) implications, while contraction operations correspond to what have been called contraction inference relations in [Bochman, 2000].

By its very design, the Ramsey test provides an operational semantics for conditionals in terms of corresponding revision operations on epistemic states. However, coupled with a specific definition of these operations in epistemic states, this operational test can be immediately transformed into a full-fledged semantic definition of conditionals. As a result, epistemic states can also serve as a representation framework for nonmonotonic inference. For instance, a definition of the revision operation employed in [Bochman, 2001] corresponds to a semantic condition according to which a conditional A/B holds in an epistemic state if and only if $A \supset B$ holds in all preferred admissible states consistent with A. This semantic definition corresponds precisely to the system P of preferential inference from [Kraus et al., 1990].

The connection between belief change and nonmonotonic inference provides us with a broader perspective on both these theories. First of all, it shows that an epistemic state is not an auxiliary, ad hoc construct serving the only purpose of representing belief change functions, but rather a versatile representation framework capable to host a broad range of our reasoning activities. In addition, the very possibility of using epistemic states in nonmonotonic reasoning provides us with further insights about potential instantiations and constructions of epistemic states themselves. Thus, as in nonmonotonic reasoning, admissible belief states of a general epistemic state can be viewed as formed by admissible combinations of *default beliefs* that we are willing to accept in the absence of evidences to the contrary. Accordingly, the models and methods of nonmonotonic reasoning acquire immediate relevance for the theory of belief change.

It is important to note, however, that a general comparison between nonmonotonic reasoning and belief change reveals also significant differences between them. To begin with, nonmonotonic conditionals cover only one-step, non-iterated belief revision, whereas one of the central tasks of a general belief change theory consists in describing iterations of belief change operations. Moreover, this apparently technical distinction is actually a by-product of more profound differences in the reasoning tasks posed by the two theories. Generally speaking, we argued elsewhere that a theory of nonmonotonic reasoning provides a formal representation for a reasoning process of *belief formation* that occurs as a result of combining facts (evidence) with default beliefs, or assumptions, represented by epistemic states. In other words, the aim of this reasoning process consists in answering the question "What should I believe if A would happen to be the case?" This process is conceptually different from traditional understanding of belief change operations in that it does not involve real changes in epistemic states; all it requires is a *suppositional change* (see [Levi, 1997]), in which the input proposition (the evidence) is temporarily assumed as true. Clearly, supposing that a proposition is true (for the purposes of argumentation, or as holding in a given situation) is different from "actually coming to believe the proposition in earnest", using Levi's phrase. In particular, changing the evidence does not necessarily lead to changing epistemic states, since the latter contain general default information that is applicable in many factual circumstances.

Despite the above mentioned differences, a most important contribution of nonmonotonic reasoning theories consists in providing a constructive view of epistemic states in terms of more basic notions. Such a constructive instantiation of epistemic states could be described in general terms as follows.

To begin with, on the 'nonmonotonic' interpretation of epistemic states, their admissible belief states are generated by allowable combinations of default beliefs. Clearly, this makes such epistemic states a natural instantiation of the base approach to belief change.

The second ingredient of general epistemic states, namely the preference relation on admissible belief states, reflects now the fact that not all combinations of defaults constitute equally preferred options for choice. For example, defaults are presumed to hold, so an admissible belief state generated by a larger set of defaults is preferred to an admissible state generated by a smaller set of defaults. In other words, the preference relation should be monotonic on default sets. This leads us to the following intermediate definition.

Definition 2. An epistemic state is *base-generated* by a set Δ of propositions with respect to a classical Tarski consequence relation Th if

- its admissible states are subsets of Δ;
- l is a function assigning each $\Gamma \subseteq \Delta$ a theory $\text{Th}(\Gamma)$;
- the preference order is monotonic on the subsets of Δ: if $\Gamma \subset \Phi$, then $\Gamma \prec \Phi$.

Now, a second important constructivization step amounts to different ways of constructing the preference order on sets of defaults from a more basic *priority relation* among individual defaults. For instance, in quantitative/probabilistic approaches to belief change, such a priority relation is a direct by-product of assigning values to all propositions of the language. In a more general relational case of partial orders, however, this task of constructing a preference relation on sets turns out to be a special case of a general problem of combining a set of preference relations into a single 'consensus' preference order. Based on the fundamental results of [Andreka et al., 2002], it has been shown in [Bochman, 2001] that a most natural and justified way of doing this is as follows.

Suppose that the set of defaults Δ is ordered by some *priority relation* $\lhd$ which will be assumed to be a strict partial order: $\alpha \lhd \beta$ will mean that α *is prior to* β. Then the resulting preference relation on sets of defaults can be defined in the following way:

$$\Gamma \prec \Phi \equiv (\forall \alpha \in \Gamma \backslash \Phi)(\exists \beta \in \Phi \backslash \Gamma)(\beta \lhd \alpha) \qquad \text{(Consensus order)}$$

$\Gamma \prec \Phi$ holds when, for each default in $\Gamma \backslash \Phi$, there is a prior default in $\Phi \backslash \Gamma$. [Lifschitz, 1985] was apparently the first to use this construction in prioritized circumscription, while [Geffner, 1992] employed it for defining preference relations among sets of default conditionals.

The construction of epistemic states from prioritized sets of defaults, sketched above, provides a basic step toward transforming epistemic states from a purely descriptive, theoretical formalism into a constructive working tool for representing and (dynamic) reasoning with beliefs. By this construction, a proper representation of both belief change and belief formation problems boils down to singling out an appropriate set of defaults and establishing priorities among them. Accordingly, the pair $(\Delta, \lhd)$ consisting of a default set and a priority relation on it constitutes on this view an ultimate information we have to possess in order to resolve these reasoning tasks. In other words, default bases can be viewed as primary informational units in the process of belief change.

And finally, the last step in the process of constructivization of epistemic states we are suggesting amounts to a particular elaboration of the notion of a default belief. The accumulated experience in nonmonotonic reasoning implies that predominant default information we use in our reasoning comes in the form of 'normality' conditionals "If A, then normally B". Furthermore, by the Ramsey test, information about such conditionals is implicit, in a sense, in our very ability to revise epistemic states in response to new information. In other words, if we know how to revise epistemic states, we should know also what are the conditionals that hold in them. All this strongly suggests that conditionals, or rules, of the form "If A, then normally B" should constitute the core of our epistemic states, namely primary informational units they are built from. Accordingly, we contend that such conditionals should serve as default beliefs in the above construction of epistemic states.

It should be noted from the outset that the restriction of default beliefs to conditionals does not reduce expressivity of our descriptions as compared with traditional models of belief change. Indeed, on most interpretations of conditionals, ordinary propositional beliefs are readily expressible as conditionals of the form t/A, where t is a truth constant. Moreover, the language of conditionals allows us even to express a distinction between such beliefs and 'hard facts', namely immutable facts (knowledge) that cannot be changed in belief revision. Such facts can be encoded using conditionals of the form $\neg A/f$, where f is a falsity constant.

2.6 Conditional Epistemic States

In order to provide a formal, yet sufficiently general, representation of epistemic states that is based on conditionals, we will extend the usual propositional language L of classical logic by adding conditionals as propositions of a new kind[2]. Formally, we add to L new propositional atoms of the form A/B, where A and B are classical propositions of L. The conditionals A/B will be viewed as propositional atoms of a new type, so nesting of conditionals will not be allowed. Still, the new propositional atoms can be freely combined with ordinary ones using the classical propositional connectives. We will denote the resulting language by L_c.

For a set Γ of conditionals, we will denote by $\vec{\Gamma}$ the corresponding set of material implications:

$$\vec{\Gamma} = \{A \supset B \mid A/B \in \Gamma\}.$$

Now, by a *conditional base* we will mean a pair $(\Delta, \lhd)$, where Δ is a set of conditionals, and $\lhd$ is a priority relation on Δ. Finally, by an epistemic state generated

---

[2] See [Kern-Isberner, 2004] for a similar construction, though in a more specific framework of quantitative, totally ordered structures.

by a conditional base $(\Delta, \lhd)$ (in short, a *conditional epistemic state*) we will mean any epistemic state $(\mathscr{S}, l, \prec)$ that satisfies the following conditions:

- Admissible belief states are subsets of Δ;
- $\prec$ is a monotonic preference relation on $\mathscr{P}(\Delta)$ that subsumes $\lhd$: if $\alpha \lhd \beta$, then $\{\beta\} \prec \{\alpha\}$;
- $l(\Gamma) = \text{Th}(\overrightarrow{\Gamma})$, for any $\Gamma \subseteq \Delta$;

The first two conditions correspond to general conditions on base-generated states, described earlier. The third condition reflects a very strong, though natural requirement that the propositional content of conditionals (i.e., the beliefs they support) amounts to the corresponding classical material implications (though this does not imply that they are reducible to the latter).

The above definition of a conditional epistemic state allows us to state now what can be seen as the main problem for both nonmonotonic reasoning and belief revision. The starting point of this problem is an observation that the above description of a conditional epistemic state does not determine, in general, a unique epistemic state for a particular conditional base. More specifically, what is missing in the above definition is a systematic and principled construction of a preference order from the source priorities among conditionals. In fact, different constructions of this kind has been suggested in the literature, corresponding to different sets of output epistemic states.

From the belief revision point of view, conditionals that hold in a generated conditional epistemic state determine one-step revisions of this epistemic state (by the Ramsey test). On the other hand, from the point of view of the theory of nonmonotonic inference, the same set of conditionals can be viewed as (nonmonotonic) consequences of the source conditional base. Consequently, a particular construction of such epistemic states determines a specific theory of *defeasible entailment* among conditionals. Moreover, it has been argued in [Bochman, 2011] that this problem is intimately related to the main problem of nonmonotonic reasoning, namely a general selection problem for defaults. Hence the problem of constructing an epistemic state from a conditional base can indeed be viewed as the main problem for both nonmonotonic inference and belief revision.

2.7 Theories of Defeasible Entailment

Daniel Lehmann, one of the authors of [Kraus et al., 1990], was also one of the first to realize that logical systems of preferential entailment are not sufficient, taken by themselves, for capturing reasoning with default conditionals and, moreover, that we cannot hope to overcome this by strengthening these systems with additional axioms or inference rules. Instead, he suggested a certain nonmonotonic semantic construction, called *rational closure*, that allows us to make appropriate default conclusions from a given set of conditionals (see [Lehmann, 1989; Lehmann and Magidor, 1992]). An essentially equivalent, though formally very different, construction has been suggested in [Pearl, 1990] and called system Z. Unfortunately, both theories have turned out to be insufficient for representing defeasible entailment, since they have not allowed to make certain intended conclusions. Hence, they have been refined in a number of ways, giving such systems as lexicographic inference [Benferhat et al., 1993; Lehmann, 1995], and similar modifications of Pearl's system. Unfortunately, these refined systems have encountered an opposite problem, namely, together with some desirable properties, they invariably produced some unwanted conclusions. A more general, and apparently more justified, approach to this problem in the framework of preferential inference has been suggested in [Geffner, 1992].

Yet another, more syntactic, approach to defeasible entailment has been pursued, in effect, in the framework of an older theory of nonmonotonic inheritance (see [Horty, 1994]). Inheritance reasoning deals with a quite restricted class of conditionals constructed from literals only. Nevertheless, in this restricted domain it has achieved a remarkably close correspondence between what is derived and what is expected intuitively. Accordingly, inheritance reasoning has emerged as an important test bed for adjudicating proposed theories.

Despite the diversity, the systems of defeasible entailment have a lot in common, and take as a starting point a few basic principles. In fact, practically all of them can be described as particular constructions of conditional epistemic states as defined above. Moreover, all such systems require that an intended epistemic state generated by a conditional base $(\Delta, \triangleleft)$ should make valid all the conditionals from Δ in accordance with the semantic definition sanctioned by the Ramsey test:

Direct Inference If $A/B \in \Delta$, then $A \supset B$ should hold in all preferred admissible states consistent with A.

A good example of such a construction, as well as the problems associated with it, is provided by Geffner's theory of conditional entailment from [Geffner, 1992]. Starting from a plain conditional base without any priorities, the theory of conditional entailment imposes prioritization by making use of the following relation among conditionals:

Definition 3. A conditional α *dominates* a set of conditionals Γ if the set of implications $\{\overrightarrow{\Gamma}, \overrightarrow{\alpha}\}$ is incompatible with the antecedent of α.

The origins of this relation can be found already in [Adams, 1975], and it has been used in practically all studies of defeasible entailment, including the notion of preemption in inheritance reasoning. A suggestive reading of dominance says that if α dominates Γ, it should have priority over at least one conditional in Γ. Among other things, this secures the above condition of Direct Inference, namely that α will hold in the resulting epistemic state. Accordingly, a priority order on the default base is called *admissible* if, for any conditional α and any set of conditionals Γ from the base, if α dominates Γ, then it is prior to at least one conditional in Γ. Then the intended models can be identified with epistemic states that are generated by all admissible priority orders on the default base (using the construction of a consensus preference order, given earlier).

Conditional entailment has shown itself as a serious candidate on the role of a general theory of defeasible entailment. Nevertheless, Geffner himself has shown that it still does not capture some desired conclusions, for which he suggested to augment it with an explicit representation of causal reasoning. The main point of the generalization was to introduce a distinction between propositions and facts that are plainly true versus facts that are *explainable* (caused) by other facts and rules. Actually, the causal generalization suggested by Geffner in the last chapters of his book has served, in part, as an inspiration for a causal theory of reasoning about actions and change (see [McCain and Turner, 1997]), a formalism that falls out of the preferential approach.

Yet another problem with conditional entailment stems from the fact that it does not capture inheritance reasoning. The main difference between the two theories is that conditional entailment is based on absolute, invariable priorities among defaults, while nonmonotonic inheritance determines such priorities in a context-dependent way, namely in presence of other defaults that provide a (preemption) link between two defaults (see [Dung and Son, 2001]). In fact, in an attempt to resolve the problem of defeasible entailment, it has been shown in [Bochman, 2001] that nonmonotonic inheritance is representable, in principle, by epistemic states ordered by *conditional* (context-dependent) priority orders. Unfortunately, the emerged construction could hardly be called simple or natural.

Speaking more generally, the above shortcomings and discrepancies arising with the theory of conditional entailment actually cast some general doubts about the preferential approach itself as a fully adequate framework for representing nonmonotonic inference, and hence belief revision. But before we turn to alternative possibilities, we should discuss how conditional epistemic states can be used for representing 'standard' and iterated revisions.

3 Evolutionary Belief Change

The main point we argued for in the introductory sections was that, in order to satisfy the required desiderata, belief change operations should preserve much more information about epistemic states than what the initial representations of belief change, namely AGM and base change, allow us to preserve. We have seen also that nonmonotonic reasoning deals primarily with suppositional belief change in which epistemic states are normally not changed by the facts. Moreover, this persistence is grounded in the very nature of epistemic states as vehicles for encoding general information about the world, and it is especially vivid in the construction of conditional epistemic states that are generated by generally applicable rules, or conditionals. Hence it comes as no surprise that a similar view on epistemic states has a long history also in the theory of belief revision. Thus, Peter Gärdenfors came very close to this view in [Gardenfors, 1988, pp. 87–88], where he argued that the relation of epistemic entrenchment can be established independently of what happens to the belief set in contractions and revisions. Though Gärdenfors did not exclude changes to the entrenchment relation, he compared such changes with scientific revolutions and paradigm shifts.

It turns out, however, that a traditional understanding of belief revision requires a certain modification of the above persistence claim about epistemic states, because the latter do change as a result of such revisions. Still, we are going to define below a broad class of belief change operations that preserve, in a sense, the original epistemic states, but only expand them with new beliefs.

3.1 Belief Revision via Base Expansions

On the standard AGM understanding of belief revision, an input proposition is viewed as a new *general belief* on a pair with the source beliefs accumulated in the current epistemic state. More precisely, it is viewed as a belief that is superior to these 'old' beliefs (in the case it is in conflict with the latter). Now, in the original, relatively poor AGM context of propositional belief sets, some of these old beliefs have to be removed in order to accommodate the new one. In the framework of conditional epistemic states, however, a more conservative, and arguably more adequate, procedure suggests itself. Namely, in order to revise a conditional epistemic state with a new belief, it is sufficient just to expand the default base of this epistemic state with this new belief, coupled with a proviso that the new belief will be prior to the old beliefs in the source base. Then, on any reasonable construction of conditional epistemic states from default bases, the new proposition will belong to all preferred admissible states, and hence will be believed in the revised epistemic state.

A formal description of the above procedure can be given as follows.

Definition 4. Let $\mathcal{E}$ be a conditional epistemic state generated by a conditional base $(\Delta, \triangleleft)$. Then a *revision* of $\mathcal{E}$ with a new belief A is a conditional epistemic state $\mathcal{E} * A$ generated by the base $(\Delta \cup \{t/A\}, \triangleleft')$, where $\triangleleft' = \triangleleft \cup \{t/A \triangleleft \alpha \mid \alpha \in \Delta\}$.

By the above definition, a revision of a conditional epistemic state with a new belief A amounts to a plain *expansion* of its conditional base Δ with a new conditional t/A, coupled with an extension of the associated priority relation that makes t/A prior to all conditionals in Δ.

The first thing we should note is that changes in propositional belief sets determined by the above definition coincide with standard AGM revisions (for partially ordered structures), and thereby with belief revision functions produced by suppositional changes. The obvious difference between them, however, is that whereas suppositional change retains the source epistemic state intact, belief revision produces a new epistemic state. Still, an important feature of the above construction is that it preserves the source epistemic state in its entirety, but only expands it with new conditionals. This modified preservation property arguably has far-reaching consequences for a general understanding of belief change. In what follows, we will call changes of epistemic states based on such revisions *evolutionary* belief changes.

Historical remark. Ideas similar to the above notion of evolutionary change can be found already in the concept of vertical revision for prioritized bases, described in [Rott, 2001]. A more elaborate concept of 'projective default epistemology' has been developed in [Weydert, 2005] using the framework of ranking measures.

3.2 Towards a General Theory: Merge Operations

As a matter of fact, the above definition of revisions makes them a very special case of a broad range of *merge operations* on epistemic states that were described in [Bochman, 2001]. More precisely, revisions in this sense are a special case of a prioritized merge of two epistemic states that can be defined as follows:

Definition 5. Let $\mathscr{E}_1$ and $\mathscr{E}_2$ be conditional epistemic states generated, respectively, by conditional bases $(\Delta_1, \triangleleft_1)$ and $(\Delta_2, \triangleleft_2)$ such that Δ_1 and Δ_2 are disjoint sets. Then a *prioritized merge* of $\mathscr{E}_1$ with $\mathscr{E}_2$ is a conditional epistemic state $\mathscr{E}_1 * \mathscr{E}_2$ that is generated by the conditional base $(\Delta_1 \cup \Delta_2, \triangleleft_0)$, where

$$\triangleleft_0 = \triangleleft_1 \cup \triangleleft_2 \cup \{\alpha \triangleleft \beta \mid \alpha \in \Delta_2, \beta \in \Delta_1\}.$$

Prioritized merge combines, in effect, a source conditional base with a new one in such a way that every conditional from the new base is prior to any conditional from the source base. Clearly, the revision operation, defined earlier, corresponds to a prioritized merge of an epistemic state $\mathscr{E}$ with a 'rudimentary' epistemic state $\mathscr{E}_A$ generated by a conditional base $(t/A, \emptyset)$ with an empty priority relation.

Example 1. Let us illustrate the above construction with a very simple example involving an addition of a new conditional belief A/B to a conditional base that contains A/C, where B is logically incompatible with C. Then both conditionals are potentially applicable in the same situations (since they have the same antecedents), but since the new rule has priority over the old one, the latter will never be applied. In other words, though both conditionals belong to the 'revised' default base, the new conditional A/B eventually disables the old one A/C. In fact, A/C could even be 'contracted' from the conditional base insofar as the new conditional A/B remains in force. Such a contraction would make, however, difference if A/B will be disabled, in turn, by subsequent revisions. According to our definition, this might lead to a situation in which A/C could be applicable again. Furthermore, such a contraction of previous conditionals appears to be even less appropriate in the case when the source conditional base contains a conflicting conditional with a different antecedent, say D/C, where, as before, B is incompatible with C, but D is compatible with A. Then again, an addition of A/B will lead to an epistemic state in which D/C will be disabled by A/B in all situations in which the latter is applicable. The contraction of D/C from the resulting conditional base, however, would disable the latter also in situations when only the latter is applicable. This would obviously be an unjustified loss of information.

The above revision operation, and even more general prioritized merge operations, were defined as operations on epistemic states, so they can be readily iterated. Moreover, iterated revisions of this kind can be described 'globally' by layered (stratified) priority structures, in which the first layer represents the source default base, while the higher layers represent subsequent revisions made to it. As a matter of fact, structures of this kind are well known in the logic programming counterpart of belief revision theory, called *dynamic logic programming* (see, e.g., [Leite, 2003]). In this theory, an iterated revision of a logic program P is defined directly as a sequence of logic programs $(P, P_1, \ldots, P_n)$, such that program rules from a layer P_j are considered as having higher priority than program rules from lower layers P_i for $i < j$. Moreover, the construction of semantic models for such sequences is guided by the *causal rejection principle*, according to which program rules are always applicable whenever not refuted by more recent rules.

In addition to the prioritized merge operation, an important role in the framework described in [Bochman, 2001] was played by a simpler operation of non-prioritized, or pure, merge that can be described as follows:

Definition 6. Let $\mathscr{E}_1$ and $\mathscr{E}_2$ be conditional epistemic states generated, respectively, by conditional bases $(\Delta_1, \triangleleft_1)$ and $(\Delta_2, \triangleleft_2)$ such that Δ_1 and Δ_2 are disjoint sets. Then a *pure merge* of $\mathscr{E}_1$ with $\mathscr{E}_2$ is a conditional state $\mathscr{E}_1 * \mathscr{E}_2$ that is generated by the conditional base $(\Delta_1 \cup \Delta_2, \triangleleft)$, where

$$\triangleleft = \triangleleft_1 \cup \triangleleft_2.$$

The pure merge operation corresponds to a natural informal understanding of information merge in which the two epistemic states are viewed as equally valuable sources of information. Accordingly, it does not stipulate any priorities among defaults from different bases. As a result, a pure merge of epistemic states leads, in general, to a 'contraction' of the belief sets associated with these epistemic states, to their intersection.

As follows from the results stated in [Bochman, 2001], the two basic merge operations, defined above, constitute, in effect, a 'functionally complete' set of operations on conditional epistemic states in the sense that any conditional base can be constructed from single conditionals using pure and prioritized merge only. More precisely, the following result can be proved[3]:

Proposition 1. *Any finite conditional epistemic state is constructible using the two basic merge operations.*

The above result can be viewed as a formal justification of the claim that conditional epistemic states, taken together with two merge operations, pure and prioritized merge, constitute a comprehensive representation framework for evolutionary belief change.

Finally, in order to situate the above notion of evolutionary change in the general theory of belief change, we will follow Peter Gärdenfors in adapting a general view

---

[3] In fact, it is an immediate consequence of Theorem 14.4.3 in [Bochman, 2001].

of scientific development from [Kuhn, 1962] to a much more specific field of belief dynamics. More precisely, we argue that conditional epistemic states and associated evolutionary belief change provide a representation for an overwhelming majority of *normal*, or regular, belief change processes. The main characteristic property of such processes is *preservation*, or accumulation: though an evolutionary revision may lead to radical changes in propositional beliefs, no information is actually lost in such changes. A conditional rule, once incorporated into a default base, remains valid forever, though it can be eventually 'disabled' by subsequent rules.

3.3 On Contractions and Revolutions

A distinctive feature of evolutionary belief change is that it does not use contractions in order to preserve consistency of resulting epistemic states; this is achieved, instead, by stipulating priority of newly introduced (conditional) beliefs over previous ones. Nevertheless, the framework of evolutionary change and associated formalism of prioritized conditional bases arguably allows us to represent the majority of ordinary belief change processes, as well as to provide a basis for a theory of nonmonotonic inference. This situation obviously requires us to reconsider the role of the traditional contraction operation in general belief change dynamics.

To begin with, there is an obvious peculiarity to contraction as a conceptually independent operation. Thus, it has often been noted in the literature that it is extremely difficult to construct natural, 'real-life' examples of contractions as stand-alone operations; in most cases contractions are discerned only as an intermediate step in more complex changes, such as revisions. In fact, it could even be argued that belief contraction is a conceptually necessary component of the original, AGM understanding of belief revision: if one belief set is changed into a new, revised one, we can always decompose this change into two parts, a removal of old beliefs that no longer hold, followed by an addition of new beliefs to the contracted set. Furthermore, since the second step, expansion, is a logically simple and uniform procedure in AGM, the essence and main variations of belief revision in AGM are reducible to properties and alternative definitions of the underlying contractions. In a sense, this makes contraction the main, defining ingredient of the AGM theory. By the same token, however, this crucial role of contractions is tightly connected to the very 'ontology' of AGM that deals with propositional belief sets, and there is no a priori guarantee that it can be freely extrapolated to other representations.

An overwhelming influence of the AGM approach has led, however, to repeated, and largely unjustified, attempts to extend the AGM decomposition of the revision operation to other areas of informational dynamics, such as belief updates, quantitative/probabilistic approaches to belief change, rule change, etc. In most cases, such attempts encountered, however, immediate difficulties in defining a reasonable counterpart of the contraction operation. From the viewpoint of the present study, a relatively modest success of these attempts strongly indicates that contraction is an essential ingredient only for quite specific and restricted processes in information dynamics.

Evolutionary belief change, as described above, is representable in the framework of conditional bases and their generated conditional epistemic states, which form a particular instantiation of the general base approach to belief change. At the beginning of this study we have argued, however, that, even taken in its full generality, the base approach does not provide an adequate representation framework for belief contractions. More precisely, an extremely simple example of contracting $A \wedge B$ from a base $\{A, B\}$, discussed earlier, shows clearly that belief contractions cannot be defined as plain operations on bases, at least if they are required to satisfy some natural desiderata. Furthermore, in order to provide a representation framework that is closed with respect to all the AGM operations, belief bases should be replaced with more complex structures, such as disjunctive combinations of bases that have been called *flocks* in [Fagin *et al.*, 1986] (see [Bochman, 2001] for details).

The main point of the above observations was not to undermine contraction itself as an intelligible operation, but rather to reassess its actual role in belief change. Belief contraction is a radically different operation when compared with evolutionary revision and merge, described earlier, in that it involves a voluntary loss, or *forgetting*, of information. Of course, such an operation seems quite natural on the level of AGM belief sets. However, taken on a deeper level of epistemic states, the role of contraction becomes much less obvious, since it has to deal with general default beliefs that tend to persist by their very definition. True, there are situations that may lead us to removal of such general beliefs. This may happen, for example, when our default base encounters unsurmountable difficulties in producing adequate beliefs for particular circumstances. Alternatively, we may intentionally abandon some more specific default beliefs in order to construct a more general theory that would be applicable to a larger class of situations. Such drastic changes can be compared, however, to what Thomas Kuhn has called scientific revolutions and paradigm shifts. Of course, a modest scale of such changes in the framework of belief revision makes the title 'revolution' a bit of an exaggeration here. Furthermore, unlike general scientific revolutions, there is a grounded hope that corresponding 'micro-revolutions' in epistemic states can also be systematically represented and studied by using, in particular, some kind of contraction operation. Still, we believe that singling out normal, evolutionary changes as a relatively independent area of research in the framework of a general theory of belief change will allow us to clarify important aspects of belief and knowledge dynamics.

3.4 Problems with Preferences

In this concluding section we will briefly describe some of the problems our suggested theory of evolutionary change have to cope with. As we will see, however, most of these problems are not specific to the suggested theory, but pertain to the general preferential approach in belief change and nonmonotonic reasoning.

Evolutionary Belief Change

We have seen earlier that practically all theories of defeasible entailment adopt the requirement of Direct Inference, which makes every conditional from the conditional base valid in the generated epistemic state. Direct Inference implies, in particular, a well known *specificity principle*, according to which more specific rules should be preferred to less specific ones in cases of conflict. For example, in the following famous 'penguins-birds' case

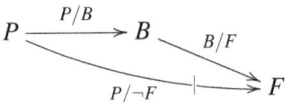

the set of conditionals $\{P/\neg F, P/B\}$ should be preferred to $\{B/F, P/B\}$ in order to support a conclusion that penguins do not fly, despite being birds. This preference can be produced by making conditional $P/\neg F$ prior to B/F. Similar considerations lead to making also P/B prior to B/F. This priority relation will be admissible in the sense of Geffner, and the corresponding generated epistemic state will satisfy Direct Inference with respect to the conditional base $\{P/B, B/F, P/\neg F\}$.

Let us look, however, at the following conditional base

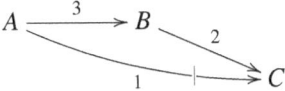

where the priority order is $1 \triangleright 2 \triangleright 3$. This priority order is not admissible, since it leads to preferring $\{2,3\}$, and this will violate both specificity and Direct Inference for the conditional $A/\neg C$. Note, however, that this priority order can be easily constructed as a result of revising a conditional base $\{A/\neg C\}$ first with B/C and then with A/B. On this construction, violation of Direct Inference for the conditional $A/\neg C$ can be viewed as its *eventual cancellation* due to revisions.

The above example clearly shows that conditional bases arising in the process of evolutionary belief change need not satisfy the principles of Direct Inference and specificity with respect to every conditional from the base. Moreover, one of the natural ways to see this situation consists in generalizing the very idea of specificity by adding a temporal dimension to it. According to this view, more recent rules added to a conditional base should be seen as more 'specific' than older ones.

It would be tempting to conclude at this point that *any* prioritized conditional base could be considered as a possible output of some evolutionary belief change process. Furthermore, a formal justification of this claim can be found in the result, stated earlier, according to which any such conditional base can be constructed using the two basic merge operations. Nevertheless, the following example shows that such a conclusion would have some problematic consequences:

In the case $3\triangleright 2\triangleright 1$, $\{1,3\}$ is preferred to $\{2\}$ (according to the consensus order), despite the fact that, given A as an evidence, only 2 and 3 are initially applicable rules, and among them 2 is preferred to 3. Furthermore, application of 2 makes 1 *inapplicable*.

The above example reveals an important specific property of default conditionals, namely that conditionals cannot be viewed as universally applicable propositions. Rather, they should be viewed as conditional *rules* that are effective (applicable) only when their antecedents hold. Accordingly, initial priorities among conditional rules should be used for establishing (stepwise) preferences among applicable conditionals only. This idea naturally leads us to the notion of a *prescriptive preference* order in which (unlike the consensus preference order) priorities are used only for constraining the order of rule applications (see [Brewka and Eiter, 2000; Delgrande et al., 2004]).

Unfortunately, even this modified construction of a preference order still does not resolve all the problems of the preferential approach. In fact, a most serious challenge to this approach is created by counterexamples indicating that the selection of intended combinations of defaults cannot always be made by establishing 'context-independent' priorities among them. The following example from [Dung and Son, 2001] illustrates this:

Example 2. Let us consider a default base consisting of the following four conditionals:

1. Students are normally young adults ($S\rightarrow Y$);
2. Young adults are adults ($Y\Rightarrow A$);
3. Adults are normally married ($A\rightarrow M$);
4. Students are normally not married ($S\rightarrow\neg M$).

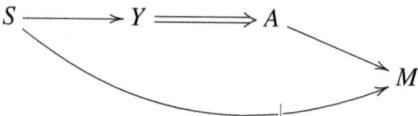

According to a natural informal understanding of this (default) information, evidence S should imply $\neg M$. In the framework of a preferential approach, this effect is achieved by making $S\rightarrow\neg M$ preferred to $A\rightarrow M$. But since such a priority stipulation is context-independent, the latter preference should hold also for a more specific evidence $S\wedge\neg Y\wedge A$, which will result in an awkward conclusion that even non-young students are not married. In contrast, a nonmonotonic inheritance theory implies for this setting a more plausible conclusion that the marital status of non-young students is not determined.

4 Belief Change and Argumentation

In order to achieve a deeper and more elaborate understanding of belief dynamics, we should take into account also the actual reasoning mechanisms used in the construction of epistemic states. This will immediately bring us to the well known connections between belief change and argumentation[4].

On a certain level of abstraction, an argumentation framework can be viewed as a structure consisting of a set of propositions, or (potential) beliefs, and a set of arguments that support such beliefs. Then the aim of argumentation theory consists in determining reasonable belief sets by selecting sustainable, or 'defensible', sets of arguments that support these beliefs. Now, this description can be directly related to the belief change theory if we will consider argumentation frameworks themselves as a particular kind of epistemic states. This will make these argumentation frameworks an instantiation of the foundational approach to belief change; changes in beliefs on this understanding will be determined by changes made to the underlying 'argument states'.

The relation between argumentation frameworks and epistemic states can be established also in the opposite direction by viewing default beliefs as *primitive arguments*. This will immediately make default bases and their associated base-generated states a particular kind of argumentation frameworks.

The above connections between the frameworks of belief revision and that of argumentation reveal, however, important differences between them, namely differences in the reasoning mechanisms employed in each of them for selecting intended argument sets. Whereas the selection mechanisms in most theories of belief change are based solely on priorities and preferences, the corresponding mechanisms in argumentation theory are based mainly on deductive inference and refutation (or attacks). In fact, the same differences constitute the main distinction between two principal approaches inside the general theory of nonmonotonic reasoning, namely preferential and explanatory approaches (see [Bochman, 2005]).

The fundamental starting point of the preferential approach to belief change amounts to the slogan *"Choice presupposes preference"* (see, e.g., [Rott, 1993]). This makes preferential approach a special case of a general methodology that is at least as old as decision theory and the theory of social choice. This approach has also been laid as a basis of all the representations that have been described so far in this study, namely preferential and conditional epistemic states. Still, we have seen that the preferential approach has problems of its own, which form an obvious incentive for trying an alternative, explanatory approach.

In the following sections we will briefly explore possibilities of replacing priorities with selection mechanisms that are provided by default logic that can be seen as a primary explanatory nonmonotonic formalism. Our explorations below should be viewed, however, only as a preliminary overview (and, hopefully, an incentive) for a more systematic study. This brief overview is purported to show that the explanatory approach and its argumentative reasoning tools provide a competitive mechanism for choosing admissible belief sets, which naturally suggests it as an alternative reasoning framework for representing and solving the problems of belief change.

---

[4] See [Falappa et al., 2011] for an overview of the relations between these two areas.

4.1 Choice and Priorities in Simple Default Theories

Explanatory nonmonotonic reasoning includes a wealth of formal systems, starting from highly abstract Dung's argumentation frameworks, and ending with specific nonmonotonic formalisms like default and modal nonmonotonic logics. A theory of belief change deals, however, mainly with a classical propositional language of beliefs, which is the main reason why we will use below default logic as a representation framework for belief formation. An additional support for this choice can be discerned from the fact that default logic constitutes, in effect, a primary instantiation of assumption-based argumentation frameworks from [Bondarenko *et al.*, 1997] for the classical language (see [Bochman, 2008a]).

It has been shown in [Bochman, 2008a] that default logic in its full generality can be reformulated in the following simple terms:

Definition 7.
- A *simple default theory* is a pair $(\vdash, \mathcal{A})$, where $\vdash$ is a supraclassical Tarski consequence relation, and $\mathcal{A}$ a distinguished subset of propositions called *default assumptions*.
- A set $\mathcal{A}_0$ of assumptions is *stable* if it is consistent and refutes any assumption not in $\mathcal{A}_0$: $\mathcal{A}_0 \vdash \neg \alpha$, for any $\alpha \in \mathcal{A} \setminus \mathcal{A}_0$.
- A set s of propositions is an *extension* of a simple default theory if $s = \text{Cn}(\mathcal{A}_0)$, for some stable set of assumptions $\mathcal{A}_0$.

According to the above description, the selection mechanism of choosing belief sets (i.e., extensions) in default logic is determined, ultimately, by establishing appropriate refutation, or cancellation, rules of the form $\mathcal{A}_0 \vdash \neg \alpha$ among defaults. Such inference rules naturally correspond to attacks among arguments in Dung's argumentation theory, and especially to the attack relations defined in assumption-based argumentation frameworks[5]. It should be clear, however, that the formalism of default logic allows us to express a much richer system of relations among defaults than just refutations, or attacks. Most importantly, it allows us to naturally express also 'positive' relations of *support* among arguments.

An obvious question that arises at this point is how the above notion of refutation, or attack, is related to the notion of priority that is central to the preferential approach.

To begin with, there is an obvious difference between these two notions. If an assumption attacks another assumption, then these two assumptions cannot belong to the same belief set. In contrast, if one assumption is prior to another one, they still may both belong to the same preferred belief set, provided this is consistent with other accepted beliefs. This means that a prior assumption 'attacks' less preferred ones only in cases when it is inconsistent with the latter.

As a first approximation, we will attempt to express the above conditional refutations in default logic using specific (non-schematic) inference rules of the form:

---

[5] Note that, since the underlying Tarski consequence relation is only supraclassical, contraposition dos not hold for it, so α may attack β, though not vice versa.

$$\frac{C,\alpha,\beta \vdash}{C,\alpha \vdash \neg\beta}$$

The above rule says, in effect, that if assumptions α and β are incompatible, given the context C, then α refutes β in this context. Accordingly, we can take this rule as a formal explication of the claim that α *is prior to* β.

It can be verified that, in most cases, the above explication of priorities provides the intended choice of assumptions. Nevertheless, this explication creates an essential side effect in making priority a *context-dependent* notion. For instance, the above inference rule is perfectly compatible with the following rule that assigns opposite priorities to α and β in a different context D:

$$\frac{D,\alpha,\beta \vdash}{D,\beta \vdash \neg\alpha}$$

In the framework of default logic, the joint use of the above two rules will have the effect that α will attack β in all extensions in which C holds, but not D, while β will attack α in all extensions in which only D holds. In the case when both C and D hold, however, we will normally have *two* extensions, one in which only α holds, another one in which only β holds.

We should recall here Example 2 (of married students) which showed the necessity of using context-dependent priorities for a more adequate theory of defeasible entailment. In this respect, the above explication of priorities in default logic suggests a plausible elaboration of the notion of (absolute) priority that forms the basis of the preferential approach. In fact, we will see in the next section that a proper representation of conditionals in the framework of default logic will provide, in particular, a more adequate representation of Example 2.

4.2 *Default Approach to Conditionals*

On a commonsense understanding, a rule "*A normally implies B*" represents a claim that A implies B, given some (unmentioned and even not fully known) conditions that are presumed to hold in normal circumstances. Thus, a default rule $TurnKey \to CarStarts$ states that if I turn the key, the car will start given the normal conditions such as there is a fuel in the tank, the battery is ok, etc. etc.

[McCarthy, 1980] has suggested a purely classical translation of such normality rules as implications of the form

$$A \wedge \neg ab \supset B,$$

where ab is a new 'abnormality' proposition serving to accumulate the conditions for violation of the source rule. Thus, *Birds fly* was translated into something like "Birds fly if they are not abnormal".

The default character of commonsense rules was captured in McCarthy's theory by a circumscription policy that minimized abnormality (and thereby maximized the acceptance of the corresponding normality claims $\neg ab$). A more elaborate version of this approach has been suggested in [McCarthy, 1986]. First, plain abnormality was replaced by *aspects* of abnormality in order to separate independent conditionals having the same antecedent. Second, priorities among (aspects of) abnormalities were introduced in order to deal with interaction of conflicting default rules. In fact, the resulting theory of prioritized circumscription can be seen as a first elaborated theory belonging to the preferential approach.

Now, the approach below is based on the idea that the default assumptions of a defeasible rule jointly function as a conditional A/B, that, once accepted, allows us to infer B from A. Accordingly, and similarly to McCarthy's representation, we may also represent such a rule as the classical implication

$$A \wedge (A/B) \supset B.$$

The default character of the inference from A to B can be captured by requiring that A/B normally holds or, more cautiously, that A normally implies A/B. We will achieve this effect, however, by representing the latter rule simply as a normal default rule in default logic.

A most important presumption behind our representation, however, will be that conditionals should have their own internal logic. Namely, we will stipulate that the conditionals A/B should satisfy the usual rules of Tarski consequence relations:

If $A \vDash B$, then A/B. \hfill (Dominance)

$$\frac{A/B \quad B/C}{A/C} \qquad \text{(Transitivity)}$$

$$\frac{A/B \quad A/C}{A/(B \wedge C)} \qquad \text{(And)}$$

Finally, we will formulate some principles of rejection for conditionals that will play a key role in our representation. Thus, the following logical principle determines how the rejection is propagated along chains of conditionals. It turns out to be suitable for representing nonmonotonic inheritance.

$$\frac{A/B \quad \neg(A/C)}{\neg(B/C)} \qquad \text{(Forward Rejection)}$$

Forward Rejection can be viewed as a partial contraposition of the basic Transitivity rule for conditionals. Note however that, since Transitivity was formulated as an inference *rule*, Forward Rejection cannot be logically derived from the latter. In fact, Forward Rejection provides a natural formalization of the principle of *forward chaining* adopted in many versions of the nonmonotonic inheritance theory.

The above logic describes the logical (monotonic) properties of arbitrary conditionals, not only default ones. The difference between the two will be reflected in the representation of normality rules in the framework of default logic, described next.

For each accepted defeasible rule *A normally implies B*, we will introduce the following two rules of default logic:

$$A : A/B \vdash A/B \qquad \text{(Default)}$$
$$A : \vdash \neg(A/\neg B). \qquad \text{(Commitment)}$$

The first rule is a normal default rule. It secures that a corresponding conditional is acceptable whenever its antecedent holds and it is not rejected. The second rule is an ordinary inference rule that reflects the following *commitment principle*: if A is known to hold, then the opposite conditional $A/\neg B$ should be rejected[6].

Note that $A/\neg B$ may be a consequence of a chain of conditionals that starts with A and ends with $\neg B$. In this case Forward Rejection dictates, in effect, that the last conditional in any such chain should also be rejected. Actually, it is this feature that allows us to reject a less specific conditional due to a commitment to a more specific one.

It has been shown in [Bochman, 2008b] that the above formalism constitutes an exact formalization of the theory of nonmonotonic inheritance. The following examples will illustrate how the resulting system works and, in particular, how it handles specificity and context-dependent priorities.

Example 3 (A Penguin-Bird story).

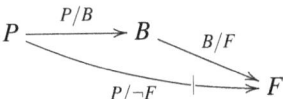

As could be expected, given the only fact B, the corresponding default theory has a unique extension that contains B and F. However, given the fact P instead, the resulting default theory also has a unique extension that includes this time P, B and $\neg F$. This happens because, given P, the commitment principle for $P|\!\!\sim\neg F$ implies $\neg(P/F)$. Taken together with P/B, this implies rejection of B/F by Forward Rejection. This is the way a more specific rule $P|\!\!\sim\neg F$ cancels a less specific rule $B|\!\!\sim F$. Note, however, that the situation is not symmetric, since the commitment to the less specific default $B|\!\!\sim F$ does not allow us to reject the more specific rule $P|\!\!\sim\neg F$. That is why, for instance, we would still have a unique extension containing $\neg F$ even if both P and B were given as facts.

The next example shows how our theory deals with context-dependent priorities, a problematic case for the preferential approach.

Example 4 (Married Students, Revisited).

---

[6] In a sense, Commitment states that A/B *is prior to* $A/\neg B$ whenever A holds.

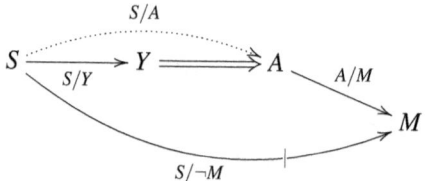

Given the only evidence S, we obtain a single extension that contains $\neg M$ and S/A, so $\neg(A/M)$ by commitment to $S \to \neg M$. However, given the evidence $S \wedge \neg Y \wedge A$, we obtain *two* extensions, one containing M, another containing $\neg M$. Consequently, (and as should be expected) the marital status of non-young students is not determined.

5 Conclusions

A general point we have argued for in this study was that a more adequate understanding of changes in beliefs can be achieved only by using a more elaborate representation of underlying epistemic states and their dynamics. Among other things, such a more elaborate representation provides us with a (much needed) broader perspective that connects belief change with other parts of our reasoning and rational activity. It should provide us also with tools for transforming highly abstract, 'postulational' belief change theories into working frameworks for representing and solving actual belief change problems in AI and related areas.

As a more specific contribution, we have suggested a general representation of epistemic states that is based on the notion of a conditional (rule) as its main ingredient. Epistemic states on this understanding are relatively stable entities that accumulate general (default) information about the world. This has naturally led us to the idea of a normal, evolutionary change of epistemic states that is representable by prioritized expansions of rules. It has been argued, in particular, that the notion of an evolutionary change requires us to reconsider the role of traditional belief change operations in belief dynamics, especially the role of contraction. Finally, we have discussed two alternative reasoning mechanisms used in selecting output belief sets, namely the traditional preferential approach and an alternative argumentative (explanatory) approach. It has been shown, in particular, that the latter approach has all the capabilities for providing a powerful reasoning framework for representing and solving the problems of belief change.

References

[Adams, 1975] E. W. Adams. *The Logic of Conditionals*. Reidel, Dordrecht, 1975.
[Alchourrón et al., 1985] C. Alchourrón, P. Gärdenfors, and D. Makinson. On the logic of theory change: Partial meet contraction and revision functions. *Journal of Symbolic Logic*, 50:510–530,

1985.

[Andreka et al., 2002] H. Andreka, M. Ryan, and P.-Y. Schobbens. Operators and laws for combining preference relations. *Journal of Logic and Computation*, 12:13–53, 2002.

[Benferhat et al., 1993] S. Benferhat, C. Cayrol, D. Dubois, J. Lang, and H. Prade. Inconsistency management and prioritized syntax-based entailment. In R. Bajcsy, editor, *Proceedings Int. Joint Conf. on Artificial Intelligence, IJCAI'93*, pages 640–645, Chambery, France, 1993. Morgan Kaufmann.

[Bochman, 2000] A. Bochman. Belief contraction as nonmonotonic inference. *Journal of Symbolic Logic*, 65:605–626, 2000.

[Bochman, 2001] A. Bochman. *A Logical Theory of Nonomonotonic Inference and Belief Change*. Springer, 2001.

[Bochman, 2005] A. Bochman. Two theories nonmonotonic reasoning. In Sergei N. Artëmov, Howard Barringer, Artur S. d'Avila Garcez, Luís C. Lamb, and John Woods, editors, *We Will Show Them! (1)*, pages 265–308. College Publications, 2005.

[Bochman, 2008a] A. Bochman. Default logic generalized and simplified. In *The Tenth International Symposium on Artificial Intelligence and Mathematics, ISAIM 2008*, Fort Lauderdale, Florida, 2008.

[Bochman, 2008b] A. Bochman. Default theory of defeasible entailment. In *Principles of Knowledge Representation and Reasoning: Proceedings of the Eleventh International Conference, KR 2008, Sydney, Australia, September 16-19, 2008*, pages 466–475. AAAI Press, 2008.

[Bochman, 2011] A. Bochman. Logic in nonmonotonic reasoning. In G. Brewka, V. W. Marek, and M. Truszczynski, editors, *Nonmonotonic Reasoning. Essays Celebrating its 30th Anniversary*, pages 25–61. College Publ., 2011.

[Bondarenko et al., 1997] A. Bondarenko, P. M. Dung, R. A. Kowalski, and F. Toni. An abstract, argumentation-theoretic framework for default reasoning. *Artificial Intelligence*, 93:63–101, 1997.

[Brewka and Eiter, 2000] G. Brewka and T. Eiter. Prioritizing default logic. In St. Hölldobler, editor, *Intellectics and Computational Logic. Papers in Honour of Wolfgang Bibel*, pages 27–45. Kluwer, 2000.

[Delgrande et al., 2004] J. Delgrande, T. Schaub, H. Tompits, and K. Wang. A classification and survey of preference handling approaches in nonmonotonic reasoning. *Computational Intelligence*, 20:308–334, 2004.

[Doyle, 1992] J. Doyle. Reason maintenance and belief revision: Foundations vs. coherence theories. In P. Gärdenfors, editor, *Belief Revision*, pages 29–51. Cambridge University Press, 1992.

[Dung and Son, 2001] P. M. Dung and T. C. Son. An argument-based approach to reasoning with specificity. *Artificial Intelligence*, 133:35–85, 2001.

[Fagin et al., 1986] R. Fagin, J. D. Ullman, G. M. Kuper, and M. Y. Vardi. Updating logical databases. In *Advances in Computing Research*, volume 3, pages 1–18, 1986.

[Falappa et al., 2011] M. A. Falappa, A. J. García, G. Kern-Isberner, and G. R. Simari. On the evolving relation between belief revision and argumentation. *Knowledge Eng. Review*, 26(1):35–43, 2011.

[Gärdenfors and Rott, 1995] P. Gärdenfors and H. Rott. Belief revision. In D. M. Gabbay et al., editor, *Handbook of Logic in Artificial Intelligence and Logic Programming. Vol. 4*, pages 35–132. Clarendon Press, 1995.

[Gardenfors, 1988] P. Gardenfors. *Knowledge in Flux: Modeling the Dynamics of Epistemic States*. Bradford Books, MIT Press, 1988.

[Gärdenfors, 1990] P. Gärdenfors. The dynamics of belief systems: Foundations vs. coherence theories. *Revue Internationale de Philosophie*, 44:24–46, 1990.

[Geffner, 1992] H. Geffner. *Default Reasoning. Causal and Conditional Theories*. MIT Press, 1992.

[Hansson, 1999] S. O. Hansson. *A Textbook of Belief Dynamics*. Kluwer, 1999.

[Horty, 1994] J. F. Horty. Some direct theories of nonmonotonic inheritance. In D. M. Gabbay, C. J. Hogger, and J. A. Robinson, editors, *Handbook of Logic in Artificial Intelligence and Logic Programming* **3**: *Nonmonotonic Reasoning and Uncertain Reasoning*. Oxford University Press, Oxford, 1994.

[Kern-Isberner, 2004] G. Kern-Isberner. A thorough axiomatization of a principle of conditional preservation in belief revision. *Ann. Math. Artif. Intell.*, 40(1-2):127–164, 2004.

[Kraus *et al.*, 1990] S. Kraus, D. Lehmann, and M. Magidor. Nonmonotonic reasoning, preferential models and cumulative logics. *Artificial Intelligence*, 44:167–207, 1990.

[Kuhn, 1962] Thomas Kuhn. *The structure of scientific revolutions*. The University of Chicago Press, Chicago London, 1962.

[Lehmann and Magidor, 1992] D. Lehmann and M. Magidor. What does a conditional knowledge base entail? *Artificial Intelligence*, 55:1–60, 1992.

[Lehmann, 1989] D. Lehmann. What does a conditional knowledge base entail? In R. Brachman and H. J. Levesque, editors, *Proc. 2nd Int. Conf. on Principles of Knowledge Representation and Reasoning, KR'89*, pages 212–222. Morgan Kaufmann, 1989.

[Lehmann, 1995] D. Lehmann. Another perspective on default reasoning. *Annals of Mathematics and Artificial Intelligence*, 15:61–82, 1995.

[Leite, 2003] J.A. Leite. *Evolving knowledge bases: specification and semantics*. Frontiers in artificial intelligence and applications. IOS Press, 2003.

[Levi, 1991] I. Levi. *The Fixation of Belief and Its Undoing*. Cambridge University Press, 1991.

[Levi, 1997] I. Levi. Contraction and Informational Value. Columbia University, fifth version, August 1997.

[Lifschitz, 1985] V. Lifschitz. Computing circumscription. In *Proc. 9th Int. Joint Conf. on Artificial Intelligence, IJCAI-85*, pages 121–127. Morgan Kaufmann, 1985.

[Makinson and Gärdenfors, 1991] D. Makinson and P. Gärdenfors. Relations between the logic of theory change and nonmonotonic logic. In A. Fuhrmann and M. Morreau, editors, *The Logic of Theory Change*, pages 185–205. Springer Verlag, 1991. Lecture Notes in AI, 465.

[McCain and Turner, 1997] N. McCain and H. Turner. Causal theories of action and change. In *Proceedings AAAI-97*, pages 460–465, 1997.

[McCarthy, 1980] J. McCarthy. Circumscription – a form of non-monotonic reasoning. *Artificial Intelligence*, 13:27–39, 1980.

[McCarthy, 1986] J. McCarthy. Applications of circumscription to formalizing common sense knowledge. *Artificial Intelligence*, 13:27–39, 1986.

[Pearl, 1990] J. Pearl. System Z: A natural ordering of defaults with tractable applications to default reasoning. In *Proceedings of the Third Conference on Theoretical Aspects of Reasoning About Knowledge (TARK'90)*, pages 121–135, San Mateo, CA, 1990. Morgan Kaufmann.

[Poole, 1988] D. Poole. A logical framework for default reasoning. *Artificial Intelligence*, 36:27–47, 1988.

[Reiter, 1980] R. Reiter. A logic for default reasoning. *Artificial Intelligence*, 13:81–132, 1980.

[Rott, 1991] H. Rott. Two methods of constructing contractions and revisions of knowledge systems. *Journal of Philosophical Logic*, 20:149–173, 1991.

[Rott, 1993] H. Rott. Belief contraction in the context of the general theory of rational choice. *Journal of Symbolic Logic*, 58:1426–1450, 1993.

[Rott, 2001] H. Rott. *Change, Choice and Inference: A Study of Belief Revision and Nonmonotonic Reasoning*. Oxford UP, 2001.

[Weydert, 2005] E. Weydert. Projective default epistemology. In G. Kern-Isberner, W. Rödder, and F. Kulmann, editors, *Conditionals, Information, and Inference*, volume 3301 of *Lecture Notes in Computer Science*, pages 65–85. Springer, 2005.

Cognitive realism in belief revision

Sven Ove Hansson

Abstract The AGM (Alchourrón-Gärdenfors-Makinson) model of belief contraction has considerable advantages but also non-negligible disadvantages, in particular with respect to cognitive realism. Based on a discussion of these disadvantages, six criteria for a more cognitively realistic model of belief contraction are presented. Six alternatives to AGM are presented and compared to these criteria, namely partial meet contraction with a finite language, base-generated partial meet contraction, specified meet contraction, perimaximal contraction, blockage contraction, and bootstrap contraction. It is concluded that among these, blockage and bootstrap contraction are in best compliance with the criteria of cognitive realism used in the evaluation.

1 Introduction

In belief change theory, a belief state is usually represented by a belief set, i.e. a logically closed set of sentences (K). The belief set is subjected to operations of change, consisting in the removal of sentences and/or the incorporation of new sentences. (Alchourrón et al 1985. Hansson 1999. Fermé and Hansson 2011.) In contraction (generically denoted ÷), a specified sentence or set of sentences is removed without any new sentence being added.

In the standard model of belief change, the AGM model (Alchourrón et al 1985), contraction is based on a selection among the inclusion-maximal subsets of K not implying the sentence to be contracted. This is a simple and elegant model, but it has been criticized for being unrealistic in important respects. In this contribution, several alternatives to AGM are evaluated in order to determine to what extent they are improvement over AGM in the relevant aspects.

Sven Ove Hansson
Department of Philosophy and the History of Technology, Royal Institute of Technology, Sweden,
e-mail: soh@kth.se

Table 1 Notation and terminology

$\mathcal{L}$	the language in which beliefs are expressed
$p, q \ldots$	sentences in $\mathcal{L}$
$\neg$	negation
&	conjunction
$\vee$	disjunction
$\rightarrow$	truth-functional ("material") implication
$\leftrightarrow$	truth-functional equivalence
$A, B, \ldots X, Y, \ldots$	sets of sentences in $\mathcal{L}$
Cn	a compact supraclassical Tarskian consequence operator that satisfies inclusion, monotony, iteration, and the deduction property
$X \vdash p$	alternative notation for $p \in \text{Cn}(X)$
$X \vdash_\exists Y$	alternative notation for $\text{Cn}(X) \cap Y \neq \emptyset$
&A	a sentence such that $\text{Cn}(A) = \text{Cn}(\{\&A\})$
$\mathcal{K}$	a logically closed set of sentences (belief set)
$X \perp p, X \perp A$	the set of inclusion-maximal subsets of X not implying p, respectively not implying any element of A (remainder set)
$X \perp\!\!\!\perp p, X \perp\!\!\!\perp A$	the set of inclusion-minimal subsets of X implying p, respectively implying some element of A (kernel set)
$\mathcal{E}_\mathcal{K}(A)$	$\{X \in \mathcal{K} \mid X \not\vdash_\exists A\}$ (the set of A-excluders in $\mathcal{K}$)
$\widehat{\mathcal{E}}_\mathcal{K}(A)$	$\{X \in \mathcal{E}_\mathcal{K}(A) \mid \neg(\exists Y)(X \subset Y \in \mathcal{E}_\mathcal{K}(A))\}$ (the set of maximal A-excluders in $\mathcal{K}$)

In Section 2 the AGM model is briefly introduced and some of its problems concerning cognitive realism are identified. In Section 3, a list of six requirements on a cognitively more realistic alternative are proposed. After that, some alternatives to AGM are presented and tested against the six requirements: partial meet contraction with a finite language (Section 4), base-generated partial meet contraction (Section 5), specified meet contraction (Section 6), perimaximal contraction (Section 7), blockage contraction (Section 8), and bootstrap contraction (Section 9). Finally, in Section 10, these operations are compared in terms of how they satisfy the requirements.

Standard notation and terminology are used, as seen in Table 1.

2 AGM, the standard approach

An operation that removes sentences is called contraction, and its specific form in AGM is *partial meet contraction*: When contracting K by a sentence p, a selection is made among the elements of $K \perp p$, the set of maximal subsets of K not implying p. This is done with a selection function γ such that for all non-tautologous p, $\gamma(K \perp p)$ is a non-empty subset of $K \perp p$. In the limiting case when p is a tautology, $\gamma(K \perp p) = \{K\}$. The partial meet contraction $\sim_\gamma$ based on γ is defined as follows:

$$K \sim_\gamma p = \bigcap \gamma(K \perp p)$$

Cognitive realism in belief revision 59

If $\gamma(K \perp p)$ is a singleton for all p, then $\sim_\gamma$ is a *maxichoice contraction*. If $\gamma(K \perp p) = K \perp p$ for all non-tautologous p, then $\sim_\gamma$ is *full meet contraction* (denoted by $\sim$). Furthermore, γ is *transitively relational* if and only if there is some transitive relation $\sqsupseteq$ such that $\gamma(K \perp p) = \{X \in K \perp p \mid (\forall Y \in K \perp p)(X \sqsupseteq Y)\}$. (See Hansson 1999 for an extensive introduction to AGM.)

The AGM model is remarkably useful, as evidenced by the many formal models that have been developed from it (Fermé and Hansson 2011). But like other highly idealized models of complex phenomena it also has limitations. Here I will focus on four such limitations that are all related to the issue of its cognitive realism.

The first of these limitations is that it requires that the following condition:

$$K \subseteq \mathrm{Cn}((K \div p) \cup \{p\})$$

is satisfied by all $p \in K$. We can call a belief p *recovering* in relation to a belief set K and an operation of contraction $\div$ if and only if this condition is satisfied. The AGM postulate Recovery states that all elements of K are recovering in relation to K. Recovery is satisfied by partial meet contraction. However, it is easy to show with examples that intuitively speaking, some but not all beliefs are recovering, and therefore the Recovery postulate appears to be too demanding. The following example has been offered to show that Recovery should not hold:

> I believed that *Cleopatra had a son* (s). Therefore I also believed that *Cleopatra had a child* (c or equivalently $s \vee d$ where d denotes *Cleopatra had a daughter*). Then I received information that made me give up my belief in c, and contract my belief set accordingly, forming $K \div c$. Soon afterwards I learned from a reliable source that Cleopatra had a child. It seems perfectly reasonable for me to then add c (i.e., $s \vee d$) to my set of beliefs without also reintroducing s. (Hansson 1991)

Recovery has been subject to extensive discussions, see for instance Gärdenfors (1982), Hansson (1991 and 1999b), Makinson (1987 and 1997), and Glaister (2000).

The second limitation concerns the cognitive realism of the process that takes us from the original belief set to the contracted one. In order to go from K to $K \div p$ we are supposed to make a choice not among potential outcomes but among elements of $K \perp p$, the set of inclusion-maximal subsets of the original belief set not implying p (and then take the intersection of the chosen such maximal sets as the contraction outcome). As was shown in Hansson (2008), even if K has a finite representation (i.e. a finite set B such that $K = \mathrm{Cn}(B)$), $K \perp p$ is an infinite set none of whose elements has a finite representation. Even in intuitively trivial cases, such as when we choose in the contraction by $p \& q$ whether to retain p, q, or neither of them, the selection process is depicted as being inserted between two logical operations, the first of which (remainder-formation) creates an infinite array of infinite entities that are eliminated in the second (intersection). From a cognitive point of view this does not seem credible as a means to take us from one finitely representable belief set to another. It seems as if the operation takes place on the wrong "level" in the logical universe, where it manipulates formal objects that a cognitive agent cannot grasp.

The third limitation is also concerned with the creation of infinite objects, but in this case not with their appearance as intermediate objects but as final outcomes of contraction. Let K be a belief set that has a finite representation. A partial meet contraction by a sentence p may very well result in a contraction outcome $K \div p$ that lacks a finite representation, i.e. there may be no finite B' with $K \div p = \text{Cn}(B')$. In other words, the standard AGM model fails to conserve finite representability. This can be most easily seen from the maxichoice case, in which $K \div p$ is never finite-based if the language is infinite and $p \in K \setminus \text{Cn}(\varnothing)$:

OBSERVATION 1 *Let the language consist of infinitely many logically independent atoms and their truth-functional combinations. Let K be a belief set that contains some non-tautology, and let $p \in K \setminus \text{Cn}(\varnothing)$. Then:*
 (1) $K \perp p$ is infinite, and
 (2) if $X \in K \perp p$, then X is not finite-based.

PROOF: See Hansson (2008, p. 33 and pp. 43-44). □

The fourth limitation concerns the plausibility of the selection mechanism. Since we do not normally reason about maximal consistent subsets of our set of beliefs, we do not have reliable intuitions on how to choose among them. Discussions on suitable formal properties of the selection function γ have therefore drawn heavily on the intuitions that that we have about choices among other types of objects such as physical objects or social states of affairs. Hans Rott (2001) has shown in considerable detail how various properties of social choice functions, when applied to the selection function γ for remainders, correspond to interesting properties of the contraction $\sim_\gamma$. However, it is not obvious that this type of choices should be based on the same principles that we have found to be plausible in the application area of social choice theory. A major reason why the intuitions in question may not be transferable to belief change is that cases when the selected set has more than one element are treated quite differently in belief contraction than in social choice. If a social choice function leaves as outcome a set with more than one element, then the choice is considered to be indeterminate; the agent may choose any of them. In belief revision, such an outcome is made deterministic by taking the intersection of the outcome set as the solution (i.e. $\cap \gamma(K \perp p)$ instead of $\gamma(K \perp p)$).

It is easy to show that the intersective solution would lead to absurd results if applied to social choice. For a simple example, consider the following offers, each of which contains two items, a vacation trip and a sum of money:

  $A = \{\text{A week for two in Florence}, €100\}$
  $B = \{\text{A week for two in Venice}, €100\}$
  $C = \{\text{A week for two in Birmingham}, €200\}$
  $D = \{\text{A week for two in Manchester}, €200\}$

Suppose that your choice set is $\{A, B\}$, i.e. you prefer these two alternatives to the other two but you do not prefer A to B or the other way around. According to the intersective approach, which we now (ungraciously) transfer from belief revision to social choice, the solution is then that you should have $A \cap B$, i.e. €100 and no travel. This is absurd, since you would probably prefer each of C, D and (most certainly) $C \cap D$ to $A \cap B$.

This would be no problem for belief contraction if it could be shown that contrary to other collections of objects, belief sets do not lose in preferability by being intersected with other equally preferable objects. However, no such argument seems to be available. To the contrary, Tor Sandqvist (2000) has shown with an ingenious counter-example that if each element of a collection $\mathcal{A}$ of belief sets is preferable to each element of the collection $\mathcal{B}$ of belief sets, it does not follow that the belief set $\bigcap \mathcal{A}$ is preferable to the belief set $\bigcap \mathcal{B}$.

3 Requirements on alternatives to AGM

The following requirements on a contraction operator $\div$ are based on the four limitations discussed in Section 2.

- *Non-recovery:* It does not hold for all $p \in K$ that $K \subseteq \mathrm{Cn}((K \div p) \cup \{p\})$.
- *Cognitive accessibility:* The contraction does not require manipulation of cognitively inaccessible objects.
- *Finite-based outcome:* If K is finite-based then so is $K \div p$ for all p.[1]
- *No meet assumption:* The preferability of belief sets is not assumed to be preserved under intersection.

Obviously, when looking for alternatives that have advantages over the standard AGM model in the above-mentioned respects, we should also attempt to retain the major advantages of that model. The following two requirements refer to advantages that may be lost in modifications of AGM that aim at increased cognitive realism:

- *Expressive power:* Beliefs and belief states that can be expressed in the AGM framework should still be expressible.
- *Parsimony:* No formal element that is complex or difficult to interpret should be added to the framework.

The last-mentioned requirement is important since little would be gained if we replaced references to the cognitively inaccessible elements of AGM by other elements that are equally problematic from a cognitive point of view.

4 Partial meet contraction with a finite language

Two of the limitations of the AGM model that were mentioned in Section 2 hinge on the assumption that the language $\mathcal{L}$ in which the belief sets are expressed is infinite. It is therefore sensible to investigate what happens if we simply restrict the framework to a finite language.

---

[1] A logically closed set A of sentences is finite-based if and only if there is some finite set A' of sentences such that $A = \mathrm{Cn}(A')$.

Non-recovery is not achieved in this way, since the Recovery postulate holds for partial meet contraction irrespective of whether the belief set is finite.

Cognitive accessibility is arguably achieved since the components of the model will all be finite. Selecting among maximal consistent subsets of the belief set is much more plausible when these are finite and therefore not obviously inaccessible to a finite mind.

Finite-based outcome is fully achieved with a finite langauge.

No meet assumption is not satisfied since the definition $K \div p = \bigcap \gamma(K \perp p)$ is still used.

The main problem with a finite language is that our requirement of retained *expressive power* is violated. As the following example shows, the resources of a finite language are insufficient to make even some fairly simple distinctions.

> For each natural number m, let p_m denote the sentence that the Roman Catholic church has fewer than m popes. Clearly, $p_m \vdash p_{m+1}$ for all m. My present beliefs concerning all these statements are characterized by $p_2 \& \neg p_1 \in K$ (from which it follows that p_m also holds for all $m > 2$. Nevertheless, whenever $m \neq m'$, my truth conditions for p_m and $p_{m'}$ are different. Therefore an infinite number of logically non-equivalent beliefs have to be represented in the belief set that represents my beliefs. This is impossible with a finite language. (Hansson 2007b)

Finally, *Parsimony* is satisfied since no extra component is added to the original AGM framework.

5 Base-generated partial meet contraction

Another way to avoid the infinity-related problems mentioned in Section 2 is to apply the operation of partial meet contraction to a finite base for K, rather than to K itself. A set B is a (belief) base for the (belief) set K if and only if $Cn(B) = K$. Base-generated partial meet contraction is defined as follows:

$$K \tilde{\sim}_\gamma p = Cn(\bigcap \gamma(B \perp p))$$

where $K = Cn(B)$ and γ is a selection function for B. (Nebel 1989. Fuhrmann 1991.) In the intended case when B is finite, $\tilde{\sim}_\gamma$ is *finitely base-generated*. Operations performed on K can be referred to as "direct" to distinguish them from the base-generated variants. For an axiomatic characterization of base-generated partial meet contraction, see Hansson (1993).

Non-recovery is achieved in this way, which was in fact an important motivation for early studies of operations on belief bases. (Hansson 1991)

The requirement *Cognitive accessibility* is satisfied by partial meet contraction on finite bases, since all operations take place on finite sets. Since the outcome is the logically closure of a subset of B, *Finite-based outcome* is also satisfied.

No meet assumption is not satisfied since we use the intersection $\bigcap \gamma(B \perp A)$ to derive the outcome. In fact, the partial meet construction may have even more problematic effects in its base-generated than in its direct variant. If all elements of

$K \perp p$ contain either q or r, then $q \vee r \in K \sim_\gamma p$, but if all elements of $B \perp p$ contain either q or r, it may nevertheless be the case that $q \vee r \notin K\tilde{\sim}_\gamma p = \text{Cn}(\bigcap \gamma(B \perp p))$.

Expressive power is satisfied since the same language that is used in classical AGM (direct partial meet contraction) can be used for base-generated operations as well. In an important way, the expressive power has even been increased, since we can interpret different belief bases with the same logical closure, such as $\{p, p \rightarrow q\}$ and $\{p, q\}$ as representing dynamically different ways to hold one and the same set of beliefs. (In the first but not in the second case we should expect belief in q to be lost after contraction by p.)

Parsimony as defined above refers to the addition of elements to the framework that are either complex or difficult to interpret. In the intended cases, when B is finite but K infinite (due to the language being infinite) the belief base is arguably not more complex than the belief set. However, it is more difficult to interpret. The belief set consists (roughly) of the beliefs an agent holds.[2] The belief base is usually assumed to consist of those elements of the belief set that are held independently of any other beliefs. This means that in order to determine whether a belief is basic we need to know precisely on what grounds it is held, which is a rather heavy epistemological demand.[3]

6 Specified meet contraction

Some of the problems with AGM that we wish to avoid depend on the manipulations on infinite sets that are required in that framework. In the two previous sections we have investigated two ways to avoid such manipulations, namely a finite language and finite belief bases. The former of these approaches restricts the expressive power of the language, thereby taking us further away from classical AGM than the second. However, it can be argued that the approach using finite belief bases also involves an unnecessarily large deviation from AGM. In order to avoid the "infinitistic problems" we have to exclude belief sets that have no finite base. However, if a belief set has a finite base it has many alternative ones. Some of the interpretative problems with the belief base approach seem to follow from the assumption that we need to choose one of these alternatives in preference to the others. This can be avoided if we instead operate in a framework with *finite-based belief sets*, so that both K and all contraction outcomes (all sets equal to $K \div p$ for some p) are finite-based but no specific base is assigned to any of them. Intuitively speaking, finite-based belief sets deviate less from AGM than belief bases (which in their turn deviate less than a finite language). It is well worth investigating whether this smaller deviation can solve the problems with AGM that we listed in Section 2 without creating new problems like what a finite language and finite belief bases tend to do. This is what

---

[2] More precisely: those that she is, in her current belief state, committed to believe in. Cf. Levi 1977 and 1991.

[3] Furthermore, since $K = \text{Cn}(B)$ it is assumed that all beliefs are deductively obtainable from beliefs that are held independently of any other beliefs, which is a far from uncontestable assumption.

we will investigate in this and the following three sections. In this section it will be combined with an alternative selection mechanism that replaces γ.

Partial meet contraction was introduced as a modification of a simpler operation that has already been mentioned, namely full meet contraction:

$$K \sim p = \bigcap (K \perp p)$$

Full meet contraction was first described by Alchourrón and Makinson (1982). It was soon realized that it removes too much from the belief set. Partial meet contraction was constructed as an improvement of full meet contraction, allowing us to retain more of the belief set. But as we have already seen, partial meet contraction has the disadvantage of operating on the elements of $K \perp p$ that are cognitively inaccessible. We do not reason about the maximal consistent subsets of our belief sets, in fact we have no idea which they are. Instead, when discussing how to contract we typically reason about sentences, for instance:

> I realized that $p \& q$ cannot be true, so I had to give up either my belief in p or my belief in q, or perhaps both. After some deliberation I chose to give up only p.

Based on examples like this it would seem much more plausible to perform the non-logical selection on sentences such as $p \& q$ and p that are clearly cognitively accessible, rather than on elements of $K \perp p$ that are not. Formally, such a selection process can be represented by a function f that takes us from the sentence to be removed to the one that is "really" eliminated. In the above example we have $f(p \& q) = p$. Such a sentential selection mechanism can be added to full meet contraction, instead of adding a selection function over remainders to it. (Full meet contraction has the important advantage of being a purely logical operation which makes it suitable as a skeleton to which non-logical mechanisms can be attached.) Formally, we define a *sentential selector* as a function from and to $\mathcal{L}$, i.e. a function that takes us from one sentence to another. We can then define an operation based on sentential selection as follows:

> An operation $\div$ on K is an operation of *specified meet contraction* if and only if there is a sentential selector f such that for all p, $K \div p = K \sim f(p)$.

The following formal results provide a strong connection between this construction and the restriction of contraction operators to finite-based belief sets:

THEOREM 1 *Let K be a finite-based belief set and let $\sim_\gamma$ be a partial meet contraction. Then $\sim_\gamma$ satisfies*

$K \div p$ *is finite-based* (Finite-based Outcome)

if and only if there is a specified meet contraction f such that $K \sim_\gamma p = K \sim f(p)$ for all p.

PROOF: For one direction, let $\sim_\gamma$ be a partial meet contraction that satisfies Finite-based Outcome. It follows from Theorem 3 in Hansson (2008, p. 36) that $\sim_\gamma$ is reconstructible as a specified meet contraction. For the other direction, let $\sim_\gamma$ be a partial meet contraction that is reconstructible as a specified meet contraction. It follows from Observation 2 in Hansson (2008, p. 34) that $\sim_\gamma$ satisfies Finite-based Outcome. □

THEOREM 2 *(Hansson 2007) Let $\div$ be a sentential operation on a finite-based belief set K. The following two conditions are equivalent:*
(1) $\div$ *satisfies:*
  $K \div p = \text{Cn}(K \div p)$ (Closure),
  $K \div p \subseteq K$ (Inclusion)*, and*
  $K \div p$ *is finite-based* (Finite-based Outcome)
(2) *There is a sentential selector f such that for all p: $f(p) \in K$ and $K \div p = K \sim f(p)$.*

PROOF: See Observation 2 in Hansson (2007, pp. 482 and 494). □

The reader is referred to Hansson (2007a, 2008) for additional formal results on specified meet contraction, including connections between postulates for the sentential selector f and various properties of the contraction operator $\div$.

As indicated in Theorem 2, specified meet contraction is a fairly general construction that subsumes not only all partial meet contractions that satisfy Finite-based Outcome but also a wide class of what Makinson (1987) calls withdrawals, i.e. operations that remove a sentence from a belief set but differ from partial meet contraction in not satisfying the recovery postulate ($K \subseteq \text{Cn}((K \div p) \cup \{p\})$). Therefore, *Non-recovery* is clearly satisfied by this construction.

The selection mechanism of specified meet contraction is a function that takes us from a single sentence to another single sentence. Therefore, *Cognitive accessibility* is satisfied. That *Finite-based outcome* is satisfied follows directly from Theorem 2.

Full meet contraction is an essential component of specified meet contraction. It makes use of the intersective method, and therefore *No meet assumption* is not satisfied by specified meet contraction.

Expressive power is satisfied since the same language can be used as in classical AGM contraction. *Parsimony* is also satisfied since the new element that we have added to the construction is a cognitively quite accessible structure, namely a sentential selector.

7 Perimaximal contraction

Can we get rid of the problematic meet construction? The obvious way to do this would be to let the selection mechanism operate directly on the set of potential contraction outcomes, rather than on cognitively more far-fetched objects such as remainders that need to be intersected in order to yield a contraction outcome.[4] In

---

[4] In many belief revision frameworks the selection is made on a set of possible worlds rather than a set of remainders. Possible worlds, as well, have to be intersected in order to deliver a contraction outcome. See for instance Grove (1988), Rott and Pagnucco (1999) and Meyer et al (2002).

this and the following two sections we are going to consider some operations that employ such direct selection among potential outcomes. A key insight when developing such directly acting selection mechanisms is that only some of the logically closed subsets of K are at all viable as contraction outcomes. (For a simple illustration of this, return to the example in Section 4. My current belief set has a logically closed subset that contains $p_{666} \& \neg p_{665}$, but due to the way my belief system is constructed no such a subset can be the outcome of contracting K by some sentence.)

We will assume, therefore, that there is a repertoire of potential outcomes. It can be conceived as consisting of those subsets of K that are stable or plausible enough to be suitable as contraction outcomes. The term *repertoire contraction* will be used to denote any operation of contraction in which the modus operandi of the selection mechanism is to select among the elements of such a repertoire. The actually chosen outcomes form the *outcome set*, $\mathcal{K} = \{K' \mid (\exists p)(K' = K \div p)\}$. (Hansson 2013a) In order to simplify the formal treatment, the repertoire and the outcome set can for most purposes be assumed to be identical.

The following example serves to show how the general principles of repertoire contraction relate to actual arguments about contraction:

> I previously believed in both the following statements:
>
> s_1: Dar es Salaam is the capital of Tanzania.
>
> s_2: Cucumbers are not fruits.
>
> When taking part in a quiz game, I find out that these statements cannot both be true. I therefore have to contract my belief set by the sentence $s_1 \& s_2$. There are three ways in which I can do this: I can remove s_1 (which I will presumably do if I consider s_2 to be clearly less dispensable than s_1). I can remove s_2 (presumably in the reverse case). Finally, I can remove both s_1 and s_2 (which I will presumably do if none of the two is clearly more dispensable for me than the other).

In the first case we can identify $K \div s_1 \& s_2$ with $K \div s_1$, in the second case with $K \div s_2$, and in the third case with $K \div \{s_1, s_2\}$, the simultaneous contraction by both s_1 and s_2. As this example shows, when choosing a contraction outcome in the repertoire of potential outcomes we have use for contractions by sets of sentences (multiple contraction, Fuhrmann and Hansson 1994). In what follows, repertoire contraction will therefore be presented in the more general format of multiple contraction, which means that the contraction inputs will be sets of sentences rather than single sentences.

Three variants of repertoire contraction will be introduced in this and the following two sections. The first of these is perimaximal contraction that is closest among the three to the AGM model. In order to explain its construction it is useful to begin with a simpler construction, *maximal contraction*. A repertoire contraction is maximal if and only if it retains as much as possible of the original belief set K. This is true if and only if it holds for all sets A of sentences that $K \div A$ is an element of $\widehat{\mathcal{E}}_\mathcal{K}(A)$, i.e. it is inclusion-maximal among those elements of the outcome set that do not imply any of the sentences in A. Maximal contraction can be seen as a generalization of the AGM notion of a maxichoice contraction. It is much more plausible than maxichoice contraction since the elements of $\widehat{\mathcal{E}}_\mathcal{K}(A)$ are by definition plausible contraction outcomes, whereas those of $K \perp A$ are notoriously implausible.

Cognitive realism in belief revision

The major problem with maximal contraction is that there may be ties between equally plausible elements of $\widehat{\mathcal{E}}_\mathcal{K}(A)$. Suppose that my original belief set K contains the two equally justified beliefs p (Pat is an honest bookkeeper) and q (Quinn is an honest bookkeeper). Then we can have $q \in K \div \{p\} \in \widehat{\mathcal{E}}_\mathcal{K}(\{p\})$ and $p \in K \div \{q\} \in \widehat{\mathcal{E}}_\mathcal{K}(\{q\})$. When I learn that some funds have been embezzled, seemingly by either Pat or Quinn, I have to contract K by $\{p\&q\}$. When doing this I hesitate between the two equally plausible options to remove p while retaining q and to remove q while retaining p. Presumably, the contraction outcome will be equal to $K \div \{p\}$ in the first case and $K \div \{q\}$ in the second. In order to be on the safe side with respect to the risk of holding false beliefs I can then let $K \div \{p\&q\}$ be a subset of both $K \div \{p\}$ and $K \div \{q\}$. In order to avoid unnecessary losses of information it should be an inclusion-maximal subset with this property. If there is a unique inclusion-maximal element of $\mathcal{K}$ that is a subset of $(K \div \{p\}) \cap (K \div \{q\})$, then it should be the choice. In other words, if there is a tie between $K \div \{p\}$ and $K \div \{q\}$, and $\bigcup \{Z \in \mathcal{K} \mid Z \subseteq (K \div \{p\}) \cap (K \div \{q\})\}$ is an element of $\mathcal{K}$, then it should be the contraction outcome.

This can be expressed in more general terms if we introduce a choice function C that for any set A selects the "best" elements of $\widehat{\mathcal{E}}_\mathcal{K}(A)$. If $C(\widehat{\mathcal{E}}_\mathcal{K}(A))$ has exactly one element, then that element is equal to $K \div A$. If $C(\widehat{\mathcal{E}}_\mathcal{K}(A))$ has more than one element then, for this contraction recipe to work, there will have to be a unique inclusion-maximal element of $\mathcal{K}$ that is included in all of them, i.e. $\bigcup\{Z \in \mathcal{K} \mid Z \subseteq \cap C(\widehat{\mathcal{E}}_\mathcal{K}(A))\}$ is an element of $\mathcal{K}$, and it is equal to $K \div A$. The operation that treats ties between maximal outcomes in this way is called *perimaximal contraction*. Obviously, a perimaximal contraction can yield a non-maximal outcome in contraction by A only if there is a unique inclusion–maximal element of $\mathcal{K}$ that is included in all elements of $C(\widehat{\mathcal{E}}_\mathcal{K}(A))$. The formal definition of perimaximal contraction is as follows:

> Let K be a belief set and $\div$ a contraction[5] on K with the outcome set $\mathcal{K}$. Then $\div$ is a *perimaximal contraction* if and only if there is a choice function C such that for all A, if $\nvdash_\exists A$ then:
>
> $K \div A = \bigcup\{Z \in \mathcal{K} \mid Z \subseteq \cap C(\widehat{\mathcal{E}}_\mathcal{K}(A))\}$,
>
> and $K \div A = K$ if $\vdash_\exists A$. (Hansson 2013c)

The reader is referred to Hansson (2013c) for an investigation of perimaximal contraction, including axiomatic characterizations.

Non-recovery is satisfied by perimaximal contraction. So is *Cognitive accessibility* since the selection takes place on finite-based potential belief states rather than on remainders or other infinite objects. Since all elements of $\mathcal{K}$ are assumed to be finite-based, *Finite-based outcome* is also satisfied. However, due to the use of $\cap C(\widehat{\mathcal{E}}_\mathcal{K}(A))$ in the construction *No meet assumption* is not satisfied. *Expressive power* is satisfied since the beliefs and belief states that are expressible in the AGM model can be expressed in this framework as well. Finally, *Parsimony* is also satisfied. Of course we have introduced a new formal element namely the choice

---

[5] An operation of contraction on a belief set K is an operation $\div$ that satisfies *Inclusion* ($K \div A \subseteq K$) and *Success* ($K \div A \nvdash_\exists A$ if $\nvdash_\exists A$) (Hansson 1999, p. 65).

function C. However, it replaces the AGM selection function γ. The main difference between the two is that C, contrary to γ, operates on cognitively accessible objects namely potential belief states.

8 Blockage contraction

Consider again the case of the two bookkeepers from the previous section. Suppose that my reluctance to give up p is equally strong as my reluctance to give up q. Since I have to give up at least one of them this results in me giving up both. This means that instead of setting $K \div \{p\&q\}$ equal to $K \div \{p\}$ or $K \div \{q\}$ I set it equal to $K \div \{p,q\}$. We can say that $K \div p$ and $K \div q$ block each other, but $K \div \{p,q\}$ is not blocked by either of them, or by anything else, and can therefore be chosen. The blocking relation will be denoted $\rightarrow$, and the intended interpretation of $X \rightarrow Y$ is:

"For all p, if X does not imply p, then $Y \neq K \div p$."

A blocking relation $\rightarrow$ on $\mathcal{K}$ generates a *blockage contraction*, whose outcome is equal to the (unique) inclusion-maximal $\rightarrow$-unblocked element of the outcome set that does not imply p. (Rather obviously X is defined as $\rightarrow$-*unblocked within* $\mathcal{X}$ if and only if $X \in \mathcal{X}$ and there is no $Y \in \mathcal{X}$ such that $Y \rightarrow X$.) Within the somewhat more general framework of multiple contraction, blockage contraction can be defined as follows:

Let $\div$ be an operation on a belief set K with sets of sentences as inputs. Let $\mathcal{K}$ be its outcome set. Then $\div$ is a *blockage contraction* if and only if there is a (blocking) relation $\rightarrow$ on $\mathcal{K}$ such that for all $A \subseteq \mathcal{L}$:

(i) If $\nvdash_\exists A$ then $K \div A$ is the unique inclusion-maximal $\rightarrow$-unblocked element within the set of elements of $\mathcal{K}$ that do not imply any element of A.
(ii) If $\vdash_\exists A$ then $K \div A = K$.

Surprisingly weak properties on $\mathcal{K}$ and $\rightarrow$ are sufficient to guarantee the existence of such a unique inclusion-maximal potential outcome.[6] Blockage contraction has been axiomatically characterized, and shown to neither include nor being included by perimaximal contraction. (Hansson 2013b)

Non-recovery, *Cognitive accessibility*, *Finite-based outcome*, and *Expressive power* are all satisfied by blockage contraction for the same reasons as for perimaximal contraction. It also satisfies *No meet assumption* since we have replaced

---

[6] Namely the following five:
$K \in \mathcal{K}$ (integrity)
$Cn(\emptyset) \in \mathcal{K}$ (depletability)
If $X \subseteq Y$ then $X \nrightarrow Y$,
If $X \nsubseteq Y \nsubseteq X$ then there is either some $X' \subseteq X$ such that $X' \rightarrow Y$ or some $Y' \subseteq Y$ such that $Y' \rightarrow X$.
If $\mathcal{X}$ is a $\mathcal{K}$-plenished subset of $\mathcal{K}$, then there is some $Y \in \mathcal{X}$ such that $(\forall X \in \mathcal{X})(X \nrightarrow Y)$.
(A subset $\mathcal{X}$ of $\mathcal{K}$ is $\mathcal{K}$-plenished if and only if it holds for all $X, Y \in \mathcal{K}$ that if $X \subseteq Y \in \mathcal{X}$ then $X \in \mathcal{X}$.)

the intersective method by blocking relations. Finally, *Parsimony* is also satisfied. Although we have introduced blocking relations that are a new formal element, they represent an improvement over the AGM selection function γ that they replace since they operate on cognitively accessible objects namely potential belief states rather than on remainders.

9 Bootstrap contraction

Our last class of contraction operators can be introduced as an improvement over specified meet contraction. The most problematic feature of that operation is its use of full meet contraction. Unfortunately, when we interpret $K \div p = K \sim f(p)$ as:

"In order to contract by p we contract by $f(p)$"

the second use of the word "contract" does not have the same meaning as the first. Whereas the first "contract" in the sentence refers to the operation of contraction that we actually use, the second refers to full meet contraction that is a highly mutilating operation. (Alchourrón and Makinson 1982) The reason why full meet contraction is used here is that the whole point of replacing a contraction operator $\div$ by a sentential selector f through the identity $K \div p = K \div' f(p)$ would seem to be lost if the replacing operator $\div'$ contains an extralogical component. No other exclusively logical contraction operator than full meet contraction seems to be available for the purpose.

However, there is a way out, provided that the contraction operator $\div$ has an exclusively logical fraction, i.e. that there are sentences q such that $K \div q$ can be determined without reference to the extralogical component of $\div$. If $f(p)$ is one of these sentences but p is not, then we can reduce the contraction $K \div p$ to $K \div f(p)$, using the same contraction operator but moving to its exclusively logical fraction with the help of the sentential selector. Now suppose that $\div$ is such that for each p there is some q such that $K \div p = K \div q$ and $K \div q$ can be obtained without reference to the extralogical component of $\div$. In such a case, we can eliminate all references to the extralogical component and replace it by a sentential selector f, without recourse to a second, less plausible contraction operator. For the formal introduction of this approach it is useful to use the more general framework of multiple contraction and to focus on classes of operations (such as all base-generated partial meet contractions or all blockage contractions on a belief set K):

> Let $\mathbb{C}$ be a set of operators for the belief set K, all of which take sets of sentences as inputs. Then $\div_\vee$, the *overlap* of $\mathbb{C}$, is the (partial) operation such that, for any $A \subseteq \mathcal{L}$:
>
> (1) If $K \div A = K \div' A$ for all $\div, \div' \in \mathbb{C}$ then $K \div_\vee A = K \div A$ for any $\div \in \mathbb{C}$.
>
> (2) Otherwise, $K \div_\vee A$ is undefined.
>
> An operation $\div \in \mathbb{C}$ is *bootstrappable* with respect to $\mathbb{C}$ if and only if for all $A \subseteq \mathcal{L}$ there is some $A' \subseteq \mathcal{L}$ such that ($K \div_\vee A'$ is defined) $K \div A = K \div_\vee A'$. A selector f, i.e. a function from and to $\wp(\mathcal{L})$, is a *bootstrapping selector* for $\div$ if and only if it holds for all $A \subseteq \mathcal{L}$ that $K \div A = K \div_\vee f(A)$.

ℂ is bootstrappable if and only if all its members are so. (Hansson 2013d)

If ℂ is a set of operations that differ only in terms of some extra-logical element, then bootstrapping replaces that element by the bootstrapping selector. If the eliminated extra-logical element is cognitively less accessible than a selector, then bootstrapping brings in an improvement in terms of cognitive realism.

A simple and quite plausible property of multiple contraction turns out to be sufficient to make bootstrapping possible, given that the outcome set is specified:

OBSERVATION 2 *Let K be a finite-based belief set and 𝒦 a set of logically closed, finite-based subsets of K. Let ℂ be a set of multiple operators on K that have 𝒦 as their outcome set and satisfy:*

If $Y \in \mathcal{K}$ and it holds for all $X \in \mathcal{K}$ that $X \subseteq Y$ if and only if $X \nvdash_\exists A$, then $Y = K \div A$. (Unique maximum)

Then ℂ is bootstrappable.

PROOF: See Hansson (2013d).

Unique maximum is one of the many belief revision postulates that codify the intuition that information should not be lost unless there are reasons to remove it. A wide range of classes of multiple operations have been shown to satisfy Unique maximum and are therefore bootstrappable:

OBSERVATION 3 *The multiple versions of base-generated partial meet contraction, perimaximal contraction, and blockage contraction are all bootstrappable.*[7]

PROOF: See Hansson (2013d).

Non-recovery, *Cognitive accessibility*, *Finite-based outcome*, and *Expressive power* are all satisfied by bootstrap contraction for the same reasons as for perimaximal and blockage contraction. Bootstrap contraction also satisfies *No meet assumption* since the intersective method for dealing with ties is not used. Finally, *Parsimony* is also satisfied. Although we have introduced a selector f as a new formal element, it represents an improvement over the AGM selection function γ that it replaces since it operates on cognitively accessible objects namely potential belief sets rather than on remainders.

10 Conclusion

The outcome of these comparisons are summarized in Table 2. In terms of the criteria used here, blockage and bootstrap contraction fare best among the evaluated classes of operators.

---

[7] Multiple direct partial meet contraction is also bootstrappable, but $f(A)$ may have to be infinite for a finite contractee A.

Cognitive realism in belief revision

Table 2 Summary of the evaluation of contraction operators in terms of cognitive realism. + denotes that the condition is satisfied, and − that it is not satisfied.

	AGM	Finite language	Base-generated	Specified meet	Peri-maximal	Blockage	Bootstrap
Non-recovery	−	−	+	+	+	+	+
Cognitive accessibility	−	+	+	+	+	+	+
Finite-based outcome	−	+	+	+	+	+	+
No meet assumption	−	−	−	−	−	+	+
Expressive power	+	−	+	+	+	+	+
Parsimony	+	+	−	+	+	+	+

The criteria that we have used can be divided into two classes that we may call "black box" and "glass box" criteria. A black box criterion is only concerned with the original belief state, the input, and the resulting new belief state. It does not refer to the mechanism by which the first two of these give rise to the third. A glass box criterion refers to the inner workings or mechanisms of the operation. Three of our criteria, namely *Non-recovery*, *Finite-based outcome*, and *Expressive power* are black box criteria. The other three, namely *Cognitive accessibility*, *No meet assumption*, and *Parsimony*, are glass box criteria.

The three black box criteria are satisfied by five of our operations, namely base-generated partial meet contraction, specified meet contraction, perimaximal contraction, blockage contraction, and bootstrap contraction. If we add the white box criteria, then only blockage and bootstrap contraction pass the test. In a search for cognitively realistic operations, the more demanding white box criteria should have a crucial role.

References

Alchourrón, Carlos and David Makinson (1982) "On the logic of theory change: Contraction functions and their associated revision functions", *Theoria* 48: 14-37.

Alchourrón, Carlos, Peter Gärdenfors, and David Makinson (1985) "On the Logic of Theory Change: Partial Meet Contraction and Revision Functions", *Journal of Symbolic Logic* 50: 510-530.

Fermé, Eduardo and Sven Ove Hansson (2011) "AGM 25 years. Twenty-Five Years of Research in Belief Change", *Journal of Philosophical Logic* 40:295-331.

Fuhrmann, André (1991) "Theory contraction through base contraction", *Journal of Philosophical Logic* 20:175-203, 1991.

Fuhrmann, André and Sven Ove Hansson (1994) "A Survey of Multiple Contraction", *Journal of Logic, Language and Information* 3:39-76.

Gärdenfors, Peter (1982) "Rules for Rational Changes of Belief", pp. 88-101 in Tom Pauli (ed) *Philosophical Essays dedicated to Lennart Åqvist on his fiftieth birthday. Philosophical studies* 34 (Published by the Philosophical Society and the Department of Philosophy). Uppsala: University of Uppsala.

Glaister, Stephen Murray (2000) "Recovery recovered", *Journal of Philosophical Logic* 29:171-206.

Grove, Adam (1988) "Two modellings for theory change", *Journal of Philosophical Logic* 17:157-170.

Hansson, Sven Ove (1991) "Belief Contraction Without Recovery", *Studia Logica* 50: 251-260.

Hansson, Sven Ove (1993) "Theory Contraction and Base Contraction Unified", *Journal of Symbolic Logic* 58: 602-625.

Hansson, Sven Ove (1999a) *A Textbook of Belief Dynamics. Theory Change and Database Updating.* Kluwer 1999.

Hansson, Sven Ove (1999b) "Recovery and epistemic residues", *Journal of Logic, Language and Information* 8:421-428,.

Hansson, Sven Ove (2007a) "Contraction Based on Sentential Selection", *Journal of Logic and Computation*, 17:479-498.

Hansson, Sven Ove (2007b) "Three Approaches to Finitude in Belief Change", Toni Rønnow-Rasmussen, Björn Petersson, Jonas Josefsson and Dan Egonsson, *Hommage à Wlodek. Philosophical Papers Dedicated to Wlodek Rabinowicz.* Lund 2007 (http://www.fil.lu.se/hommageawlodek)

Hansson, Sven Ove (2008) "Specified Meet Contraction", *Erkenntnis* 69:31-54.

Hansson, Sven Ove (2013a) "Outcome level analysis of belief contraction", *Review of Symbolic Logic* 6:183-204.

Hansson, Sven Ove (2013b) "Blockage contraction", *Journal of Philosophical Logic* 42:415-442, 2013.

Hansson, Sven Ove (2013c) "Maximal and Perimaximal Contraction", *Synthese* 190: 3325-3348.

Hansson, Sven Ove (2013d) "Bootstrap Contraction", *Studia Logica* 101:1013-1029.

Levi, Isaac (1977) "Subjunctives, dispositions and chances", *Synthese* 34:423-455.

Levi, Isaac (1991) *The Fixation of Belief and Its Undoing.* Cambridge, MA.: Cambridge University Press.

Makinson, David (1987) "On the Status of the Postulate of Recovery in the Logic of Theory Change", *Journal of Philosophical Logic* 16: 383-394.

Makinson, David (1997) "On the force of some apparent counterexamples to recovery", pp. 475-481 in E. Garzón Valdéz et al. (eds.) *Normative Systems in Legal and Moral Theory: Festschrift for Carlos Alchourrón and Eugenio Bulygin*, Berlin: Duncker & Humblot.

Meyer, Thomas, Johannes Heidema, Willem Labuschagne, and Louise Leenen (2002) "Systematic Withdrawal", *Journal of Philosophical Logic* 31: 415-443.

Nebel, B. (1989) "A knowledge level analysis of belief revision". In *Proceedings of the 1st International Conference of Principles of Knowledge Representation and Reasoning*, pp. 301–311. Morgan Kaufmann.

Rott, Hans (2001) *Change, choice and inference. A study of belief revision and nonmonotonic reasoning*, Oxford University Press. Volume 29, Number 2

Rott, Hans and Maurice Pagnucco (1999) "Severe Withdrawal (and Recovery)", *Journal of Philosophical Logic* 28:501-547.

Sandqvist, Tor (2000) "On Why the Best Should Always Meet", *Economics and Philosophy* 16:287-313.

Characterizing change in abstract argumentation systems

Pierre Bisquert, Claudette Cayrol, Florence Dupin de Saint-Cyr, and Marie-Christine Lagasquie-Schiex

Abstract An argumentation system can undergo changes (addition/removal of arguments/interactions). At an abstract level, we propose a typology to classify the different properties describing a change operation. This typology reflects the evolution of three features:

- the set of extensions in Dung's sense (*e.g.*, the set of extensions is empty before the change and not empty after the change),
- the sets of accepted arguments (*e.g.*, all the arguments skeptically accepted before the change are still skeptically accepted after the change) and
- the status of some given argument (*e.g.*, an accepted argument may become rejected after the change).

Then, an important issue is to provide characterizations for these properties: *i.e.* conditions on the argumentation system and on the change operation that are necessary or sufficient to guarantee that the properties are satisfied.

So, in this paper, we present this typology and the characterization results obtained either directly or by using an approach based on a notion of duality. Our results are twofold, they can be considered as a guide for selecting the change operation to perform in order to obtain a desired property on an argumentation system and they may also be used as a tool for predicting the result of a change operation in a given context.

1 Introduction

The main feature of argumentation framework is the ability to deal with incomplete and / or contradictory information, especially for reasoning [Dung, 1995;

Pierre Bisquert · Claudette Cayrol · Florence Dupin de Saint-Cyr · Marie-Christine Lagasquie-Schiex
IRIT laboratory, Toulouse, France,
e-mail: bisquert,ccayrol, bannay, lagasq@irit.fr

Amgoud and Cayrol, 2002]. Moreover, argumentation can be used to formalize dialogues between several agents by modeling the exchange of arguments in, *e.g.*, negotiation between agents [Amgoud *et al.*, 2000]. An argumentation system (AS) consists of a collection of arguments interacting with each other through a relation reflecting conflicts between them, called *attack*. The issue of argumentation is then to determine "acceptable" sets of arguments (*i.e.*, sets able to defend themselves collectively while avoiding internal attacks), called *"extensions"*, and thus to reach a coherent conclusion. Another form of analysis of an AS is the study of the particular status of each argument, this status is based on membership (or non-membership) of the extensions. Formal frameworks have greatly eased the modeling and study of AS. In particular, the framework of [Dung, 1995] allows to completely abstract the "concrete" meaning of the arguments and relies only on binary interactions that may exist between them. This approach enables the user to focus on other aspects of argumentation, including its dynamic side. Indeed, in the course of a discussion or due to the acquisition of new pieces of information, an AS can undergo changes such as the addition of a new argument or the removal of an argument considered as illegal. Thus, it is interesting to study these changes and to characterize them by giving properties describing a change operation and by providing conditions under which these properties hold. Moreover, the study of the links between addition and removal through the concept of duality is a way to complete the characterization of removal through the work previously done on addition, and conversely. The following example shows that some knowledge about duality could help to benefit from known results:

> *Mr Pink knows that one simple argument could defeat Mr White's argumentation, but this argument is lacking. Another way to win could be to remove one of Mr White's arguments (e.g. by doing an objection). Unfortunately, Mr Pink does not know the consequences of this removal.*

Although the research on dynamics of AS is growing [Paglieri and Castelfranchi, 2005; 2006; Boella *et al.*, 2009a; 2009b; Baumann and Brewka, 2010; Moguillansky *et al.*, 2010; 2013; Liao *et al.*, 2011], the removal of argument has so far been little considered. A realistic example of the use of removal may nevertheless be found in [Bisquert *et al.*, 2011] and shows that studying argument removal is at least as important as studying argument addition. *A fortiori*, the relationship between addition and removal of argument has not, to our knowledge, been treated so far.

This paper presents a synthesis about change characterization based on already published papers [Cayrol *et al.*, 2010; Bisquert *et al.*, 2012b; 2012a] and including new results. A brief background is given in Sect. 2. Sect. 3 displays properties of a change operation reflecting possible modifications of an AS. A direct characterization of these properties is given in Sect. 4 to 6. Then various notions of duality are presented in Sect. 7 and are used for enriching the characterization (see Sect. 8 to 9). Sect. 10.1 describes the possible use of these characterization results. Finally, Sect. 10.3 concludes and suggests perspectives of our work. In Appendix 11, the reader will find tables synthesizing all characterization results and the proofs of the new direct results are given in Appendix 12.

2 Background

We give here some background concerning argumentation systems (Sect. 2.1) as well as change operations (Sect. 2.2).

2.1 Dung's abstract argumentation system

The work presented in this paper uses the framework of [Dung, 1995]:

Def. 1 (Argumentation System) *An* argumentation system (AS) *is a pair* $\langle \mathbf{A}, \mathbf{R} \rangle$, *where* $\mathbf{A}$ *is a finite nonempty set of arguments and* $\mathbf{R}$ *is a binary relation on* $\mathbf{A}$, *called* attack relation. *Let* $A, B \in \mathbf{A}$, ARB *means that A attacks B.* $\langle \mathbf{A}, \mathbf{R} \rangle$ *will be represented by an* argumentation graph $\mathscr{G}$ *whose vertices are the arguments and whose edges correspond to* $\mathbf{R}$[1].

Let $A \in \mathbf{A}, B \in \mathbf{A}$, A *indirectly attacks* B iff[2] there exists an odd-length path from A to B in $\mathscr{G}$. In this paper, we also use the following notions based on the attack relation, namely the attack of an argument to - and from - a set:

Def. 2 (Attack from and to a set) *Let* $A \in \mathbf{A}$ *and* $\mathscr{S} \subseteq \mathbf{A}$. $\mathscr{S}$ *attacks A iff* $\exists X \in \mathscr{S}$ *s.t.*[3] XRA. A *attacks* $\mathscr{S}$ *iff* $\exists X \in \mathscr{S}$ *s.t.* ARX.

The acceptable sets of arguments ("extensions") are determined according to a given semantics which is usually based on the following concepts:

Def. 3 (Conflict-freeness, defense, admissibility) *Let* $A \in \mathbf{A}$ *and* $\mathscr{S} \subseteq \mathbf{A}$. $\mathscr{S}$ *is* conflict-free *iff there does not exist* $A, B \in \mathscr{S}$ *s.t.* ARB. $\mathscr{S}$ defends *an argument A iff each attacker of A is attacked by an argument of* $\mathscr{S}$. *The set of the arguments defended by* $\mathscr{S}$ *is denoted by* $\mathscr{F}(\mathscr{S})$; $\mathscr{F}$ *is called the* characteristic function *of* $\langle \mathbf{A}, \mathbf{R} \rangle$. *More generally,* $\mathscr{S}$ indirectly defends A *iff* $A \in \bigcup_{i \geq 1} \mathscr{F}^i(\mathscr{S})$. $\mathscr{S}$ *is an* admissible set *iff it is conflict-free and it defends all its elements.*

The set of extensions of $\langle \mathbf{A}, \mathbf{R} \rangle$ is denoted by $\mathbf{E}$ (with $\mathscr{E}_1, \ldots, \mathscr{E}_n$ standing for the extensions). In this article, we restrict our study to the most traditional semantics proposed by [Dung, 1995]:

Def. 4 (Acceptability semantics) *Let* $\mathscr{E} \subseteq \mathbf{A}$, $\mathscr{E}$ *is a* preferred extension *iff* $\mathscr{E}$ *is a maximal admissible set (with respect to set inclusion* $\subseteq$*).* $\mathscr{E}$ *is the only* grounded extension *iff* $\mathscr{E}$ *is the least fixed point (with respect to* $\subseteq$*) of the characteristic function* $\mathscr{F}$. $\mathscr{E}$ *is a* stable extension *iff* $\mathscr{E}$ *is conflict-free and attacks any argument not belonging to* $\mathscr{E}$.

---

[1] In this paper, we use freely $\langle \mathbf{A}, \mathbf{R} \rangle$ or $\mathscr{G}$ to refer to an AS. Similarly, if there is no ambiguity, we use without distinction $\mathbf{A}$ and $\mathscr{G}$.

[2] iff = if and only if.

[3] s.t. = such that.

The status of an argument is determined by its membership of the extensions of the selected semantics: *e.g.*, an argument A can be "skeptically accepted" (resp. "credulously") if $A \in$ to all the extensions (resp. at least to one extension) and be "rejected" if $A \notin$ to any extension.

Prop. 1 *[Dung, 1995]*
1. *There is at least one preferred extension, always a unique grounded extension, while there may be zero, one or several stable extensions.*
2. *Each admissible set is included in a preferred extension.*
3. *Each stable extension is a preferred extension, the converse is false.*
4. *The grounded extension is included in each preferred extension.*
5. *Each argument which is not attacked belongs to the grounded extension (hence to each preferred and to each stable extension).*
6. *If $\mathbf{R}$ is finite, then the grounded extension can be computed by iteratively applying the function $\mathscr{F}$ from the empty set.*
7. *If $\mathbf{A}$ is non empty, then a stable extension is always non empty.*

Prop. 2 *[Dunne and Bench-Capon, 2001; 2002]*
1. *If $\mathscr{G}$ contains no cycle, then $\langle \mathbf{A}, \mathbf{R} \rangle$ has a unique preferred extension, which is also the grounded extension and the unique stable extension.*
2. *If $\{\}$ is the unique preferred extension of $\langle \mathbf{A}, \mathbf{R} \rangle$, then $\mathscr{G}$ contains an odd-length cycle.*
3. *If $\langle \mathbf{A}, \mathbf{R} \rangle$ has no stable extension, then $\mathscr{G}$ contains an odd-length cycle.*
4. *If $\mathscr{G}$ contains no odd-length cycle, then preferred and stable extensions coincide.*
5. *If $\mathscr{G}$ contains no even-length cycle, then $\langle \mathbf{A}, \mathbf{R} \rangle$ has a unique preferred extension.*

2.2 Dynamics in argumentation systems

We rely on the work of [Cayrol *et al.*, 2010] which have distinguished four change operations; in this paper, we only use the operations of addition and removal of an argument and its interactions:

Def. 5 (Change operations) *Let $\langle \mathbf{A}, \mathbf{R} \rangle$ be an AS, Z be an argument and $\mathscr{I}_z$ be a set of interactions concerning Z*
- *Adding $Z \notin \mathbf{A}$ and $\mathscr{I}_z \subseteq (\mathbf{A} \cup \{Z\}) \times (\mathbf{A} \cup \{Z\})$ is a change operation, denoted by $\oplus$, providing a new AS s.t. $\langle \mathbf{A}, \mathbf{R} \rangle \oplus (Z, \mathscr{I}_z) = \langle \mathbf{A} \cup \{Z\}, \mathbf{R} \cup \mathscr{I}_z \rangle$.*
- *Removing $Z \in \mathbf{A}$ and $\mathscr{I}_z \subseteq \mathbf{R}$ is a change operation, denoted by $\ominus$, providing a new system[4] s.t. $\langle \mathbf{A}, \mathbf{R} \rangle \ominus (Z, \mathscr{I}_z) = \langle \mathbf{A} \setminus \{Z\}, \mathbf{R} \setminus \mathscr{I}_z \rangle$.*

We denote by $\mathscr{O}$ a change operation ($\oplus$ or $\ominus$)[5]. The new AS $\langle \mathbf{A}', \mathbf{R}' \rangle$ obtained by the application of $\mathscr{O}$ will be represented by the argumentation graph $\mathscr{G}' = \mathscr{O}(\mathscr{G})$. Moreover, we assume that Z does not attack itself and $\forall (X, Y) \in \mathscr{I}_z$, we have either

---

[4] It is the subgraph of $\langle \mathbf{A}, \mathbf{R} \rangle$ induced by $\mathbf{A} \setminus \{Z\}$.

[5] The symbols $\oplus$ and $\ominus$ used here correspond to the symbols $\oplus_I^a$ and $\ominus_I^a$ of [Cayrol *et al.*, 2010], where *a* stands for "argument" and *I* for "interactions", meaning that the operation concerns an argument and its interactions.

Characterizing change in abstract argumentation systems

$(X = Z$ and $Y \neq Z$, $Y \in \mathbf{A})$ or $(Y = Z$ and $X \neq Z$, $X \in \mathbf{A})$. In case of removing, let us note that $\mathscr{I}_z$ is the set of all the interactions concerning Z in $\langle \mathbf{A}, \mathbf{R} \rangle$.

The set of extensions of $\langle \mathbf{A}', \mathbf{R}' \rangle$ is denoted by $\mathbf{E}'$ (with $\mathscr{E}'_1, \ldots, \mathscr{E}'_n$ standing for the extensions). In this chapter, we assume that the semantics remains the same before and after any change operation. Note that if the removal operation was only described by the name of the removed argument then removal could be viewed as a non injective application (while surjectivity holds due to Def. 5 since $\forall \mathscr{G}$, $\mathscr{G}' = \mathscr{O}(\mathscr{G})$ is unique even if the removed attacked (the set $\mathscr{I}_z$) associated to $\mathscr{O}$ are not described).

Ex. 1 *The following 3 systems can be changed into the same system by the removal of Z (see Table 1 which also gives the grounded extension of each system), however each change corresponds to a distinct operation since $\mathscr{I}_z$ is different:*

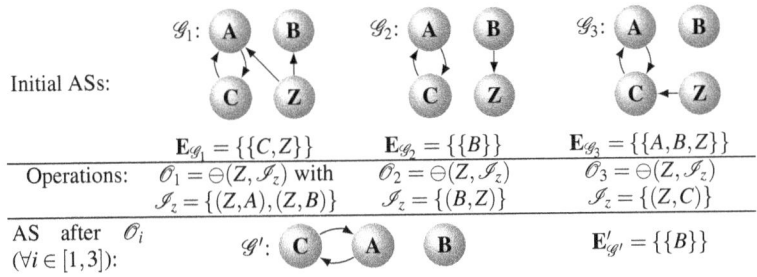

Table 1 Illustration of the removal operation.

The impact of a change operation will be studied through the notion of change property. A change property $\mathscr{P}$ can be seen as a set of pairs $(\mathscr{G}, \mathscr{G}')$, where $\mathscr{G}$ and $\mathscr{G}'$ are argumentation graphs:

Ex. 1 (cont'd) *Let $\mathscr{P}$ be the property defined by "$\mathscr{P}(\mathscr{G}, \mathscr{G}')$ holds iff any extension of $\mathscr{G}'$ is included in at least one extension of $\mathscr{G}$". Thus, $\mathscr{P}(\mathscr{G}_1, \mathscr{G}')$ does not hold while $\mathscr{P}(\mathscr{G}_2, \mathscr{G}')$ and $\mathscr{P}(\mathscr{G}_3, \mathscr{G}')$ hold.*

Def. 6 (Operation satisfying a property) *A change operation $\mathscr{O}$ satisfies a property $\mathscr{P}$ iff $\forall \mathscr{G}$, $\mathscr{P}(\mathscr{G}, \mathscr{O}(\mathscr{G}))$ holds.*

Ex. 1 (cont'd) *$\mathscr{P}(\mathscr{G}_1, \mathscr{G}')$ does not hold. Thus, $\mathscr{O}_1$ does not satisfy $\mathscr{P}$.*

3 Properties of change operations

For a change operation, there exist three classes of properties concerning:
- either the evolution of the set of extensions,
- or the evolution of the acceptability of a set of arguments,
- or the evolution of the status of a given argument.

3.1 Properties about the set of extensions

Change properties express structural modifications of an AS that are caused by a change operation. In this section, we focus on these modifications in order to obtain a clear and accurate classification. For that purpose, a new partition, inspired by the work of [Cayrol *et al.*, 2010] and based on three possible cases of evolution of the set of extensions, has been defined:
- the *extensive* case, in which the number of extensions increases,
- the *restrictive* case, in which the number of extensions decreases,
- the *constant* case, in which the number of extensions remains the same.

For each case, numerous sub-cases are proposed and denoted by a letter (*e* for the *extensive* case, *r* for the *restrictive* case and *c* for the *constant* case) subscripted by the expression $\gamma - \gamma'$, where γ (resp. γ') describes the set of extensions before (resp. after) the change. Thus γ and γ' can be:
- $\emptyset$: the set of extensions is empty,
- $1e$: the set of extensions is reduced to one empty extension,
- $1ne$: the set of extensions is reduced to one non-empty extension,
- k (resp. j): the set of extensions contains k (resp. j) extensions s.t. $1 < k$ (resp. $1 < j < k$: note that the symbol j is used only if the symbol k belongs also to the expression $\gamma - \gamma'$).

For instance, the notation $e_{\emptyset - 1ne}$ means that the change increases the number of extensions (so it is an *extensive* case), with no initial extension ($\emptyset$) and one non-empty final extension ($1ne$).

Nevertheless, some special sub-cases of the *constant* case are denoted by another method since they are based on notions distinct from the emptiness or the number of the extensions; for these sub-cases, the subscript is replaced by a qualifier. For instance, the *c-conservative* case describes the case where the extensions remain unchanged after the change.

Note also that for the sake of clarity, we say that a change satisfying a property $\mathscr{P}$ is a "$\mathscr{P}$ change"; for example, a change that satisfies the *constant* property is said *constant* change.

Here is the formal definition of these changes. First, we study the case in which a change increases the number of extensions, called *extensive* change.

Def. 7 (Extensive change) *The change from $\mathscr{G}$ to $\mathscr{G}'$ is extensive iff $|\mathbf{E}| < |\mathbf{E}'|$. The sub-cases of extensive changes from $\mathscr{G}$ to $\mathscr{G}'$ are:*
1. $e_{\emptyset - 1ne}$ iff $|\mathbf{E}| = 0$ and $|\mathbf{E}'| = 1$, with $\mathscr{E}' \neq \emptyset$.
2. $e_{\emptyset - k}$ iff $|\mathbf{E}| < |\mathbf{E}'|$, $|\mathbf{E}| = 0$ and $|\mathbf{E}'| > 1$.
3. e_{1e-k} iff $|\mathbf{E}| < |\mathbf{E}'|$ and $|\mathbf{E}| = 1$, with $\mathscr{E} = \emptyset$.
4. e_{1ne-k} iff $|\mathbf{E}| < |\mathbf{E}'|$ and $|\mathbf{E}| = 1$, with $\mathscr{E} \neq \emptyset$.
5. e_{j-k} iff $1 < |\mathbf{E}| < |\mathbf{E}'|$.

The *restrictive* change, in which a change decreases the number of extensions, is defined symmetrically to the *extensive* change.

Def. 8 (Restrictive change) *The change from $\mathscr{G}$ to $\mathscr{G}'$ is restrictive iff $|\mathbf{E}| > |\mathbf{E}'|$. The sub-cases of restrictive changes from $\mathscr{G}$ to $\mathscr{G}'$ are:*

1. $r_{1ne-\varnothing}$ iff $|\mathbf{E}| = 1$, with $\mathcal{E} \neq \varnothing$, and $|\mathbf{E}'| = 0$.
2. $r_{k-\varnothing}$ iff $|\mathbf{E}| > |\mathbf{E}'|$, $|\mathbf{E}| > 1$ and $|\mathbf{E}'| = 0$.
3. r_{k-1e} iff $|\mathbf{E}| > |\mathbf{E}'|$ and $|\mathbf{E}'| = 1$, with $\mathcal{E}' = \varnothing$.
4. r_{k-1ne} iff $|\mathbf{E}| > |\mathbf{E}'|$ and $|\mathbf{E}'| = 1$, with $\mathcal{E}' \neq \varnothing$.
5. r_{k-j} iff $1 < |\mathbf{E}'| < |\mathbf{E}|$.

The *constant* change corresponds to the case where the number of extensions remains unchanged and its sub-cases depend on the inclusions between the various possible extensions ($\mathcal{G}$ to $\mathcal{G}'$ and vice versa), emptiness of these extensions, ...

Def. 9 (Constant change) *The change from $\mathcal{G}$ to $\mathcal{G}'$ is constant iff $|\mathbf{E}| = |\mathbf{E}'|$. The sub-cases of constant changes from $\mathcal{G}$ to $\mathcal{G}'$ are:*

1. **c-conservative** *iff* $\mathbf{E} = \mathbf{E}'$.
2. c_{1e-1ne} *iff* $\mathbf{E} = \{\{\}\}$ *and* $\mathbf{E}' = \{\mathcal{E}'\}$, *with* $\mathcal{E}' \neq \varnothing$.
3. c_{1ne-1e} *iff* $\mathbf{E} = \{\mathcal{E}\}$, *with* $\mathcal{E} \neq \varnothing$ *and* $\mathbf{E}' = \{\{\}\}$.
4. **c-expansive** *iff* $\mathbf{E} \neq \varnothing$ *and* $|\mathbf{E}| = |\mathbf{E}'|$ *and* $\forall \mathcal{E}_i \in \mathbf{E}, \exists \mathcal{E}'_j \in \mathbf{E}', \varnothing \neq \mathcal{E}_i \subset \mathcal{E}'_j$ *and* $\forall \mathcal{E}'_j \in \mathbf{E}', \exists \mathcal{E}_i \in \mathbf{E}, \varnothing \neq \mathcal{E}_i \subset \mathcal{E}'_j$.
5. **c-narrowing** *iff* $\mathbf{E} \neq \varnothing$ *and* $|\mathbf{E}| = |\mathbf{E}'|$ *and* $\forall \mathcal{E}_i \in \mathbf{E}, \exists \mathcal{E}'_j \in \mathbf{E}', \varnothing \neq \mathcal{E}'_j \subset \mathcal{E}_i$ *and* $\forall \mathcal{E}'_j \in \mathbf{E}', \exists \mathcal{E}_i \in \mathbf{E}, \varnothing \neq \mathcal{E}'_j \subset \mathcal{E}_i$.
6. **c-altering** *iff* $|\mathbf{E}| = |\mathbf{E}'|$ *and it is neither* c-conservative, *nor* c_{1e-1ne}, *nor* c_{1ne-1e}, *nor* c-expansive, *nor* c-narrowing.

Def. 9.1, 9.2, 9.3 and 9.6 are fairly straightforward. Def. 9.4 states that a *c-expansive* change is a change where all the extensions of $\mathcal{G}$, which are not initially empty, are increased by some arguments. A *c-narrowing* change, according to Def. 9.5, is a change where all the extensions of $\mathcal{G}$ are reduced by some arguments without becoming empty.

3.2 Properties about the acceptability of a set of arguments

A change can also have an impact on the acceptability of sets of arguments. For instance, in a dialog, it would be interesting to know if the addition or the removal of an argument modifies the acceptability of the arguments previously accepted. We speak of "monotony from $\mathcal{G}$ to $\mathcal{G}'$" when every argument accepted *before* the change is still accepted *after* the change, *i.e.*, no accepted argument is lost and there is a (not necessarily strict) *expansion* of acceptability. A second case, referred as "monotony from $\mathcal{G}'$ to $\mathcal{G}$", occurs when every argument accepted *after* the change was already accepted *before* the change, *i.e.*, no new accepted argument appears and there is a (not necessarily strict) *restriction* of acceptability.

Def. 10 (Simple monotony)
1. *The change from $\mathcal{G}$ to $\mathcal{G}'$ satisfies the property of* simple expansive monotony *iff* $\forall \mathcal{E}_i \in \mathbf{E}, \exists \mathcal{E}'_j \in \mathbf{E}', \mathcal{E}_i \subseteq \mathcal{E}'_j$.
2. *The change from $\mathcal{G}$ to $\mathcal{G}'$ satisfies the property of* simple restrictive monotony *iff* $\forall \mathcal{E}'_j \in \mathbf{E}', \exists \mathcal{E}_i \in \mathbf{E}, \mathcal{E}'_j \subseteq \mathcal{E}_i$.

These properties are refined into *credulous expansive monotony* and *credulous restrictive monotony* when acceptability is restricted to credulous acceptability, and into *skeptical expansive monotony* and *skeptical restrictive monotony* when skeptical acceptability is considered:

Def. 11 (Credulous and skeptical monotonies)

1. The change from $\mathcal{G}$ to $\mathcal{G}'$ satisfies the property of credulous expansive monotony iff: $\bigcup_{1 \leq i \leq |\mathbf{E}|} \mathcal{E}_i \subseteq \bigcup_{1 \leq j \leq |\mathbf{E}'|} \mathcal{E}'_j$.

2. The change from $\mathcal{G}$ to $\mathcal{G}'$ satisfies the property of credulous restrictive monotony iff: $\bigcup_{1 \leq j \leq |\mathbf{E}'|} \mathcal{E}'_j \subseteq \bigcup_{1 \leq i \leq |\mathbf{E}|} \mathcal{E}_i$.

3. The change from $\mathcal{G}$ to $\mathcal{G}'$ satisfies the property of skeptical expansive monotony iff: $\bigcap_{1 \leq i \leq |\mathbf{E}|} \mathcal{E}_i \subseteq \bigcap_{1 \leq j \leq |\mathbf{E}'|} \mathcal{E}'_j$.

4. The change from $\mathcal{G}$ to $\mathcal{G}'$ satisfies the property of skeptical restrictive monotony iff $\bigcap_{1 \leq j \leq |\mathbf{E}'|} \mathcal{E}'_j \subseteq \bigcap_{1 \leq i \leq |\mathbf{E}|} \mathcal{E}_i$.

Some links between these properties are given in Prop 6, 7, Sect. 4.

3.3 Properties about the status of a given argument

A change operation about an argument Z can of course have an influence on Z, but also on other given arguments (particularly arguments that are attacked or defended by Z). Considering the influence on Z, there are only three possible cases that concern the establishment of the acceptability of Z (credulous, skeptical or not established at all) and they occur only when there is an *addition* of Z (if we remove Z, its acceptability is of course modified but it is obvious and irrelevant):

Def. 12 (Priority to recency) *The change from $\mathcal{G}$ to $\mathcal{G}'$ which adds Z satisfies priority to recency iff*

- (credulous-only priority to recency) $\exists \mathcal{E}'_j, \mathcal{E}'_l \in \mathbf{E}', Z \in \mathcal{E}'_j$ and $Z \notin \mathcal{E}'_l$.
- (skeptical priority to recency) $\forall \mathcal{E}'_j \in \mathbf{E}', Z \in \mathcal{E}'_j$.

Considering the influence on a given argument X distinct from Z (so $X \in \mathcal{G} \cap \mathcal{G}'$), we identify several properties expressing the modification of the status of X when a change operation is done on the AS.

Def. 13 (Acceptability establishment) *Let* $X \in \mathcal{G} \cap \mathcal{G}'$. *The change from $\mathcal{G}$ to $\mathcal{G}'$ establishes acceptability for X iff* $\forall \mathcal{E}_i \in \mathbf{E}, X \notin \mathcal{E}_i$ *and*

- (credulous-only acceptability establishment) $\exists \mathcal{E}'_j, \mathcal{E}'_l \in \mathbf{E}', X \in \mathcal{E}'_j$ and $X \notin \mathcal{E}'_l$.
- (skeptical acceptability establishment) $\forall \mathcal{E}'_j \in \mathbf{E}', X \in \mathcal{E}'_j$.

Def. 14 (Acceptability removal) *Let* $X \in \mathcal{G} \cap \mathcal{G}'$. *The change from $\mathcal{G}$ to $\mathcal{G}'$ removes acceptability for X iff* $\forall \mathcal{E}'_j \in \mathbf{E}', X \notin \mathcal{E}'_j$ *and*

- (credulous-only acceptability removal) $\exists \mathcal{E}_i, \mathcal{E}_k \in \mathbf{E}, X \in \mathcal{E}_i$ and $X \notin \mathcal{E}_k$.
- (skeptical acceptability removal) $\forall \mathcal{E}_i \in \mathbf{E}, X \in \mathcal{E}_i$.

The modification of the status of X can be less "drastic", *e.g.*, an argument can belong to some extension before the change, and become member of every extension after it:

Def. 15 (General diffusion of acceptability) *Let $X \in \mathscr{G} \cap \mathscr{G}'$. The change from $\mathscr{G}$ to $\mathscr{G}'$ is a general diffusion of acceptability for X iff $\exists \mathscr{E}_i \in \mathbf{E}, X \in \mathscr{E}_i, \exists \mathscr{E}_k \in \mathbf{E}, X \notin \mathscr{E}_k$ and $\forall \mathscr{E}'_j \in \mathbf{E}', X \in \mathscr{E}'_j$.*

Def. 16 (Partial degradation of acceptability) *Let $X \in \mathscr{G} \cap \mathscr{G}'$. The change from $\mathscr{G}$ to $\mathscr{G}'$ partially degrades acceptability for X iff $\forall \mathscr{E}_i \in \mathbf{E}, X \in \mathscr{E}_i, \exists \mathscr{E}'_j \in \mathbf{E}', X \in \mathscr{E}'_j$ and $\exists \mathscr{E}'_l \in \mathbf{E}', X \notin \mathscr{E}'_l$.*

And lastly, the acceptability of X can also remain unchanged (note that the following definition refines the *partial monotony* property defined by [Cayrol *et al.*, 2010]).

Def. 17 (Status preservation) *Let $X \in \mathscr{G} \cap \mathscr{G}'$. The change from $\mathscr{G}$ to $\mathscr{G}'$ preserves the status of X iff*
- *(preserves credulous-only acceptability)* $\exists \mathscr{E}_i, \mathscr{E}_k \in \mathbf{E}, X \in \mathscr{E}_i, X \notin \mathscr{E}_k, \exists \mathscr{E}'_j, \mathscr{E}'_l \in \mathbf{E}', X \in \mathscr{E}'_j$ and $X \notin \mathscr{E}'_l$.
- *(preserves skeptical acceptability)* $\forall \mathscr{E}_i \in \mathbf{E}, X \in \mathscr{E}_i$ and $\forall \mathscr{E}'_j \in \mathbf{E}', X \in \mathscr{E}'_j$.
- *(preserves the rejected status)* $\forall \mathscr{E}_i \in \mathbf{E}, X \notin \mathscr{E}_i$ and $\forall \mathscr{E}'_j \in \mathbf{E}', X \notin \mathscr{E}'_j$.

Note that (credulous-only or skeptical) *preservation of acceptability* for X does not mean that arguments that were accepted together with X remain accepted after the change (this differs from the monotony property presented in Sect. 3.2). Nevertheless, some links exist (see Prop. 7, Sect. 4).

4 Characterizing change operations: Preliminary results

In this section, we give some general results about the characterization of addition and removal in argumentation under some semantics (some of them are taken from [Cayrol *et al.*, 2010; Bisquert *et al.*, 2012b]). The first result is due to the uniqueness of the grounded extension (Prop. 1.1).

Prop. 3 *Under the grounded semantics, a change (addition or suppression) is never $e_{\varnothing-1ne}$, nor $e_{\varnothing-k}$, nor $r_{k-\varnothing}$, nor $r_{1ne-\varnothing}$, nor e_{1e-k}, nor e_{1ne-k}, nor e_{j-k}, nor r_{k-j}, nor r_{k-1e}, nor r_{k-1ne}.*

The second result follows simply from the fact that there always exists a preferred extension (Prop. 1.1):

Prop. 4 *Under the preferred semantics, a change (addition or suppression) is never $e_{\varnothing-1ne}$, nor $e_{\varnothing-k}$, nor $r_{k-\varnothing}$, nor $r_{1ne-\varnothing}$.*

The third result concerns the stable semantics and follows simply from Prop. 1.7 and from the assumption that the set of arguments is not empty:

Prop. 5 *Under the stable semantics, a change (addition or suppression) is never c_{1e-1ne}, nor c_{1ne-1e}, nor e_{1e-k}, nor r_{k-1e}.*

The following proposition is also simply due to the uniqueness of the grounded extension:

Prop. 6
- *Under the grounded semantics,* skeptical expansive monotony *and* credulous expansive monotony *both correspond to* simple expansive monotony.
- *Under the grounded semantics,* skeptical restrictive monotony *and* credulous restrictive monotony *both correspond to* simple restrictive monotony.

According to monotony definitions given in Sect. 3.2, the following proposition holds for each semantics studied in this paper:

Prop. 7
- *Simple expansive monotony implies* credulous expansive monotony.
- *Simple restrictive monotony implies* credulous restrictive monotony.
- *Simple expansive monotony implies* preservation of acceptability *and* preservation of credulous-only acceptability.
- *Skeptical expansive monotony implies* preservation of skeptical acceptability.

The next propositions and notations will be used for establishing propositions given in Sect. 5 to 9.

Nota. 1 *Let* $Z \in \mathcal{G}$, $\mathcal{U}_Z = \{X \in \mathcal{G} \text{ s.t. } X \text{ is not attacked by } \mathcal{G} \setminus \{Z\}\}$.

Lem. 1 *When removing an argument Z under the grounded semantics, Z does not attack $\mathcal{E}'$ in $\mathcal{G}$ iff $\forall X \in \mathcal{G}'$, if Z attacks X in $\mathcal{G}$ then (X is attacked by $\mathcal{G} \setminus \{Z\}$ and X is not indirectly defended by $\mathcal{U}_Z$ in $\mathcal{G} \setminus \{Z\}$).*

Lem. 2 *When removing an argument Z under the grounded semantics, if Z does not attack $\mathcal{E}'$, then the following equivalence holds: $Z \in \bigcup_{i \geq 1} \mathcal{F}^i(\mathcal{U}_Z)$ iff $\mathcal{E}'$ defends Z in $\mathcal{G}$.*

The following lemma uses the fact that the argument Z is the only one argument which is added or removed; moreover, by assumption, Z is not a self-attacking argument:

Lem. 3
1. *When adding an argument Z, Z does not attack $\mathcal{G}'$ (resp. is not attacked by $\mathcal{G}'$) iff Z does not attack $\mathcal{G}$ (resp. is not attacked by $\mathcal{G}$) in $\mathcal{G}'$.*
2. *When removing an argument Z, Z does not attack $\mathcal{G}'$ (resp. is not attacked by $\mathcal{G}'$) in $\mathcal{G}$ iff Z does not attack $\mathcal{G}$ (resp. is not attacked by $\mathcal{G}$).*

The following new lemma is straightforward:

Lem. 4
- *When removing an argument Z from $\mathcal{G}$, let $X \neq Z$ be an argument of $\mathcal{G}$. If X is attacked in $\mathcal{G}'$ then X is also attacked in $\mathcal{G}$ by an argument distinct from Z. If X is attacked in $\mathcal{G}$ then X is either attacked in $\mathcal{G}'$ or X is attacked only by Z in $\mathcal{G}$.*
- *When adding an argument Z to $\mathcal{G}$, let $X \neq Z$ be an argument of $\mathcal{G}$. If X is attacked in $\mathcal{G}$ then X is also attacked in $\mathcal{G}'$ by an argument distinct from Z. If X is attacked in $\mathcal{G}'$ then X is either attacked in $\mathcal{G}$ or X is attacked only by Z in $\mathcal{G}$.*

5 Characterizing argument addition: direct results

Among the results given here, one is taken from [Bisquert et al., 2012b], two others are new and the others are taken from [Cayrol et al., 2010] (sometimes simplified: useless conditions have been removed). They only concern two semantics (the grounded and the preferred one). The proofs of the new propositions are given in Appendix 12.

5.1 Results for the grounded semantics

Prop. 8 *Under the grounded semantics, if X belongs to $\mathcal{E}$, and Z does not indirectly attack X, then $\oplus$ preserves the acceptability status for X (i.e. X belongs to $\mathcal{E}'$).*

Prop. 9 *Under the grounded semantics, if Z is not attacked by $\mathcal{G}$, then $\oplus$ satisfies priority to recency (i.e. Z belongs to $\mathcal{E}'$).*

Prop. 10 *Under the grounded semantics, in the case of addition,*
1. *if $\mathcal{E} = \{\}$ then it holds that: $\mathcal{E}' = \{\}$ iff Z is attacked by $\mathcal{G}$, moreover,*
2. *if $\mathcal{E} = \{\}$ and Z is not attacked by $\mathcal{G}$, then $\mathcal{E}' = \{Z\} \cup \bigcup_{i \geq 1} \mathcal{F}'^i(\{Z\})$.*

Note that Prop. 10.1 implies that "if Z is attacked by $\mathcal{G}$ and $\mathcal{E} = \{\}$ then $\mathcal{E}' = \{\}$". Thus "Z is attacked by $\mathcal{G}$ and $\mathcal{E} = \{\}$" is a sufficient condition for having a *c-conservative* change. The following two propositions are extensions of propositions of [Cayrol et al., 2010] to cover the case $\mathcal{E} = \emptyset$.

Prop. 11 *Under the grounded semantics, if Z does not attack $\mathcal{E}$, then $\oplus$ satisfies simple expansive monotony (i.e. $\mathcal{E} \subseteq \mathcal{E}'$).*

Prop. 12 *Under the grounded semantics, if Z does not attack $\mathcal{E}$, we have:*
1. *if $\mathcal{E}$ does not defend Z, then $\mathcal{E}' = \mathcal{E}$. (The change $\oplus$ is c-conservative).*
2. *if $\mathcal{E}$ defends Z, then $\mathcal{E}' = \mathcal{E} \cup \{Z\} \cup \bigcup_{i \geq 1} \mathcal{F}'^i(\{Z\})$.*
3. *Moreover, if $\mathcal{E}$ defends Z and Z does not attack $\mathcal{G}$ then $\mathcal{E}'$ reduces to $\mathcal{E} \cup \{Z\}$. (The change $\oplus$ is c-expansive if $\mathcal{E} \neq \{\}$, otherwise it is c_{1e-1ne}).*

Let $X \in \mathcal{G} \cap \mathcal{G}'$, a consequence of Prop. 12 gives a characterization (sufficient condition) of *acceptability establishment* for X:

Conseq. 1 *In case of addition under the grounded semantics, if Z does not attack $\mathcal{E}$ and $\mathcal{E}$ defends Z and Z indirectly defends X and $X \notin \mathcal{E}$ then $X \in \mathcal{E}'$.*

Prop. 13 *Under the grounded semantics, if $\mathcal{E} \neq \emptyset$ and Z attacks each unattacked argument of $\mathcal{G}$ and Z is attacked by $\mathcal{G}$ then the change $\oplus$ is c_{1ne-1e}; the converse also holds.*

Let $X \in \mathcal{G} \cap \mathcal{G}'$, the *preservation of the rejected status* for X is characterized by:

Prop. 14 *When adding an argument Z under the grounded semantics, $\forall X \in \mathcal{G}$, if $X \notin \mathcal{E}$ and Z does not indirectly defend X, then $X \notin \mathcal{E}'$.*

Let $X \in \mathcal{G} \cap \mathcal{G}'$, the following results are new and give respectively a characterization of *acceptability establishment* for X, then a characterization of *acceptability removal* for X:

Prop. 15 *When adding an argument Z under the grounded semantics, if Z is not attacked by $\mathcal{G}$ and Z indirectly defends X and $X \notin \mathcal{E}$ then $X \in \mathcal{E}'$.*

Prop. 16 *When adding an argument Z under the grounded semantics, if $\mathcal{E} \setminus \{X\}$ does not attack Z and Z attacks X and $X \in \mathcal{E}$ then $X \notin \mathcal{E}'$.*

5.2 Results for the preferred semantics

Prop. 17 *Under the preferred semantics, if Z is not attacked by $\mathcal{G}$, then $\oplus$ satisfies skeptical priority to recency (i.e. Z belongs to each $\mathcal{E}'_i$).*

Prop. 18 *When adding an argument Z under the preferred semantics,*
1. *if Z does not attack $\mathcal{E}_i$, then $\mathcal{E}_i$ remains admissible in $\mathcal{G}'$;*
2. *if Z does not attack $\mathcal{E}_i$ and $\mathcal{E}_i$ defends Z in $\mathcal{G}'$, then $\mathcal{E}_i \cup \{Z\}$ is admissible in $\mathcal{G}'$.*

Prop. 19 *When adding an argument Z under the preferred semantics, if $\mathbf{E} = \{\{\}\}$ and Z is not attacked by $\mathcal{G}$ and there is no even-length cycle in $\mathcal{G}$ then $\mathbf{E}' = \{\mathcal{E}'\}$ and Z belongs to $\mathcal{E}'$ (so, $\oplus$ is c_{1e-1ne}).*

Prop. 20 *When adding an argument Z under the preferred semantics, if Z attacks no argument of $\mathcal{G}$ and $\mathbf{E} = \{\{\}\}$, then $\mathbf{E}' = \{\{\}\}$; or equivalently, if $\mathbf{E} = \{\{\}\}$ the change $\oplus$ by Z is c_{1e-1ne} only if Z attacks $\mathcal{G}$.*

Note that the previous proposition gives a sufficient condition for having a *c-conservative* change: "Z attacks no argument of $\mathcal{G}$ and $\mathbf{E} = \{\{\}\}$".

Prop. 21 *When adding an argument Z under the preferred semantics, if Z attacks no argument of $\mathcal{G}$ and $\mathbf{E} \neq \{\{\}\}$, then for each i:*
1. *if $\mathcal{E}_i$ defends Z, then $\mathcal{E}_i \cup \{Z\}$ is an extension of $\mathcal{G}'$;*
2. *if $\mathcal{E}_i$ does not defend Z, then $\mathcal{E}_i$ is an extension of $\mathcal{G}'$;*

moreover, $\mathcal{G}$ and $\mathcal{G}'$ have the same number of extensions (so the change is constant).

Note again that Prop. 21.2 gives a sufficient condition for having a *c-conservative* change: "Z attacks no argument of $\mathcal{G}$ and $\mathbf{E} \neq \{\{\}\}$ and for each i, $\mathcal{E}_i$ does not defend Z".

Prop. 22 *Under the preferred semantics, if Z attacks no extension of $\mathcal{G}$ then the change $\oplus$ satisfies simple expansive monotony.*

Prop. 23 *Under the preferred semantics, assume that $\mathcal{G}$ contains no controversial argument[6]. If Z does not attack $\bigcap_{i \geq 1} \mathcal{E}_i$, then the change $\oplus$ satisfies skeptical expansive monotony, that is $\bigcap_{i \geq 1} \mathcal{E}_i \subseteq \bigcap_{i \geq 1} \mathcal{E}'_i$.*

Prop. 24 *Under the preferred semantics, if $\mathbf{E} \neq \{\{\}\}$ and there is no even-length cycle in $\mathcal{G}'$ and each unattacked argument of $\mathcal{G}$ is attacked in $\mathcal{G}'$ and Z is attacked in $\mathcal{G}'$ then the change $\oplus$ is c_{1ne-1e}.*

---

[6] An argument A is controversial iff there exists at least an argument B s.t. A is a defender (direct or indirect) and an attacker (direct or indirect) of B.

6 Characterizing argument removal: direct results

6.1 Results concerning the three semantics

These results can be found in [Bisquert *et al.*, 2012a] and concern the three well-known semantics (grounded, stable and preferred).

Prop. 25 *When removing an argument Z,*
1. *If $\mathcal{E}$ is a preferred extension of $\mathcal{G}$ and $Z \notin \mathcal{E}$ then $\mathcal{E}$ is admissible in $\mathcal{G}'$ and so there exists a preferred extension $\mathcal{E}'$ of $\mathcal{G}'$ s.t. $\mathcal{E} \subseteq \mathcal{E}'$.*
2. *If $\mathcal{E}$ is a stable extension of $\mathcal{G}$ and $Z \notin \mathcal{E}$ then $\mathcal{E}$ is stable in $\mathcal{G}'$.*
3. *If $\mathcal{E}$ is the grounded extension of $\mathcal{G}$ and $Z \notin \mathcal{E}$ then $\mathcal{E} \subseteq \mathcal{E}'$ where $\mathcal{E}'$ is the grounded extension of $\mathcal{G}'$.*

The following result gives a necessary and sufficient condition for *simple expansive monotony*.

Prop. 26 *When removing an argument Z under preferred, stable or grounded semantics, it holds that: $(\forall \mathcal{E} \in \mathbf{E}, Z \notin \mathcal{E})$ iff $(\forall \mathcal{E} \in \mathbf{E}, \exists \mathcal{E}' \in \mathbf{E}'$ s.t. $\mathcal{E} \subseteq \mathcal{E}')$.*

The following proposition concerns the notion of "weak" simple expansive monotony (i.e. a kind of monotony in which Z is not taken into account):

Prop. 27 *When removing an argument Z, if Z attacks no argument in $\mathcal{G}$, then*
1. *$\forall \mathcal{E}$ preferred extension of $\mathcal{G}$, $\mathcal{E} \setminus \{Z\}$ is admissible in $\mathcal{G}'$ and so $\exists \mathcal{E}'$ preferred extension of $\mathcal{G}'$ s.t. $\mathcal{E} \setminus \{Z\} \subseteq \mathcal{E}'$.*
2. *$\forall \mathcal{E}$ stable extension of $\mathcal{G}$, $\mathcal{E} \setminus \{Z\}$ is a stable extension of $\mathcal{G}'$.*
3. *If $\mathcal{E}$ is the grounded extension of $\mathcal{G}$, $\mathcal{E} \setminus \{Z\}$ is the grounded extension of $\mathcal{G}'$.*

Prop. 28 *When removing an argument Z under preferred, stable or grounded semantics, if Z attacks no argument in $\mathcal{G}$, then for any extension $\mathcal{E}$ of $\mathcal{G}$ s.t. $Z \notin \mathcal{E}$, $\mathcal{E}$ is an extension of $\mathcal{G}'$.*

Prop. 29 *When removing an argument Z under preferred, stable or grounded semantics, if the change is c-narrowing, then there exists an extension $\mathcal{E}$ of $\mathcal{G}$ s.t. $Z \in \mathcal{E}$.*

6.2 Results concerning only some semantics

Three new results are given here (proofs in Appendix 12), the other propositions being taken from [Bisquert *et al.*, 2012a].

Prop. 30 *When removing an argument Z under the preferred semantics, if Z attacks no argument in $\mathcal{G}$ then, for each extension $\mathcal{E}_i$ of $\mathcal{G}$,*
1. *If $Z \notin \mathcal{E}_i$ then $\mathcal{E}_i$ is a preferred extension of $\mathcal{G}'$.*
2. *If $Z \in \mathcal{E}_i$ then $\mathcal{E}_i \setminus \{Z\}$ is a preferred extension of $\mathcal{G}'$.*

Moreover, $|\mathbf{E}| = |\mathbf{E}'|$ (so the change is constant).

Prop. 31 *When removing an argument under the stable semantics, a change cannot be c-expansive.*

Prop. 32 *When removing an argument Z under the preferred or the grounded semantics, if this change is c-expansive, then*
1. *Z belongs to no extension of $\mathcal{G}$ and*
2. *Z attacks at least one element of $\mathcal{G}$.*

Prop. 33 *When removing an argument Z under the preferred or the grounded semantics, if Z attacks no argument of $\mathcal{G}$ and $\forall \mathcal{E}, Z \in \mathcal{E}$, then the change is c-narrowing.*

Let $X \in \mathcal{G} \cap \mathcal{G}'$, the three following propositions are new. The first one gives a characterization of *acceptability establishment* for X:

Prop. 34 *When removing an argument Z under the grounded semantics, if $X \notin \mathcal{E}$ and Z is the unique attacker of X then $X \in \mathcal{E}'$.*

The previous proposition is weak, but many examples can be easily found for illustrating the fact that an argument X attacked by several arguments and s.t. $X \notin \mathcal{E}$ cannot be reinstated by the removal of only one argument. For instance, with the attack relation $\{(a,b), (b,c), (c,b), (c,d), (d,c), (b,e)\}$, for reinstating b, both a and c must be removed, and for reinstating c, both b and d must be removed.

Let $X \in \mathcal{G} \cap \mathcal{G}'$, a characterization of *acceptability removal* for X is given by the two following propositions:

Prop. 35 *When removing an argument Z under the grounded semantics, if $X \neq Z$ and $X \in \mathcal{E}$ and $Z \in \mathcal{E}$ and $X \in \bigcup_{i \geq 1} \mathcal{F}^i(\{Z\})$ and (there exists Y attacker of X s.t. each odd-length path from $\mathcal{F}(\emptyset)$ to Y contains Z at an even place) and (there exists no odd-length path from Z to X), then $X \notin \mathcal{E}'$.*

Prop. 36 *When removing an argument Z under the grounded semantics, if $X \in \mathcal{E}$ and X is attacked in $\mathcal{G}$ and $\forall S$ s.t. $X \in \mathcal{F}(S), Z \in S$ then $X \notin \mathcal{E}'$.*

7 Duality

As far as we know, the problem of removing an argument and, *a fortiori*, the link between addition and removal of an argument have been little discussed. However, it can be worthy to use the links between these operations in order to study the properties characterizing the changes that may impact an AS. For that purpose, the notion of duality seems pertinent.

7.1 Two definitions of duality

We focus on two concepts of duality: first, duality at the level of change operations, *based on the notion of inverse*, expressing the opposite nature of two operations,

Characterizing change in abstract argumentation systems

then duality at the level of change properties, *based on the notion of symmetry*, conveying a correspondence between two properties.

Def. 18 (Duality based on the notion of inverse) *Two change operations $\mathcal{O}$ and $\mathcal{O}'$ are the inverse of each other iff:* "$\forall \mathcal{G}, \forall \mathcal{G}',\ \mathcal{O}(\mathcal{G}) = \mathcal{G}'$ iff $\mathcal{O}'(\mathcal{G}') = \mathcal{G}$".

Obviously, following Def. 18, it is clear that addition and removal operations defined in Sect. 2.2 are the inverse of each other:

Ex. 1 (cont'd) *Let $\mathcal{O}_1 = \ominus(Z, \mathcal{I}_z)$ with $\mathcal{I}_z = \{(Z,A),(Z,B)\}$ and $\mathcal{O}'_1 = \oplus(Z, \mathcal{I}_z)$ with $\mathcal{I}_z = \{(Z,A),(Z,B)\}$, $\mathcal{O}_1$ and $\mathcal{O}'_1$ are the inverse of each other.*

Def. 19 (Duality based on the notion of symmetry) *Two properties $\mathcal{P}$ and $\mathcal{P}'$ are symmetric iff:* "$\forall \mathcal{G},\ \forall \mathcal{G}',\ \mathcal{P}'(\mathcal{G}', \mathcal{G})$ holds iff $\mathcal{P}(\mathcal{G}, \mathcal{G}')$ holds".

From these definitions, it is straightforward to draw the following condition on properties that hold for inverse change operations:

Prop. 37 *Let $\mathcal{O}$ and $\mathcal{O}'$ be two inverse change operations and $\mathcal{P}$ and $\mathcal{P}'$ be two symmetric properties. $\mathcal{O}$ satisfies $\mathcal{P}$ iff $\mathcal{O}'$ satisfies $\mathcal{P}'$.*

Both concepts of duality enable us to straightforwardly obtain the following results about change properties:

Prop. 38
1. *A change is restrictive iff the inverse change is extensive.*
2. *A change is $r_{1ne-\varnothing}$ iff the inverse change is $e_{\varnothing - 1ne}$.*
3. *A change is $r_{k-\varnothing}$ iff the inverse change is $e_{\varnothing - k}$.*
4. *A change is r_{k-1e} iff the inverse change is e_{1e-k}.*
5. *A change is r_{k-1ne} iff the inverse change is e_{1ne-k}.*
6. *A change is r_{k-j} iff the inverse change is e_{j-k}.*

Prop. 39
1. *A change is constant iff the inverse change is also constant.*
2. *A change is c_{1ne-1e} iff the inverse change is c_{1e-1ne}.*
3. *A change is c-conservative iff the inverse change is also c-conservative.*
4. *A change is c-narrowing iff the inverse change is c-expansive.*
5. *A change is c-altering iff the inverse change is also c-altering.*

Prop. 40
1. *A change satisfies* simple restrictive monotony *iff the inverse change satisfies* simple expansive monotony.
2. *A change satisfies* credulous restrictive monotony *iff the inverse change satisfies* credulous expansive monotony.
3. *A change satisfies* skeptical restrictive monotony *iff the inverse change satisfies* skeptical expansive monotony.

Prop. 41 *Let $X \in \mathcal{G} \cap \mathcal{G}'$.*
1. *A change* establishes credulous-only acceptability *for X iff the inverse change* removes credulous-only acceptability *of X.*
2. *A change* establishes skeptical acceptability *for X iff the inverse change* removes skeptical acceptability *of X.*

Prop. 42 *Let* $X \in \mathscr{G} \cap \mathscr{G}'$. *A change is a* general diffusion of acceptability *for X iff the inverse change* partially degrades acceptability *for X*.

Prop. 43 *Let* $X \in \mathscr{G} \cap \mathscr{G}'$.
1. *A change* preserves credulous-only acceptability *for X iff the inverse change* preserves credulous-only acceptability *for X*.
2. *A change* preserves skeptical acceptability *for X iff the inverse change* preserves skeptical acceptability *for X*.
3. *A change* preserves the rejected status *for X iff the inverse change* preserves the rejected status *for X*.

7.2 Methodology for Using Duality

This part describes how to use duality in order to obtain new propositions for the operation of removal, starting from propositions concerning addition[7]. Note first that we restrict our study to the grounded semantics. Let us describe this methodology using Prop. 14. In order to clarify the presentation, the graphs and the extensions are renamed by adding two capital letters in subscripts - IA, OA, IR and OR - representing respectively the **I**nput system for **A**ddition, the **O**utput system for **A**ddition, the **I**nput system for **R**emoval and the **O**utput system for **R**emoval. Thus, Prop. 14 can be rewritten as follows:

Prop. 14.1 *When adding an argument Z under the grounded semantics, if* $X \notin \mathscr{E}_{IA}$ *and Z does not indirectly defend X, then* $X \notin \mathscr{E}_{OA}$.

Let $\mathscr{P}$ be a property and $\mathscr{P}^{-1}$ its symmetric. Due to Prop. 37, it holds that: $\oplus$ *satisfies* $\mathscr{P}$ *iff* $\ominus$ *satisfies* $\mathscr{P}^{-1}$.

And due to Def. 6, we know that a change operation $\mathscr{O}$ satisfies $\mathscr{P}$ iff $\forall \mathscr{G}$, it holds that $\mathscr{P}(\mathscr{G}, \mathscr{O}(\mathscr{G}))$. Hence:

$\forall \mathscr{G}_{IA}, \mathscr{P}(\mathscr{G}_{IA}, \oplus(\mathscr{G}_{IA}))$ holds iff $\forall \mathscr{G}_{IR}, \mathscr{P}^{-1}(\mathscr{G}_{IR}, \ominus(\mathscr{G}_{IR}))$ holds.

Moreover, due to Def. 19, we have:

$\forall \mathscr{G}_{IR}, \mathscr{P}^{-1}(\mathscr{G}_{IR}, \ominus(\mathscr{G}_{IR}))$ holds iff $\mathscr{P}(\ominus(\mathscr{G}_{IR}), \mathscr{G}_{IR})$ holds.

And so, we have:

$\forall \mathscr{G}_{IA}, \forall \mathscr{G}_{IR}, \mathscr{P}(\mathscr{G}_{IA}, \oplus(\mathscr{G}_{IA}))$ holds iff $\mathscr{P}(\ominus(\mathscr{G}_{IR}), \mathscr{G}_{IR})$ holds.

Let $\mathscr{G}_{OA} = \oplus(\mathscr{G}_{IA})$ and $\mathscr{G}_{OR} = \ominus(\mathscr{G}_{IR})$. Since we know that $\mathscr{P}$ holds for the operation of addition, we can rewrite it for removal:

Prop. 14.2 *When removing an argument Z under the grounded semantics, if* $X \notin \mathscr{E}_{OR}$ *and Z does not indirectly defend X, then* $X \notin \mathscr{E}_{IR}$.

Which is equivalent to:

Prop. 14.3 *When removing an argument Z under the grounded semantics, if* $X \in \mathscr{E}_{IR}$ *and Z does not indirectly defend X, then* $X \in \mathscr{E}_{OR}$.

Thus, for the operation of removal, we obtain a proposition analogous to Prop. 14 denoted by Prop. $14^{\ominus}$; in the remainder of this article, the exponent ($\oplus$ or $\ominus$) will rep-

---

[7] This methodology can also be used the other way round from removal to addition.

resent the correspondence between a proposition and the one obtained by applying the duality methodology:

Prop. 14$^{\ominus}$ *When removing an argument Z under the grounded semantics, if $X \in \mathcal{E}$ and Z does not indirectly defend X, then $X \in \mathcal{E}'$.*

In the next sections, we use this methodology on the propositions given in Sect. 5 and 6 together with the lemmas given in Sect. 4 in order to obtain new results.

8 Characterizing argument addition thanks to duality

The results given here are new, each proposition is obtained by using the methodology described in the previous section (Sect. 7.2).

8.1 Results for the three semantics

Prop. 25$^{\oplus}$ *When adding an argument Z,*
1. *If $\mathcal{E}'$ is a preferred extension of $\mathcal{G}'$ and $Z \notin \mathcal{E}'$ then $\mathcal{E}'$ is admissible in $\mathcal{G}$, hence there exists a preferred extension $\mathcal{E}$ of $\mathcal{G}$ s.t. $\mathcal{E}' \subseteq \mathcal{E}$.*
2. *If $\mathcal{E}'$ is a stable extension of $\mathcal{G}'$ and $Z \notin \mathcal{E}'$ then $\mathcal{E}'$ is stable in $\mathcal{G}$.*
3. *If $\mathcal{E}'$ is the grounded extension of $\mathcal{G}'$ and $Z \notin \mathcal{E}'$ then $\mathcal{E}' \subseteq \mathcal{E}$ where $\mathcal{E}$ is the grounded extension of $\mathcal{G}$.*

Prop. 26$^{\oplus}$ *When adding an argument Z under the preferred, grounded or stable semantics, $(\forall \mathcal{E}' \in \mathbf{E}', Z \notin \mathcal{E}')$ iff $(\forall \mathcal{E}' \in \mathbf{E}', \exists \mathcal{E} \in \mathbf{E} \text{ s.t. } \mathcal{E}' \subseteq \mathcal{E})$.*

Prop. 27$^{\oplus}$ *When adding an argument Z, if Z attacks no argument in $\mathcal{G}$, then*
1. *$\forall \mathcal{E}'$ preferred extension of $\mathcal{G}'$, $\mathcal{E}' \setminus \{Z\}$ is admissible in $\mathcal{G}$ and so $\exists \mathcal{E}$ preferred extension of $\mathcal{G}$ s.t. $\mathcal{E}' \setminus \{Z\} \subseteq \mathcal{E}$.*
2. *$\forall \mathcal{E}'$ stable extension of $\mathcal{G}'$, $\mathcal{E}' \setminus \{Z\}$ is a stable extension of $\mathcal{G}$.*
3. *If $\mathcal{E}'$ is the grounded extension of $\mathcal{G}'$, $\mathcal{E}' \setminus \{Z\}$ is the grounded extension of $\mathcal{G}$.*

Prop. 27$^{\oplus}$ completes the results given by Prop. 20, 21 (for the preferred semantics) and 12 (for the grounded semantics).

Prop. 28$^{\oplus}$ *When adding an argument Z under the preferred, grounded or stable semantics, if Z attacks no argument in $\mathcal{G}$, then for any extension $\mathcal{E}'$ of $\mathcal{G}'$ s.t. $Z \notin \mathcal{E}'$, $\mathcal{E}'$ is an extension of $\mathcal{G}$.*

Prop. 29$^{\oplus}$ *When adding an argument Z under the preferred, grounded or stable semantics, if the change is c-expansive, then there exists an extension $\mathcal{E}'$ of $\mathcal{G}'$ s.t. $Z \in \mathcal{E}'$.*

8.2 Results concerning only some semantics

Prop. 30$^{\oplus}$ *When adding an argument Z under the preferred semantics, if Z attacks no argument in $\mathscr{G}$ then, for each extension $\mathscr{E}'_i$ of $\mathscr{G}'$,*
1. *If $Z \notin \mathscr{E}'_i$ then $\mathscr{E}'_i$ is a preferred extension of $\mathscr{G}$.*
2. *If $Z \in \mathscr{E}'_i$ then $\mathscr{E}'_i \setminus \{Z\}$ is a preferred extension of $\mathscr{G}$.*

Moreover, $|\mathbf{E}'| = |\mathbf{E}|$.

Prop. 30$^{\oplus}$ completes the results given by Prop. 21.

Prop. 31$^{\oplus}$ *When adding an argument under the stable semantics, a change cannot be c-narrowing.*

Prop. 32$^{\oplus}$ *When adding an argument Z under preferred or grounded semantics, if the change is c-narrowing, then*
1. *Z belongs to no extension of $\mathscr{G}'$ and*
2. *Z attacks at least one element of $\mathscr{G}'$.*

Prop. 33$^{\oplus}$ *When adding an argument Z under preferred or grounded semantics, if Z attacks no argument in $\mathscr{G}'$ and belongs to each extension of $\mathscr{G}'$, then the change is c-expansive.*

Note that the duality-based translation of Prop. 34 is not interesting because the resulting proposition produces a contradictory condition ($X \notin \mathscr{E}$ and Z is the only attacker of X, Z being the added argument).

Similarly, Prop. 35 is not translated since the resulting condition is contradictory ($X \in \mathscr{E}$ and there exists an attacker Y of X s.t. each odd-length path from $\mathscr{F}'(\varnothing)$ to Y contains Z at an even place and there exists no odd-length path from Z to X).

Let $X \in \mathscr{G} \cap \mathscr{G}'$, the following proposition gives a characterization of *acceptability removal* for X. However, even if the proposed condition is not contradictory, it is not easy to check.

Prop. 36$^{\oplus}$ *When adding an argument Z under the grounded semantics, if $X \in \mathscr{E}$, X is attacked in $\mathscr{G}'$, and $\forall S$ s.t. $X \in \mathscr{F}'(S), Z \in S$, then $X \notin \mathscr{E}'$.*

9 Characterizing removal thanks to duality

9.1 Results for the grounded semantics

These results can be found in [Bisquert et al., 2012b]. Let $X \in \mathscr{G} \cap \mathscr{G}'$, the first proposition characterizes the *preservation of the rejected status* for X:

Prop. 8$^{\ominus}$ *When removing an argument Z under the grounded semantics, if $X \notin \mathscr{E}$ and Z does not indirectly attack X, then $X \notin \mathscr{E}'$.*

Note that the duality-based translation of Prop. 9 would give a trifling result under the grounded semantics ("$Z \notin \mathscr{E}$ implies Z is attacked by $\mathscr{G}$").

Prop. 10.1$^{\ominus}$ *When removing an argument Z under the grounded semantics, if $\mathscr{E} \neq \varnothing$ and Z is attacked by $\mathscr{G}$, then $\mathscr{E}' \neq \varnothing$.*

Prop. 10.2$^{\ominus}$ *When removing an argument Z under the grounded semantics, if $\mathcal{E} \neq \{Z\} \cup \bigcup_{i \geq 1} \mathcal{F}^i(\{Z\})$ and Z is not attacked by $\mathcal{G}$, then $\mathcal{E}' \neq \emptyset$.*

Corol. 1 *When removing an argument Z under the grounded semantics, if the change is c_{1ne-1e}, then Z is not attacked by $\mathcal{G}$ and $\mathcal{E} = \{Z\} \cup \bigcup_{i \geq 1} \mathcal{F}^i(\{Z\})$.*

Prop. 11$^{\ominus}$ *When removing an argument Z under the grounded semantics, if $\forall X \in \mathcal{G}$, if Z attacks X then (X is attacked by $\mathcal{G} \setminus \{Z\}$ and X is not indirectly defended by $\mathcal{U}_Z$ in $\mathcal{G} \setminus \{Z\}$), then $\mathcal{E}' \subseteq \mathcal{E}$.*

Prop. 12.1$^{\ominus}$ *When removing an argument Z under the grounded semantics, if (1) $\forall X \in \mathcal{G}$, if Z attacks X then (X is attacked by $\mathcal{G} \setminus \{Z\}$ and X is not indirectly defended by $\mathcal{U}_Z$ in $\mathcal{G} \setminus \{Z\}$) and (2) $Z \notin \bigcup_{i \geq 1} \mathcal{F}^i(\mathcal{U}_Z)$, then $\mathcal{E} = \mathcal{E}'$.*

The following proposition is a translation by duality of Prop.12.2, but this result is not easy to exploit in a prescriptive purpose since the condition concerns the system after the change.

Prop. 12.2$^{\ominus}$ *When removing an argument Z under the grounded semantics, if Z does not attack $\mathcal{E}'$ and $\mathcal{E}'$ defends Z, then $\mathcal{E} = \mathcal{E}' \cup \{Z\} \cup \bigcup_{i \geq 1} \mathcal{F}^i(\{Z\})$.*

Prop. 12.3$^{\ominus}$ *When removing an argument Z under the grounded semantics, if (1) $Z \in \bigcup_{i \geq 1} \mathcal{F}^i(\mathcal{U}_Z)$ and (2) Z does not attack $\mathcal{G} \setminus \{Z\}$, then $\mathcal{E}' = \mathcal{E} \setminus \{Z\}$.*

The next proposition gives a sufficient and necessary condition for a c_{1e-1ne} change:

Prop. 13$^{\ominus}$ *When removing an argument Z under the grounded semantics, if $\mathcal{E}' \neq \emptyset$ and Z attacks each unattacked argument of $\mathcal{G} \setminus \{Z\}$ and Z is attacked by $\mathcal{G} \setminus \{Z\}$ then $\mathcal{E} = \emptyset$. And the converse also holds.*

Let $X \in \mathcal{G} \cap \mathcal{G}'$, the following proposition (used for illustrating our methodology in Sect. 7.2) characterizes *preservation of acceptability* for X:

Prop. 14$^{\ominus}$ *When removing an argument Z under the grounded semantics, if $X \in \mathcal{E}$ and Z does not indirectly defend X, then $X \in \mathcal{E}'$.*

Note that the application of duality on Prop. 15 and 16 gives impossible conditions (for Prop. 15: "Z is not attacked by $\mathcal{G}$, Z indirectly defends X and $X \notin \mathcal{E}$", and for Prop. 16: "$\mathcal{E}' \setminus \{X\}$ does not attack Z, Z attacks X and $X \in \mathcal{E}$").

9.2 Results for the preferred semantics

The results given in this section are new, each proposition is obtained by using the methodology described in Sect. 7.2. Note that the duality-based translation of Prop. 17 would give a trifling result under the preferred semantics ("Z not attacked by $\mathcal{G}$ implies Z belongs to each $\mathcal{E}'_i$").

Prop. 18$^{\ominus}$ *When removing Z under the preferred semantics,*
1. *if Z does not attack $\mathcal{E}'_i$, then $\mathcal{E}'_i$ is admissible in $\mathcal{G}$;*
2. *if Z does not attack $\mathcal{E}'_i$ and $\mathcal{E}'_i$ defends Z, then $\mathcal{E}'_i \cup \{Z\}$ is admissible in $\mathcal{G}$.*

This result is related to Prop. 27 to 30 which are specific cases of this proposition when we restrict to preferred semantics.

Prop. 19$^\ominus$ *When removing Z under the preferred semantics, if $\mathbf{E}' = \{\{\}\}$ and Z is not attacked by $\mathcal{G}$ and there is no even-length cycle in $\mathcal{G}'$ then $\mathbf{E} = \{\mathcal{E}\}$ and Z belongs to $\mathcal{E}$ (so, $\ominus$ is c_{1ne-1e}).*

Prop. 20$^\ominus$ *When removing Z under the preferred semantics, if Z attacks no argument of $\mathcal{G}$ and $\mathbf{E}' = \{\{\}\}$, then $\mathbf{E} = \{\{\}\}$; or equivalently, if $\mathbf{E}' = \{\{\}\}$ the change $\ominus$ by Z is c_{1ne-1e} only if Z attacks $\mathcal{G}'$.*

The following proposition completes Prop. 30:

Prop. 21$^\ominus$ *When removing Z under the preferred semantics, if Z attacks no argument of $\mathcal{G}$, and $\mathbf{E}' \neq \{\{\}\}$, then for each i:*
1. *if $\mathcal{E}'_i$ defends Z, then $\mathcal{E}'_i \cup \{Z\}$ is an extension of $\mathcal{G}$;*
2. *if $\mathcal{E}'_i$ does not defend Z, then $\mathcal{E}'_i$ is an extension of $\mathcal{G}$;*

moreover, $\mathcal{G}'$ and $\mathcal{G}$ have the same number of extensions.

Prop. 22$^\ominus$ *Under the preferred semantics, if Z attacks no extension of $\mathcal{G}'$ then the change $\ominus$ satisfies simple restrictive monotony.*

Prop. 23$^\ominus$ *Under the preferred semantics, assume that $\mathcal{G}'$ contains no controversial argument. If Z does not attack $\bigcap_{i\geq 1} \mathcal{E}'_i$, then the change $\ominus$ satisfies skeptical restrictive monotony, that is $\bigcap_{i\geq 1} \mathcal{E}'_i \subseteq \bigcap_{i\geq 1} \mathcal{E}_i$.*

Note that the following proposition is not easy to exploit in a prescriptive purpose since the condition concerns the system after the change:

Prop. 24$^\ominus$ *When removing Z under the preferred semantics, if $\mathbf{E}' \neq \{\{\}\}$, and there is no even-length cycle in $\mathcal{G}$ and each unattacked argument of $\mathcal{G}'$ is attacked in $\mathcal{G}$ and Z is attacked in $\mathcal{G}$ then $\mathbf{E} = \{\{\}\}$ (and so the change $\ominus$ is c_{1e-1ne}).*

10 Discussion

10.1 How to use these change properties? A road-map

Among these properties, the user may wonder how to select the useful properties. For this purpose, three criteria may be taken into account:

- the kind of change concerned: some changes may be considered as useful according to the role of the user, *e.g.* a debate moderator may be interested in focusing or enlarging the dialog depending on the remaining time, while an orator may have dialog strategies and may want to focus on particular arguments. Let us review some of these properties:
  - A "decisive" (*e.g.* c_{1e-1ne}) change is useful to lower ignorance since after this change one and only one extension remains. It can be used by a moderator for concluding the debate.
  - An "expansive" (*e.g. c-expansive*) change increases the accepted arguments while conserving those already accepted, it can also be used by a moderator or

Characterizing change in abstract argumentation systems

by an orator in order to convince a larger audience about the current view of the debate.
- A "conservative" (*e.g. c-conservative*) change may be a more neutral attitude that can be adopted by a moderator or an orator that does not want to deliver new information but wants to participate (very useful political waffle).
- "Monotony" and "priority to recency" allow to focus on some particular arguments and may be used strategically by an orator.
- "Questioning" (*e.g.* e_{j-k}) and "destructive" (*e.g.* r_{k-1e}) changes are increasing ignorance either by augmenting the possible views or by destroying any coherent view, they may be used desperately by a strategical orator/manager that wants to forbid any decision to be made.
- An "altering" (*e.g. c-altering*) change allows to completely change the point of view, it may also be done to reverse the course of the debate.
• The nature of the characterization obtained in terms of computational time required to check its condition (*e.g.*, checking if an argument attacks no other argument is easier than checking if it belongs to an extension).
• The nature of the characterization obtained in terms of typicality: is the condition often realized in usual AS or is this condition scarcely encountered?

10.2 How to use the characterization results?

Beyond the theoretical point of view, the characterization results may be used to address computational issues concerning change. In recent work about goal-driven change in argumentation (see [Bisquert *et al.*, 2013]), we consider an agent that can act on an argumentation system (the target system) in order to achieve some desired goal. We developed a software for computing the change operations that the agent should perform in order to accomplish her goal. This software uses the properties characterizing change as follows. Given a goal to achieve, for each type of operation (*i.e.* addition/removal), if there exists a characterization corresponding to this type of operation and this goal, and if its conditions are satisfied, then the precise operation is provided to the agent (with the corresponding characterization as an explanation).

10.3 Conclusion and future works

In this paper, we have presented a comprehensive study of change in argumentation (addition or removal of an argument and its interactions). The first step of this study is the definition of change properties that describe the impact of change on argumentation systems. The second step is to characterize these properties by giving (sufficient or necessary) conditions under which they hold. Some of the characterization results are obtained by using duality between addition and removal of an argument.

Let us come back to the example given in Sect. 1. For Mister Pink, adding a new argument attacking a specific argument of Mister White without threatening his own accepted arguments corresponds to Prop. 8. Prop. 14$^{\ominus}$, on the other hand, allows him to ensure that the removal of his opponent's argument achieves the same result if this argument is not giving assistance to any of his own accepted arguments. Thereby, instead of using Prop. 8, Mister Pink can benefit from Prop. 14$^{\ominus}$ thanks to our methodology (see *"preserves acceptability"* lines of the tables in Appendix 11).

Our work deals with a facet of the argumentation theory that has not been studied so far. Hence, many points are to be deepened or explored further; here are some issues that seem to be of short-term importance:

- In this work, we have studied only two of the four operations of [Cayrol *et al.*, 2010]. A first issue is to extend our work to the two missing operations, addition and removal of an interaction.
- Moreover, we could consider the addition or removal of a set of arguments. The study of these special operations (seen as a sequence of change operations) seems essential in order to approach minimal change problems.

References

[Amgoud and Cayrol, 2002] L. Amgoud and C. Cayrol. A reasoning model based on the production of acceptable arguments. *Annals of Mathematics and Artificial Intelligence*, 34:197–216, 2002.

[Amgoud *et al.*, 2000] L. Amgoud, N. Maudet, and S. Parsons. Modelling dialogues using argumentation. In *Proc. of ICMAS*, pages 31–38, 2000.

[Baumann and Brewka, 2010] R. Baumann and G. Brewka. Expanding argumentation frameworks: Enforcing and monotonicity results. In *Proc. of COMMA*, pages 75–86. IOS Press, 2010.

[Bisquert *et al.*, 2011] P. Bisquert, C. Cayrol, F. Dupin de Saint-Cyr, and MC. Lagasquie-Schiex. Change in argumentation systems: exploring the interest of removing an argument. In *Proc. of SUM (LNCS 6929)*, pages 275–288. Springer, 2011.

[Bisquert *et al.*, 2012a] P. Bisquert, C. Cayrol, F. Dupin de Saint-Cyr, and MC. Lagasquie-Schiex. Changement dans un système d'argumentation : suppression d'un argument. *Revue d'Intelligence Artificielle*, 6:225–254, 2012.

[Bisquert *et al.*, 2012b] P. Bisquert, C. Cayrol, F. Dupin de Saint-Cyr, and MC. Lagasquie-Schiex. Duality between addition and removal: a tool for studying change in argumentation. In *Proc. of IPMU*, pages 219–229, 2012.

[Bisquert *et al.*, 2013] P. Bisquert, C. Cayrol, F. Dupin de Saint-Cyr, and M.-C. Lagasquie-Schiex. Goal-driven changes in argumentation: A theoretical framework and a tool. Technical Report RR–2013-33–FR, IRIT-UPS, 2013.

[Boella *et al.*, 2009a] G. Boella, S. Kaci, and L. van der Torre. Dynamics in argumentation with single extensions: Abstraction principles and the grounded extension. In *Proc. of ECSQARU (LNAI 5590)*, pages 107–118, 2009.

[Boella *et al.*, 2009b] G. Boella, S. Kaci, and L. van der Torre. Dynamics in argumentation with single extensions: Attack refinement and the grounded extension. In *Proc. of AAMAS*, pages 1213–1214, 2009.

[Cayrol *et al.*, 2010] C. Cayrol, F. Dupin de Saint-Cyr, and MC. Lagasquie-Schiex. Change in abstract argumentation frameworks: adding an argument. *Journal of Artificial Intelligence Research*, 38:49–84, 2010.

[Dung, 1995] P. M. Dung. On the acceptability of arguments and its fundamental role in nonmonotonic reasoning, logic programming and n-person games. *Artificial Intelligence*, 77:321–357, 1995.

[Dunne and Bench-Capon, 2001] P. Dunne and T. Bench-Capon. Complexity and combinatorial properties of argument systems. Tech. report, U.L.C.S., 2001.

[Dunne and Bench-Capon, 2002] P. Dunne and T. Bench-Capon. Coherence in finite argument system. *Artificial Intelligence*, 141(1-2):187–203, 2002.

[Liao et al., 2011] B. Liao, L. Jin, and R. C. Koons. Dynamics of argumentation systems: A division-based method. *Artificial Intelligence*, 175(11):1790 – 1814, 2011.

[Moguillansky et al., 2010] M. O. Moguillansky, N. D. Rotstein, M. A. Falappa, A. J. García, and G. R. Simari. Argument theory change through defeater activation. In *Proc. of COMMA 2010*, pages 359–366. IOS Press, 2010.

[Moguillansky et al., 2013] M. O. Moguillansky, N. D. Rotstein, M. A. Falappa, A. J. Garcia, and G. R. Simari. Dynamics of knowledge in delp through argument theory change. *Theory and Practice of Logic Programming*, pages 1–65, 2013.

[Paglieri and Castelfranchi, 2005] F. Paglieri and C. Castelfranchi. Revising beliefs through arguments: Bridging the gap between argumentation and belief revision in MAS. In *Argumentation in Multi-Agent Systems*, pages 78–94. Springer, 2005.

[Paglieri and Castelfranchi, 2006] F. Paglieri and C. Castelfranchi. The toulmin test: Framing argumentation within belief revision theories. In *Arguing on the Toulmin model*, pages 359–377. Springer, 2006.

11 Synthesis

All the characterization results given in this paper are synthesized in the following tables. It should be noted that some CS (Sufficient Condition) or CN (Necessary Condition) obtained by the application of duality are not really useful because they relate in general to the output system and it is very difficult to translate them in terms of conditions on the input system, whether it is for the addition or the removal (these CS and CN are denoted with *CS* and *CN* in the tables).

For each change operation ($\ominus$ and $\oplus$), two tables are given, the first long table giving the CS and CN found for the properties defined in Sect. 3 and the second short one giving some additional propositions.

Properties of the change $\oplus$	Grounded semantics	Preferred semantics	Stable semantics
$e_{\emptyset-1ne}$, $e_{\emptyset-k}$, $r_{k-\emptyset}$, $r_{1ne-\emptyset}$	Never (Prop. 3)	Never (Prop. 4)	
c_{1e-1ne}	CNS: Prop. 10.2	CS: Prop. 19 CN: Prop. 20	Never (Prop. 5)
r_{k-1ne}	Never (Prop. 3)		
r_{k-j}	Never (Prop. 3)	CN: $\exists$ even-length cycle in $\mathscr{G}$ + Prop. 18	
e_{1ne-k}	Never (Prop. 3)	CN: $\exists$ even-length cycle in $\mathscr{G}'$ + Prop. 20, Prop. 21	
e_{1e-k}	Never (Prop. 3)	CN: $\exists$ even-length cycle in $\mathscr{G}'$ + Prop. 20, Prop. 21	Never (Prop. 5)

e_{j-k}	Never (Prop. 3)	CN: $\exists$ even-length cycle in $\mathcal{G}'$ + Prop. 20, Prop. 21	
c_{1ne-1e}	CNS: Prop. 13	CS: Prop. 24 CN: Prop. 1.5, 2.2	Never (Prop. 5)
r_{k-1e}	Never (Prop. 3)	CN: Prop. 1.5, 2.2	Never (Prop. 5)
c-expansive	CS: Prop. 12.2 + 12.3 *CN* : Prop. 29$^{\oplus}$ *CS* : Prop. 33$^{\oplus}$	CS: Prop. 21.1 *CN* : Prop. 29$^{\oplus}$ *CS* : Prop. 33$^{\oplus}$	*CN* : Prop. 29$^{\oplus}$
c-conservative	CS: Prop. 10.1 CS: Prop. 12.1	CS: Prop. 20 CS: Prop. 21.2	
c-narrowing	CN: Prop. 11 *CN* : Prop. 32$^{\oplus}$	CN: Prop. 18 *CN* : Prop. 32$^{\oplus}$	Never (Prop. 31$^{\oplus}$)
c-altering	CN: Prop. 11	CN: Prop. 18	
simple expansive monotony	CS: $\mathscr{E} = \{\}$ CS: Prop. 11	CS: Prop. 22	
simple restrictive monotony	*CNS* : Prop. 26$^{\oplus}$	*CNS* : Prop. 26$^{\oplus}$	*CNS* : Prop. 26$^{\oplus}$
weak expansive monotony			
weak restrictive monotony	*CNS* : Prop. 27.3$^{\oplus}$	*CNS* : Prop. 27.1$^{\oplus}$	*CNS* : Prop. 27.2$^{\oplus}$
priority to recency	CS: Prop. 9 CS: Prop. 12	CS: Prop. 17 CS: Prop. 19	
skeptical expansive monotony	cf. simple expansive monotony	CS: Prop. 23	
skeptical restrictive monotony	cf. simple restrictive monotony		
credulous expansive monotony	cf. simple expansive monotony	cf. simple expansive monotony	cf. simple expansive monotony
credulous restrictive monotony	cf. simple restrictive monotony	cf. simple restrictive monotony	cf. simple restrictive monotony
preserves the rejected status	CS: Prop. 14		
preserves acceptability	CS: Prop. 8 and also cf. simple expansive monotony	cf. simple expansive monotony	cf. simple expansive monotony
preserves skeptical acceptability	CS: Prop. 8	cf. skeptical expansive monotony	cf. skeptical expansive monotony
preserves credulous-only acceptability	CS: Prop. 8	cf. simple expansive monotony	cf. simple expansive monotony
establishes acceptability	CS: Conseq.1 CS: Prop. 15		
removes acceptability	CS: Prop. 16 *CS*: Prop. 36$^{\oplus}$		

Characterizing change in abstract argumentation systems

Prop. number	For $\oplus$, corresponds to	Semantics
Prop. $28^{\oplus}$	*CS* for the "preservation of a final extension" (*i.e.* in which case a final extension was already an initial extension)	S, P, G
Prop. $30^{\oplus}$	*CS* for the preservation (eventually weak) of a final extension (*i.e.* in which case a final extension without Z was already an initial extension)	P

Properties of the change $\ominus$	Grounded semantics	Preferred semantics	Stable semantics
$e_{\varnothing-1ne}$, $e_{\varnothing-k}$, $r_{k-\varnothing}$, $r_{1ne-\varnothing}$	Never (Prop. 3)	Never (Prop. 4)	
c_{1e-1ne}	CNS: Prop. $13^{\ominus}$	*CS*: Prop. $24^{\ominus}$ CN: Prop. 1.5 + Prop. 2.2	Never (Prop. 5)
r_{k-1ne}	Never (Prop. 3)	*CN*: $\exists$ even-length cycle in $\mathscr{G}$ + Prop. $20^{\ominus}$, Prop. $21^{\ominus}$	
r_{k-j}	Never (Prop. 3)	*CN*: $\exists$ even-length cycle in $\mathscr{G}$ + Prop. $20^{\ominus}$, Prop. $21^{\ominus}$	
e_{1ne-k}	Never (Prop. 3)		
e_{1e-k}	Never (Prop. 3)	CN: Prop. 1.5 + Prop. 2.2	Never (Prop. 5)
e_{j-k}	Never (Prop. 3)	*CN*: $\exists$ even-length cycle in $\mathscr{G}'$ + Prop. $18^{\ominus}$	
c_{1ne-1e}	CN : Prop. $10.1^{\ominus}$ CN : Prop. $10.2^{\ominus}$ CN : Corollary 1	*CS*: Prop. $19^{\ominus}$ *CN*: Prop. $20^{\ominus}$	Never (Prop. 5)
r_{k-1e}	Never (Prop. 3)	*CN*: $\exists$ even-length cycle in $\mathscr{G}$ + Prop. $20^{\ominus}$, Prop. $21^{\ominus}$	Never (Prop. 5)
c-expansive	CN : Prop. 32	CN : Prop. 32 *CN*: Prop. $18^{\ominus}$	Never (Prop. 31)
c-conservative	CS : Prop. $12.1^{\ominus}$	*CS*: Prop. $20^{\ominus}$ *CS*: Prop. $21.2^{\ominus}$	
c-narrowing	CN : Prop. 29 CS : Prop. 33 *CS* : Prop. $12.2^{\ominus}$ CS : Prop. $12.3^{\ominus}$	CN : Prop. 29 CS : Prop. 33 *CS*: Prop. $21.1^{\ominus}$	CN : Prop. 29
c-altering	CN: Prop. $11^{\ominus}$	*CN*: Prop. $18^{\ominus}$	
simple expansive monotony	CNS: Prop. 26	CNS : Prop. 26	CNS : Prop. 26
simple restrictive monotony	CS: $\mathscr{E}' = \{\}$ CS: Prop. $11^{\ominus}$	*CS*: Prop. $22^{\ominus}$	
weak expansive monotony	CNS : Prop. 27.3	CNS : Prop. 27.1	CNS : Prop. 27.2
weak restrictive monotony			
priority to recency	Never (def. priority to recency)	Never (def. priority to recency)	Never (def. priority to recency)
skeptical expansive monotony	(*cf.* simple expansive monotony)		

skeptical restrictive monotony	(cf. simple restrictive monotony)	*CS*: Prop. 23$^\ominus$	
credulous expansive monotony	(cf. simple expansive monotony)	(cf. simple expansive monotony)	(cf. simple expansive monotony)
credulous restrictive monotony	(cf. simple restrictive monotony)	(cf. simple restrictive monotony)	(cf. simple restrictive monotony)
preserves the rejected status	CS: Prop. 8$^\ominus$		
preserves acceptability	CS: Prop. 14$^\ominus$ and also cf. simple expansive monotony	cf. simple expansive monotony	cf. simple expansive monotony
preserves skeptical acceptability	cf. skeptical expansive monotony	cf. skeptical expansive monotony	cf. skeptical expansive monotony
preserves credulous-only acceptability	cf. simple expansive monotony	cf. simple expansive monotony	cf. simple expansive monotony
establishes acceptability	CS: Prop. 34		
removes acceptability	CS: Prop. 35 CS: Prop. 36		

Prop. number	For $\ominus$, corresponds to	Semantics
Prop. 28	CS for the "preservation of an initial extension" (*i.e.* in which case an initial extension is always a final extension)	S, P, G
Prop. 30	CS for the preservation (eventually weak) of an initial extension (*i.e.* in which case an initial extension with eventually the adding of Z is always a final extension)	P

12 Proofs

This chapter gathers some results coming from several papers together with some new results. Table 4 gives the correspondence between the numbering used here and the numbering used in the initial papers where the proofs can be found. The proofs of the new propositions are given below while the proofs of Prop. 4, 5, 6, 7, 37-43, $25^\oplus$-$33^\oplus$, $36^\oplus$, $12.2^\ominus$, $18^\ominus$-$24^\ominus$ are not detailed (brief explanations were given in the main part).

Proof of Prop. 15: If Z is not attacked in $\mathscr{G}$, then $Z \in \mathscr{E}'$. So $\bigcup_{i \geq 1} \mathscr{F}'^i(\{Z\}) \subseteq \mathscr{E}'$ since the function $\mathscr{F}'$ is monotonic. Since Z indirectly defends X, $X \in \bigcup_{i \geq 1} \mathscr{F}'^i(\{Z\})$, so $X \in \mathscr{E}'$. □

Proof of Prop. 16: For proving this proposition, the following lemmas are used:
Lem. 5 *If $S \subseteq \mathscr{G}$ and $\mathscr{F}'(S) \subseteq \mathscr{G}$, then $\mathscr{F}'(S) \subseteq \mathscr{F}(S)$.*

Proof of Lem. 5: Let $X \in \mathscr{F}'(S)$. By assumption, $\mathscr{F}'(S) \subseteq \mathscr{G}$ so $X \in \mathscr{G}$, and by definition S defends X in $\mathscr{G}'$. Let assume that X is attacked by Y in $\mathscr{G}$. Then Y also attacks X in $\mathscr{G}'$. So, S attacks Y in $\mathscr{G}'$. But, $S \subseteq \mathscr{G}$ and $Y \in \mathscr{G}$, so S also attacks Y in $\mathscr{G}$. Thus S defends X in $\mathscr{G}$, i.e. $X \in \mathscr{F}(S)$. □

Lem. 6 *If $Z \notin \mathscr{E}'$, then $\forall i \geq 1, \mathscr{F}'^i(\varnothing) \subseteq \mathscr{F}^i(\varnothing)$.*

Characterizing change in abstract argumentation systems

Table 4 Numbering correspondence with existing results (Pr = Proposition, Lem = Lemma, Cor = Corollary, ext = extended, rest = restricted)

Correspondence with [Cayrol et al., 2010]		Correspondence with [Bisquert et al., 2012a]		Correspondence with [Bisquert et al., 2012b]	
New nb.	Old nb.	New nb.	Old nb.	New nb.	Old nb.
Pr.3	Pr.12 ext.	Pr.25	Pr.1	Lem.1	Lem.1
Pr.8	Pr.7	Pr.26	Pr.2	Lem.2	Lem.2
Pr.9	Pr.8	Pr.27	Pr.3	Pr.14	Pr.3
Pr.10	Pr.9	Pr.28	Pr.4	Pr.8$^\ominus$	Pr.1.1$^\ominus$
Pr.11	Pr.10 ext.	Pr.29	Pr.8	Pr.10.1$^\ominus$	Pr.1.2$^\ominus$
Pr.12	Pr.11 ext.	Pr.30	Pr.5	Pr.10.2$^\ominus$	Pr.1.3$^\ominus$
Pr.13	Pr.13	Pr.31	Pr.6	Cor.1	Cor.1
Pr.17	Pr.14	Pr.32	Pr.7	Pr.11$^\ominus$	Pr.2.1$^\ominus$ (v2)
Pr.18	Pr.15	Pr.33	Pr.9	Pr.12.1$^\ominus$	Pr.2.2$^\ominus$ (v2)
Pr.19	Pr.16			Pr.12.3$^\ominus$	Pr.2.4$^\ominus$ (v2)
Pr.20	Pr.17			Pr.13$^\ominus$	Pr.2.5$^\ominus$ rest.
Pr.21	Pr.18			Pr.14$^\ominus$	Pr.3$^\ominus$
Pr.22	Pr.19				
Pr.23	Pr.20				
Pr.24	Pr.21				

Proof of Lem. 6: by induction on i and using Lemma 5.
First note that, if $Z \notin \mathscr{E}'$, then $\mathscr{E}' \subseteq \mathscr{G}$ and so $\forall i \geq 1, \mathscr{F}'^i(\varnothing) \subseteq \mathscr{G}$.
Case $i = 1$: since $Z \notin \mathscr{E}'$, Z is attacked by $\mathscr{G}$ in $\mathscr{G}'$. The elements of $\mathscr{F}'(\varnothing)$ are not equal to Z and by definition they are unattacked in $\mathscr{G}'$. Since no attack is removed, they are also unattacked arguments in $\mathscr{G}$. So $\mathscr{F}'(\varnothing) \subseteq \mathscr{F}(\varnothing)$. Let assume that $\mathscr{F}'^k(\varnothing) \subseteq \mathscr{F}^k(\varnothing)$ and let consider $\mathscr{F}'^{(k+1)}(\varnothing) = \mathscr{F}'(\mathscr{F}'^k(\varnothing))$. Let $S = \mathscr{F}'^k(\varnothing)$. $S \subseteq \mathscr{G}$ and $\mathscr{F}'(S) \subseteq \mathscr{G}$. Using Lemma 5, it holds that $\mathscr{F}'^{(k+1)}(\varnothing) \subseteq \mathscr{F}(\mathscr{F}'^k(\varnothing))$. Using the induction assumption, $\mathscr{F}'^k(\varnothing) \subseteq \mathscr{F}^k(\varnothing)$. And using the monotony of $\mathscr{F}$, and the transitivity of the set-inclusion, it holds that $\mathscr{F}'^{(k+1)}(\varnothing) \subseteq \mathscr{F}(\mathscr{F}^k(\varnothing))$, i.e. $\mathscr{F}'^{(k+1)}(\varnothing) \subseteq \mathscr{F}^{(k+1)}(\varnothing)$. □

Let $X \in \mathscr{E}$ s.t. Z attacks X and $\mathscr{E} \setminus \{X\}$ does not attack Z. Assume that $X \in \mathscr{E}' = \bigcup_{i \geq 1} \mathscr{F}'^i(\varnothing)$. As Z attacks X, $X \notin \mathscr{F}'(\varnothing)$. Let i be the smallest index ≥ 2 s.t. $X \in \mathscr{F}'^i(\varnothing)$. So $X \in \mathscr{F}'(\mathscr{F}'^{(i-1)}(\varnothing))$ and $X \notin \mathscr{F}'^{(i-1)}(\varnothing)$. Z attacks X and so there exists $Y \in \mathscr{F}'^{(i-1)}(\varnothing)$ s.t. Y attacks Z in $\mathscr{G}'$. Since $X \notin \mathscr{F}'^{(i-1)}(\varnothing)$, it holds that $Y \neq X$. Moreover, Z attacks X and we assume that $X \in \mathscr{E}$, so $Z \notin \mathscr{E}'$ and thus $\mathscr{E}' \subseteq \mathscr{G}$. Using Lemma 6, we can infer that $\mathscr{F}'^{(i-1)}(\varnothing) \subseteq \mathscr{F}^{(i-1)}(\varnothing)$. Thus $Y \in \mathscr{E}$. So we have found $Y \in \mathscr{E} \setminus \{X\}$ which attacks Z, that is contradictory with the assumption. □

Proof of Prop. 34: X is never attacked in $\mathscr{G}'$. □

Proof of Prop. 35: The proof uses the following lemma:
Lem. 7 *Let $\mathscr{G}$ be an argumentation graph and $\mathscr{F}$ be the associated characteristic function. If $X \in \bigcup_{i \geq 1} \mathscr{F}^i(\varnothing)$, then for each T attacker of X, there exists an odd-length path from $\mathscr{F}(\varnothing)$ to T.*

Proof of Lem. 7: By induction on $i \geq 1$, one proves that, if $X \in \mathscr{F}^i(\varnothing)$, then for each T attacker of X, there exists an odd-length path from $\mathscr{F}(\varnothing)$ to T. For the case $i = 1$, the result obviously holds ($X \in \mathscr{F}(\varnothing)$ implies that X is unattacked). For the other cases, assume that the property holds at the rank $p \geq 1$ and that $X \in \mathscr{F}^{p+1}(\varnothing)$. Let T be an attacker of X. Thus T is attacked by an argument $W \in \mathscr{F}^p(\varnothing)$. If W is unattacked in $\mathscr{G}$, then $W \in \mathscr{F}(\varnothing)$, so there exists a path whose length equals 1 from $\mathscr{F}(\varnothing)$ to T. Otherwise, there exists U attacker of W. Using the assumption of the induction on W, there exists an odd-length path from $\mathscr{F}(\varnothing)$ to U. Then this path is augmented with the attack from U to W and the attack from W to T. So we obtain a new odd-length path from $\mathscr{F}(\varnothing)$ to T. □

First, note that $Y \neq Z$. Indeed, by assumption, $X \in \mathscr{E}$ and $Z \in \mathscr{E}$, so Z cannot attack X. So $Y \in \mathscr{G}'$. Moreover, $X \neq Z$ is also an assumption of the proposition. Using a reductio ad absurdum, assume that $X \in \mathscr{E}'$, i.e. $X \in \bigcup_{i \geq 1} \mathscr{F}'^i(\emptyset)$. Following the assumption on X, there exists Y attacker of X in $\mathscr{G}$ s.t. each odd-length path from $\mathscr{F}(\emptyset)$ to Y in $\mathscr{G}$ contains Z at an even place. Note that $X \in \mathscr{G}'$ and $Y \in \mathscr{G}'$. So, Lemma 7 can be applied on $\mathscr{G}'$ and $\mathscr{F}'$. Y attacking X in $\mathscr{G}'$, there exists an odd-length path from $\mathscr{F}'(\emptyset)$ to Y. Let $T \in \mathscr{F}'(\emptyset)$ be the root argument of this path to Y. This path only contains arguments of $\mathscr{G}'$, so it cannot contain Z so T cannot belong to $\mathscr{F}(\emptyset)$. This means that T is attacked in $\mathscr{G}$ and not attacked in $\mathscr{G}'$, so T is only attacked by Z. Thus there exists an even-length from Z to Y (via T) and so there exists an odd-length path from Z to X, that contradicts the last assumption of the proposition. □

Proof of Prop. 36: First note that $Z \in \mathscr{E}$. Indeed, $\mathscr{E} = \mathscr{F}(\mathscr{E})$ so $X \in \mathscr{F}(\mathscr{E})$ and so, with the assumption, $Z \in \mathscr{E}$. So there is no attack between Z and X, and the attackers of X in $\mathscr{G}$ are also the attackers of X in $\mathscr{G}'$. Using a reductio ad absurdum, assume that $X \in \mathscr{E}'$. We will show that $\mathscr{E}'$ defends X in $\mathscr{G}$. Let Y be an attacker of X in $\mathscr{G}$. Then Y attacks X in $\mathscr{G}'$. As $X \in \mathscr{E}'$ and $\mathscr{E}' = \mathscr{F}'(\mathscr{E}')$, it holds that $\mathscr{E}'$ defends X in $\mathscr{G}'$. So $\mathscr{E}'$ attacks Y in $\mathscr{G}'$ and also in $\mathscr{G}$. Thus $X \in \mathscr{F}(\mathscr{E}')$. Using the assumption given the proposition, it holds that $Z \in \mathscr{E}'$ that is impossible since Z has been removed. □

Argumentation and Belief Updating in Social Networks: A Bayesian Model

George Masterton and Erik J. Olsson

Abstract Human rationality involves arguing for one's beliefs as well as revising those beliefs in the face of good reasons to do so. An important problem in computer science and philosophical logic is to find models that combine argumentation and belief updating in one formal framework. In this article, we present the Laputa model which attempts to do precisely this. Laputa differs from some other frameworks with similar goals (a) in being Bayesian, (b) in taking *pro et contra* argumentation as the paradigm case, and (c) in taking the persuasive effect of an argument, rather than its strict acceptability, to be the more fundamental notion. This article provides a general introduction to the model for a computer science audience as a way of integrating argumentation and belief updating. It features detailed derivations of how inquirers in a social network update their credences and trust assessments in response to inquiry and argumentation.

1 Introduction

Human inquiry has many facets of which argumentation and belief revision stand out as particularly salient. Unsurprisingly, both have been studied extensively in various areas of scientific research. Computer science and philosophical logic are no exception to the rule. Both argumentation and belief revision have received extensive treatment by computer scientists and logicians alike. Yet, in standard treatments these aspects of rational inquiry are investigated separately rather than in conjunction. For example, most models of belief revision leave argumentation out of the picture; which is true, for instance, of the influential AGM model (Alchourrón et al., 1985). This is so for a good methodological reason: both areas are quite complex in themselves and it makes good methodological sense to adopt a simplified picture for

---

George Masterton · Erik J. Olsson
Department of Philosophy, Lund University, Sweden,
e-mail: George.Masterton,Erik_J.Olsson@fil.lu.se

the time being. Still, the ultimate goal must surely be a model which incorporates both these phenomena. How this could be accomplished is the subject of this article.

There are many ways in which one could incorporate a role for arguments in belief revision. Belief revisionists realised early on that not all beliefs have the same status. Rather, some beliefs are held because of other beliefs: the latter provide the reasons for the former. A straightforward way to capture this phenomenon in a belief revision framework is to introduce a belief base into the model (e.g. Furhmann (1991), Hansson (1991)). The belief base of an agent contains all the beliefs that the agent holds independently of other beliefs. The total set of beliefs can be understood as the logical closure of the belief base. Another early theory which focused on storing arguments is the Truth Maintenance System (Doyle, 1979). An elaborate new development in the same general category is Tennant's dependency network approach (Tennant, 2012).

However important as a research program in its own right, finding ways to represent the argumentative structure of a belief system is still not accounting for argumentation, i.e., the process of giving and taking arguments. Argumentation, in this sense, is a social practice involving more than one participant. Understanding argumentation in a belief revision framework presupposes a multi-agent, or social network, approach to belief revision. One promising point of departure for a possible marriage between belief revision and argumentation is the influential view on argumentation presented in Dung (1995). Dung is concerned with the acceptability of arguments, the fundamental principle being that the one who has the last word laughs best. Dung gives the following examples:

Mary: My government cannot negotiate with your government because your government doesnt even recognize my government.
John: Your government doesn't recognize my government either.
Mary: But your government is a terrorist government.

Assuming (somewhat unrealistically) that the debate stopped there, Mary won the argument since she had the last word. The general idea is that a statement is believable if it can be argued successfully against attacking arguments, and so whether or not a rational agent believes in a statement depends on whether or not the argument supporting this statement can be successfully defended against the counterarguments (Dung, 1995, p. 323). It is easy to imagine how this model could be used to study belief revision as arising from the dynamics of argumentation.

Formally, an argumentation framework in Dung's sense is a pair $\langle AR, attacks \rangle$, where AR is a set of arguments and attacks is a binary relation on AR. In Dung's framework an argument is viewed as an abstract entity whose role is determined by its relation to other arguments. No special attention is paid to the internal structure of arguments. We will follow Dung in this respect. However, we will focus on a type of argumentation that does not immediate fit into the last word wins category. Consider the following debate on the future of the euro:

Mary: I believe the Eurozone will survive because the German Bundesbank has strongly committed itself the future of the euro.

John: I doubt you are right considering the strong political divisions that exist within the Eurozone.

Mary: I grant this, but it should also be pointed out that in the most recent meeting among the Eurozone leaders a strong statement was issued in favor of increased political integration and unity.

Here we wouldn't say that Mary won the argument in the strict sense, although she did have the last word. Rather, both parties have contributed to the argumentation process by putting various pro and con considerations on the table, all of which deserve to be taken into account and none of which suffices by itself to establish anything beyond reasonable doubt.

What we will be concerned with, then, is not so much the strict acceptability of an argument as its more general persuasive effect. This effect has been studied extensively in social psychology within the influential and empirically robust Persuasive Argument Theory (PAT) tradition (e.g. Vinokur and Burnstein (1978), Isenberg (1986)). According to PAT, an individual's position on an issue is a function of the number and persuasiveness of pro and con arguments that the person recalls from memory when formulating his or her own position. Thus in assessing the guilt or innocence of an accused in trial, jurors come to predeliberation decisions on the basis of the relative number and persuasiveness of arguments favoring guilt or innocence. Argumentation, or group deliberation as we will also call it, will cause an individual to shift in a given direction to the extent that the discussion exposes that individual to persuasive arguments favoring that direction rather than to arguments favoring the opposite direction. How persuasive an argument is to a given individual is determined by the validity and novelty of the argument. One factor, among several, affecting perceived validity is the extent to which the argument fits into the person's previous views. Novelty has to do with how new and unusual the argument is to the person in question. Everything else equal, a novel argument has a greater persuasive force than a common place argument.

Our model, which we call Laputa (e.g. (Angere, forthcoming), (Olsson, 2011), (Vallinder and Olsson, 2012)), is similar in spirit to PAT. Laputa is also based on the assumption that the persuasive effect of an argument depends essentially on two factors: its perceived validity (including the trustworthiness of the presenter) and novelty. There are some differences, though. For instance, Laputa is more specific than PAT in assuming that individual inquirers update their degrees of belief in a particular way; namely, that dictated by Bayesianism. PAT, as such, does not postulate any more specific updating mechanism, let alone a Bayesian one. Laputa assumes, in addition, that individuals' degrees of trust are dynamically updated in a Bayesian fashion.

Furthermore, inquirers in Laputa engaging in group deliberation update their credences in a piecemeal, or sequential, fashion. The presentation of a novel argument, or collection of arguments, will normally affect the receiving inquirer's credence in the conclusion. As PAT is normally formulated, inquirers are supposed to collect in memory all the arguments they are presented with during group deliberation, postponing their own verdict on the matter until deliberation has come to an end. When the deliberation has ended the inquirer takes a stand on the basis of a holistic assessment of the number and merits of the pro and con arguments retained in memory. This holistic aspect of PAT is not unproblematic in the light of experiments indicating that the order in which arguments are presented will affect the conclusion reached. Thus, Kaplan (1977) found that subjects tend to recall persuasive arguments that they had been exposed to most recently rather than the ones they had been exposed to first.

The paper shall proceed by first informally introducing Laputa as a theory of argumentation in a group. Subsequently, the dynamical formula of Laputa are derived from accepted Bayesian principles. The paper concludes by briefly describing how this theoretical framework has been implemented in a computer program baring the same name, followed by a very brief summary of its applications to date.

2 An informal introduction to Laputa

Traditional belief revision has focused on the single inquirer setting. We wish to study belief updating in a social (network) context. An interesting complication is that inquirers in a social network not only update their degrees of belief but also their degrees of trust in their interlocutors. Ideally, we would like to have a model featuring both a rich language and a rich cognitive state representation and dynamics. However, as the matter is already quite complex— especially the proper handling of trust—some sort of compromise is necessary at the present state of investigation. One compromise would involve having a rich language but simplifying the state representation and dynamics. We will choose the opposite strategy by adopting a simple language but a rich cognitive state representation and dynamics.

In fact, the Laputa model has an extremely simple language consisting of only two propositions: p, not-p. The proposition p can stand for *The eurozone will disintegrate in 2012* or *John was at the party last night* or anything else with a truth value. Thus p and not-p can be seen as the two potential answers to the socially debated question: Whether p is the case?. While the language is simple, the cognitive state representation and dynamics will be quite complex. Our formal framework of choice will be Bayesianism, where belief states are represented as credences/subjective-probabilities and the basic method of belief updating is conditionalization. A social network is conceived as a set of inquirers with links between them. If there is a link from inquirer A to inquirer B that means that A can send a message to B. All inquirers focus on answering the question whether p is true. Each inquirer assigns to p at time t a certain credence, $C^t(p)$ (subjective probability). The messages inquirers can send are either "p" or "not-p". The preferred interpretation is the following:

- "σ sends the message p" means "σ gives a reason/argument for p"
- "σ sends the message not-p" means "σ gives a reason/argument against p"

Under this interpretation, the model is a model of deliberation/argumentation where reasons/arguments are, as in the Dung model, treated as black boxes. Inquirers also have a private signal they can listen to representing contributions to the deliberative process from external sources. In addition, each inquirer assigns to each information source a certain trust at t.

Now there are two main problems that need to be solved in order to make this model work: The Credence Problem and The Trust Problem. The former concerns how to update an inquirer's credence in p given new information, while the latter concerns how to update an inquirer's trust in a given source in response to arguments from that source. Being good Bayesians, we want to solve these two problems by means of conditionalization on new evidence. For the credence problem this means computing $C^{t+1}(p) = C^t(p|\sigma \text{ says that } p)$ or $C^{t+1}(p) = C^t(p|\sigma \text{ says that not-}p)$ depending on the incoming evidence. But how do we compute the right hand side of these equations? Our new credence in p after having listened to σ will depend on how much trust we placed in σ. Hence, already the credence problem requires that we also model epistemic trust; but how?

The proposal is that trust is also a form of credence; namely, a credence in the reliability of the source. This idea goes back to the Scandinavian School of Evidentiary Value (e.g. Hansson (1983), and it has been used extensively in the literature on epistemic coherence (e.g. Bovens and Olsson (2000), Olsson (2002) and Olsson (2005)).

Definition: By a source σ's degree of reliability with respect to p we shall mean the (objective) probability that σ says that p given (i) that σ says anything at all on p and (ii) that p is true.

In modeling trust Laputa takes into account every possible degree of reliability and every possible degree of unreliabilityby assigning a credence to the proposition expressing that the source is reliable/unreliable to that degree. For example, an inquirer's trust function assigns a certain credence to the proposition that the source is 75 percent reliable. Thus, trust values are here seen as second order probabilities: subjective probabilities about objective probabilities.

Returning to the credence problem, our strategy will be to proceed in two steps:

Step 1: Addressing the credence problem for one source.
Step 2: Extending this solution to a solution to the credence problem for n sources.

We need a few more assumptions before any useful work can be done:

Source Symmetry: σ's reliability with respect to p equals σ's reliability with respect to not-p.
Principal Principle: An inquirer α assigns credence ρ to the proposition that σ will report that p on the assumptions that (i) σ reports anything at all (ii) p is true and (iii) σ is reliable to degree ρ[1].
Communication Independence: Whether a source σ says something at all is independent of whether p is true as well as of σ's degree of reliability.

---

[1] A question mark about whether this use of the Principal Principle is valid in a context where α's total evidence includes that σ has reported p has been raised (see Meacham (2010, p. 411-413)). Here, the principle is assumed as an expediency to be justified empirically on the basis of how well Laputa models deliberation and debate.

Once these assumptions are in place, the rest is pure mathematics, as we shall see in the next section. What about the case of n sources? Here the new credence would be calculated as $C^{t+1}(p) = C^t(p|\sigma_1 \text{ says } p, \sigma_2 \text{ says not-}p, ...)$. To facilitate the calculations of the new credence in this case, we need to make a further assumption:

Source Independence: Inquirers take information they receive from other sources to be independent evidence.

Source Independence can be expressed in a standard way as a form of conditional independence: The credence assigned to the proposition that source σ_1 will report that p is independent of the credence assigned to the proposition that source σ_2 will report that p, and so on, conditional on the truth/falsity of p. This assumption is often used in the literature on epistemic coherence and in artificial intelligence (e.g. Pearl (1988)). As we shall see, given Source Independence the general credence problem has a simple mathematical solution.

Let us finally now turn to the trust problem: the problem of stating how to update an inquirer's trust function in the light of new evidence. Fortunately, no additional assumptions are needed to solve the trust problem (and we don't need Source Independence). In the case where the source says that p:

$$\text{Trust in } \sigma \text{ as a source on } p = C_\alpha^{t+1}(\sigma \text{ is reliable to degree } \rho)$$
$$= C_\alpha^t(\sigma \text{ is reliable to degree } \rho | \sigma \text{ says that } p),$$

where the latter is a function of (i) ρ, (ii) $C_\alpha^t(p)$, and (iii) the inquirer's trust function for σ at t (or rather its expected value). Now that we have developed a sense of how this works, let us move on to the formal details.

3 Derivation of the Update Functions

This section follows the structure of (Angere, forthcoming), though it deviates from his presentation in its explicit incorporation of the present interpretation of efficacious testimony as the receipt of a novel argument/reason for, or against, p. In Bayesian fashion, the epistemic state of an individual α at time t is given by a *credence function* $C_\alpha^t : \mathscr{L} \to [0,1]$, where we can take $\mathscr{L}$ to be a classical propositional language. The expression $C_\alpha^t(p) = x$ should be read as "Agent α's credence in proposition p at time t is x.". Let t be the time just prior to the first round of debate, $t+1$ be the time just prior to the 2nd, and so on up to $t+n$; the time just after deliberation concludes. Laputa works by determining the value of $C_\alpha^{t+1}(p)$ from C_α^t on the basis of any novel argument α receives, or private inquiry α makes, in the period t to $t+1$ for all α partaking in the debate, then subsequently determining the value of $C_\alpha^{t+2}(p)$ from C_α^{t+1} for all α on the same basis in the period $t+1$ to $t+2$, and so on all the way up to the determination of $C_\alpha^{t+n}(p)$ for all α. This sequential nature of such updates is why Laputa simulates, rather than models, debates.

Argumentation and Belief Updating in Social Networks: A Bayesian Model

Bayesian epistemology includes a principle—the principle of conditionalization—which allows $C_\alpha^{t+j}(p)$ to be determined from C_α^{t+i}, where $j = i+1$. Let E_α^{t+i} represent all inquiry conducted, and novel arguments received, by α in the period $t+i$ to $t+j$ regarding p, then the principle of conditionalization states that $C_\alpha^{t+j}(p) = C_\alpha^{t+i}(p|E_\alpha^{t+i})$. The problem becomes one of finding an expression for $C_\alpha^{t+i}(p|E_\alpha^{t+i})$ that allows its value to be computed from information available at $t+i$.

Relative to some inquirer α and time t, other inquirers in the network can be considered as *sources* of novel (for α at t) arguments for, or against, p. Let S be an argument either for, or against, p, let $S_{\sigma\alpha}$ be the proposition 'S is received by α from σ', let S_α^t be the proposition 'S is novel for α at t and received by α in the period t to $t+1$'. Let $S_{\sigma\alpha}^t$ be the proposition $S_{\sigma\alpha} \wedge S_\alpha^t$ and let S^+ and S^- be the propositions 'S is for p.' and 'S is against p.', respectively. Note also that an argument against p is an argument for not-p.

But what of each inquirer's private inquiry? It is not reasonable to interpret such inquiry as the receipt of an argument. For instance, how can a first person empirical investigation result in receipt of an argument for p? What such investigations can give w.r.t. p are novel *reasons* to believe p. All arguments for p are reasons to believe p, but not all reasons to believe p are arguments. Let i be the "own inquiry source" of reasons to believe p such that each α participant has such a source. Let $S_{i\alpha}$ be the proposition 'S is received by α from i'. Where this proposition holds, then S is to be interpreted as a reason to believe either p or $\neg p$, let S_α^t be the proposition 'S is novel for α at t and received by α in the period t to $t+1$'. Let $S_{i\alpha}^t$ be the proposition $S_{i\alpha} \wedge S_\alpha^t$.

We can then define the *reliability* of σ as a source of arguments on p for α as:

$$R_{\sigma\alpha}^\pm =_{\text{df.}} P(S^+|S_{\sigma\alpha} \wedge p) = P(S^-|S_{\sigma\alpha} \wedge \neg p),$$

where $P(S^+ | S_{\sigma\alpha} \wedge p)$ is the objective probability with which a p-argument is an argument for p, when α receives that argument from σ and p is true and similarly for $P(S^- | S_{\sigma\alpha} \wedge \neg p)$. This probability might be interpreted as a (conceptual) relative frequency, or as a propensity. It is also useful to define

$$R_{\sigma\alpha}^\mp =_{\text{df.}} P(S^+|S_{\sigma\alpha} \wedge \neg p) = P(S^-|S_{\sigma\alpha} \wedge p)$$

Plainly, $R_{\sigma\alpha}^\pm = 1 - R_{\sigma\alpha}^\mp$ and note that the reliability of a source for an inquirer is assumed to be constant through time.

Similarly, we can define the reliability of an agent's own inquiry as:

$$R_{i\alpha}^\pm =_{\text{df.}} P(S^+|S_{i\alpha} \wedge p) = P(S^-|S_{i\alpha} \wedge \neg p),$$

where $P(S^+ | S_{i\alpha} \wedge p)$ is the objective probability with which a p-reason is a reason to believe p, when α receives that reason from their own inquiry and p is true. Again, it is also useful to define

$$R_{i\alpha}^\mp =_{\text{df.}} P(S^+|S_{i\alpha} \wedge \neg p) = P(S^-|S_{i\alpha} \wedge p).$$

Now we assume that at any time t, α has a credence distribution over the reliability of σ as a source of novel arguments/reasons for them on p. Because such reliabilities can take any real value between 0 and 1, this distribution must be represented by a continuous density function $\tau_{\sigma\alpha}^{t}{}^{2}$. The credence at t that α has that the reliability of σ for them lies in the interval $[a,b]$ is given by the integral of $\tau_{\sigma\alpha}^{t+}$ between these limits:

$$C_\alpha^t(R_{\sigma\alpha}^\pm \in [a,b]) = \int_a^b \tau_{\sigma\alpha}^{t+}(\rho)\,d\rho.$$

Given α's credence distribution over σ's reliability we can also determine the reliability of σ that α should expect. Let $\langle \tau_{\sigma\alpha}^{t+} \rangle$ be this expected value, then

$$\langle \tau_{\sigma\alpha}^{t+} \rangle = \int_0^1 \rho \tau_{\sigma\alpha}^{t+}(\rho)\,d\rho.$$

It is also useful to define $\tau_{\sigma\alpha}^{t-}$ so that:

$$C_\alpha^t(R_{\sigma\alpha}^\mp \in [a,b]) = \int_a^b \tau_{\sigma\alpha}^{t-}(\rho)d\rho = \int_{1-b}^{1-a} \tau_{\sigma\alpha}^{t+}(\rho)d\rho = C_\alpha^t(R_{\sigma\alpha}^\pm \in [1-b,1-a]),$$

and consequently, that the expected value of $R_{\sigma\alpha}^\mp$ is $\int_0^1 \rho \tau_{\sigma\alpha}^{t-}(\rho)d\rho = \langle \tau_{\sigma\alpha}^{t-} \rangle = 1 - \langle \tau_{\sigma\alpha}^{t+} \rangle$. In Laputa, $\tau_{\sigma\alpha}^{t+}$ is referred to as α's *trust* (in σ at t) function. The extent to which trust understood in this way corresponds to what we typically mean by trusting a source is debatable, but it does seem to capture at least part of what trust in a source is about.

3.1 The credence update function

Now we consider the effect on α's credence of receiving a positive novel argument/reason from a single source σ. By conditionalization we have:

$$C_\alpha^{t+1}(p) = C_\alpha^t(p|S_{\sigma\alpha}^t \wedge S^+).$$

In words, α's credence at $t+1$ in p equals α's credence at t in p, given that α has received a novel argument/reason S on p from σ in the period t to $t+1$ and S is for p. Similarly, conditionalization gives $C_\alpha^{t+1}(p) = C_\alpha^t(p|S_{\sigma\alpha}^t \wedge S^-)$, so we have an expression that allows us to calculate the effect of α receiving a novel argument/reason from σ on α's credence in p whether that argument/reason is for

---

[2] The derivation proceeds on this rigorous basis, but in fact the computer program Laputa approximates continuous $\tau_{\sigma\alpha}^t$ by a discrete distribution over the following set of values for $R_{\sigma\alpha}^\pm$: $\{0, \frac{1}{40}, \frac{2}{40}, \ldots, 1\}$. Laputa does this to make evaluation of the integrals needed to calculate the expected reliabilities—which are what is required for updating credence and trust (see later)—computationally tractable. It was found by trial and error that 40-step discrete distribution offered the best balance between accuracy and required computing time.

(S^+), or against (S^-), p.

$$C_\alpha^{t+1}(p) = C_\alpha^t(p|S_{\sigma\alpha}^t \wedge S^\pm)$$

By Bayes theorem and the theorem of total probability, this gives:

$$C_\alpha^{t+1}(p^+) = \frac{C_\alpha^t(p^+)C_\alpha^t(S_{\sigma\alpha}^t \wedge S^\pm|p^+)}{C_\alpha^t(p^+)C_\alpha^t(S_{\sigma\alpha}^t \wedge S^\pm|p^+) + C_\alpha^t(p^-)C_\alpha^t(S_{\sigma\alpha}^t \wedge S^\mp|p^-)}, \quad (1)$$

where p^+ is p and p^- is not-p. $C_\alpha^t(p^+)$ is given and $C_\alpha^t(p^-) = 1 - C_\alpha^t(p^+)$, so $C_\alpha^{t+1}(p^+)$ is assuredly determined in the model if values for $C_\alpha^t(S_{\sigma\alpha}^t \wedge S^+|p^+)$, $C_\alpha^t(S_{\sigma\alpha}^t \wedge S^-|p^+)$, $C_\alpha^t(S_{\sigma\alpha}^t \wedge S^-|p^-)$ and $C_\alpha^t(S_{\sigma\alpha}^t \wedge S^+|p^-)$ are determined in the model. Each of these credences can be expanded using the continuous version of the conditional total probability theorem.

$$C_\alpha^t(S_{\sigma\alpha}^t \wedge S^\pm|p^\pm) = \int_0^1 C_\alpha^t(S_{\sigma\alpha}^t \wedge S^\pm|R_{\sigma\alpha}^\pm = \rho \wedge p^\pm)C_\alpha^t(R_{\sigma\alpha}^\pm = \rho|p^\pm)d\rho$$

$$C_\alpha^t(S_{\sigma\alpha}^t \wedge S^\mp|p^\pm) = \int_0^1 C_\alpha^t(S_{\sigma\alpha}^t \wedge S^\mp|R_{\sigma\alpha}^\mp = \rho \wedge p^\pm)C_\alpha^t(R_{\sigma\alpha}^\mp = \rho|p^\pm)d\rho$$

Laputa then assumes that in every C and for all ρ; $S_{\sigma\alpha}^t$, $R_{\sigma\alpha}^\pm = \rho$, $R_{\sigma\alpha}^\mp = \rho$ and p are independent of each other (the communication independence assumption). This allows the above to be manipulated into the following forms by using the definition of conditional probability, cancelling terms and noting that where $C_\alpha^t(R_{\sigma\alpha}^\pm = \rho)$ and $C_\alpha^t(R_{\sigma\alpha}^\mp = \rho)$ appear in the integral they stand for the density functions $\tau_{\sigma\alpha}^{t+}(\rho)$ and $\tau_{\sigma\alpha}^{t-}(\rho)$, respectively:

$$C_\alpha^t(S_{\sigma\alpha}^t \wedge S^\pm|p^\pm) = C_\alpha^t(S_{\sigma\alpha}^t)\int_0^1 C_\alpha^t(S^\pm|S_{\sigma\alpha}^t \wedge p^\pm \wedge R_{\sigma\alpha}^\pm = \rho)\tau_{\sigma\alpha}^{t+}(\rho)d\rho \quad (2)$$

$$C_\alpha^t(S_{\sigma\alpha}^t \wedge S^\mp|p^\pm) = C_\alpha^t(S_{\sigma\alpha}^t)\int_0^1 C_\alpha^t(S^\mp|S_{\sigma\alpha}^t \wedge p^\pm \wedge R_{\sigma\alpha}^\mp = \rho)\tau_{\sigma\alpha}^{t-}(\rho)d\rho. \quad (3)$$

By the aforementioned Principal Principal we have:

$$C_\alpha^t(S^\pm|S_{\sigma\alpha}^t \wedge p^\pm \wedge R_{\sigma\alpha}^\pm = \rho) = \rho \quad (4)$$
$$C_\alpha^t(S^\mp|S_{\sigma\alpha}^t \wedge p^\pm \wedge R_{\sigma\alpha}^\mp = \rho) = \rho \quad (5)$$

Using (4) and (5) to make substitutions into (2) and (3) we have:

$$C_\alpha^t(S_{\sigma\alpha}^t \wedge S^\pm|p^\pm) = C_\alpha^t(S_{\sigma\alpha}^t)\int_0^1 \rho\tau_{\sigma\alpha}^{t+}(\rho)d\rho = C_\alpha^t(S_{\sigma\alpha}^t)\langle\tau_{\sigma\alpha}^{t+}\rangle, \quad (6)$$

$$C_\alpha^t(S_{\sigma\alpha}^t \wedge S^\mp|p^\pm) = C_\alpha^t(S_{\sigma\alpha}^t)\int_0^1 \rho\tau_{\sigma\alpha}^{t-}(\rho)d\rho = C_\alpha^t(S_{\sigma\alpha}^t)\langle\tau_{\sigma\alpha}^{t-}\rangle. \quad (7)$$

Finally, substitution of (6) and (7) back into (1) gives:

$$C_\alpha^{t+1}(p^+) = C_\alpha^t(p^+|S_{\sigma\alpha}^t \wedge S^\pm) = \frac{C_\alpha^t(p^+)\langle \tau_{\sigma\alpha}^{t\pm}\rangle}{C_\alpha^t(p^+)\langle \tau_{\sigma\alpha}^{t\pm}\rangle + C_\alpha^t(p^-)\langle \tau_{\sigma\alpha}^{t\mp}\rangle}. \quad (8)$$

E.g. if the novel (for α at t) p-argument/reason S, received by α from σ in the period t to $t+1$ is for p, then we read the top line of plus/minus signs in (8) to give:

$$C_\alpha^{t+1}(p^+) = C_\alpha^t(p^+|S_{\sigma\alpha}^t \wedge S^+) = \frac{C_\alpha^t(p^+)\langle \tau_{\sigma\alpha}^{t+}\rangle}{C_\alpha^t(p^+)\langle \tau_{\sigma\alpha}^{t+}\rangle + C_\alpha^t(p^-)\langle \tau_{\sigma\alpha}^{t-}\rangle}.$$

In any period α might receive a novel argument/reason from any one of its sources. Let Σ_α^{t+} be the set of sources from which α receives novel arguments/reasons for p at t, Σ_α^{t-} be the set of sources from which α receives novel arguments/reasons for p at t and $\Sigma_\alpha^t = \Sigma_\alpha^{t+} \cup \Sigma_\alpha^{t-}$. Then, again by conditionalization, Bayes theorem and the law of total probability, we have:

$$C_\alpha^{t+1}(p^+) = C_\alpha^{t+1}\left(p^+ \bigg| \bigwedge_{\sigma \in \Sigma_\alpha^t} S_{\sigma\alpha}^t \wedge S^\pm \right)$$

$$= \frac{C_\alpha^t(p^+) C_\alpha^t\left(\bigwedge_{\sigma \in \Sigma_\alpha^t} S_{\sigma\alpha}^t \wedge S^\pm | p^+\right)}{C_\alpha^t(p^+) C_\alpha^t\left(\bigwedge_{\sigma \in \Sigma_\alpha^t} S_{\sigma\alpha}^t \wedge S^\pm | p^+\right) + C_\alpha^t(p^-) C_\alpha^t\left(\bigwedge_{\sigma \in \Sigma_\alpha^t} S_{\sigma\alpha}^t \wedge S^\mp | p^-\right)},$$

Laputa now makes the assumption that all α's sources are independent of one another for α; hence that:

$$C_\alpha^t\left(\bigwedge_{\sigma \in \Sigma_\alpha^t} S_{\sigma\alpha}^t \wedge S^\pm | p^\pm\right) = \prod_{\sigma \in \Sigma_\alpha^t} C_\alpha^t(S_{\sigma\alpha}^t \wedge S^\pm | p^\pm) = \prod_{\sigma \in \Sigma_\alpha^t} C_\alpha^t(S_{\sigma\alpha}^t)\langle \tau_{\sigma\alpha}^{t+}\rangle$$

$$C_\alpha^t\left(\bigwedge_{\sigma \in \Sigma_\alpha^t} S_{\sigma\alpha}^t \wedge S^\mp | p^\pm\right) = \prod_{\sigma \in \Sigma_\alpha^t} C_\alpha^t(S_{\sigma\alpha}^t \wedge S^\mp | p^\pm) = \prod_{\sigma \in \Sigma_\alpha^t} C_\alpha^t(S_{\sigma\alpha}^t)\langle \tau_{\sigma\alpha}^{t-}\rangle$$

By substitution and cancellation of terms into the above this gives:

$$C_\alpha^{t+1}(p^+) = C_\alpha^t \left(p^+ \mid \bigwedge_{\sigma \in \Sigma_\alpha^t} S_{\sigma\alpha}^t \wedge S^\pm \right)$$

$$= \frac{C_\alpha^t(p^+) \prod_{\sigma \in \Sigma_\alpha^t} C_\alpha^t(S_{\sigma\alpha}^t)\langle \tau_{\sigma\alpha}^{t\pm}\rangle}{C_\alpha^t(p^+) \prod_{\sigma \in \Sigma_\alpha^t} C_\alpha^t(S_{\sigma\alpha}^t)\langle \tau_{\sigma\alpha}^{t\pm}\rangle + C_\alpha^t(p^-) \prod_{\sigma \in \Sigma_\alpha^t} C_\alpha^t(S_{\sigma\alpha}^t)\langle \tau_{\sigma\alpha}^{t\mp}\rangle}$$

$$= \frac{C_\alpha^t(p^+) \prod_{\sigma \in \Sigma_\alpha^t} C_\alpha^t(S_{\sigma\alpha}^t) \prod_{\sigma \in \Sigma_\alpha^t} \langle \tau_{\sigma\alpha}^{t\pm}\rangle}{C_\alpha^t(p^+) \prod_{\sigma \in \Sigma_\alpha^t} C_\alpha^t(S_{\sigma\alpha}^t) \prod_{\sigma \in \Sigma_\alpha^t} \langle \tau_{\sigma\alpha}^{t\pm}\rangle + C_\alpha^t(p^-) \prod_{\sigma \in \Sigma_\alpha^t} C_\alpha^t(S_{\sigma\alpha}^t) \prod_{\sigma \in \Sigma_\alpha^t} \langle \tau_{\sigma\alpha}^{t\mp}\rangle}$$

$$= \frac{C_\alpha^t(p^+) \prod_{\sigma \in \Sigma_\alpha^t} \langle \tau_{\sigma\alpha}^{t\pm}\rangle}{C_\alpha^t(p^+) \prod_{\sigma \in \Sigma_\alpha^t} \langle \tau_{\sigma\alpha}^{t\pm}\rangle + C_\alpha^t(p^-) \prod_{\sigma \in \Sigma_\alpha^t} \langle \tau_{\sigma\alpha}^{t\mp}\rangle}$$

$$C_\alpha^{t+1}(p^+) = \frac{C_\alpha^t(p^+) \prod_{\sigma \in \Sigma_\alpha^{t+}} \langle \tau_{\sigma\alpha}^{t+}\rangle \prod_{\sigma \in \Sigma_\alpha^{t-}} \langle \tau_{\sigma\alpha}^{t-}\rangle}{C_\alpha^t(p^+) \prod_{\sigma \in \Sigma_\alpha^{t+}} \langle \tau_{\sigma\alpha}^{t+}\rangle \prod_{\sigma \in \Sigma_\alpha^{t-}} \langle \tau_{\sigma\alpha}^{t-}\rangle + C_\alpha^t(p^-) \prod_{\sigma \in \Sigma_\alpha^{t+}} \langle \tau_{\sigma\alpha}^{t-}\rangle \prod_{\sigma \in \Sigma_\alpha^{t-}} \langle \tau_{\sigma\alpha}^{t+}\rangle}. \tag{9}$$

As $C_\alpha^t(p)$, $C_\alpha^t(\neg p)$, $\langle \tau_{\sigma\alpha}^{t+}\rangle$ and $\langle \tau_{\sigma\alpha}^{t\pm}\rangle$ are known quantities for all σ and α at t, $C_\alpha^{t+1}(p)$ can be calculated for all α's in the network of concern on the basis of a record of the novel arguments/reasons they received from their sources and whether these were for, or against, p in the chosen period.

3.2 The trust update function

α's trust at t in σ—α's credence distribution at t over the reliability of σ as a source of novel arguments/reasons—also calls for updating. Let $\tau_{\sigma\alpha}^{t+1(\pm)}$ be α's credence distribution over σ's reliability as a source of novel p-arguments/reasons after receiving an argument/reason for ($\tau_{\sigma\alpha}^{t+1(+)}$), or against ($\tau_{\sigma\alpha}^{t+1(-)}$) p in the period t to $t+1$ from σ. Then by the principle of conditionalization we have:

$$\tau_{\sigma\alpha}^{t+1(\pm)}(\rho) = \{C_\alpha^{t+1}(R_{\sigma\alpha}^\pm = \rho) : \rho \in [0,1]\} = \{C_\alpha^t(R_{\sigma\alpha}^\pm = \rho \mid S_{\sigma\alpha}^t \wedge S^\pm) : \rho \in [0,1]\} \tag{10}$$

Then for each ρ we have by the definition of conditional probability and the expansion theorem:

$$C_\alpha^t(R_{\sigma\alpha}^\pm = \rho | S_{\sigma\alpha}^t \wedge S^\pm) =$$

$$\left(\frac{1}{C_\alpha^t(S_{\sigma\alpha}^t \wedge S^\pm)}\right) C_\alpha^t(S^\pm | R_{\sigma\alpha}^\pm = \rho \wedge S_{\sigma\alpha}^t \wedge p^\pm) C_\alpha^t(R_{\sigma\alpha}^\pm = \rho \wedge S_{\sigma\alpha}^t \wedge p^\pm)$$
$$+ C_\alpha^t(S^\pm | R_{\sigma\alpha}^\pm = \rho \wedge S_{\sigma\alpha}^t \wedge p^\mp) C_\alpha^t(R_{\sigma\alpha}^\pm = \rho \wedge S_{\sigma\alpha}^t \wedge p^\mp)$$

Using (4) and (5), together with the fact that $C_\alpha^t(S^\pm | R_{\sigma\alpha}^\pm = \rho \wedge S_{\sigma\alpha}^t \wedge p^\mp) = 1 - C_\alpha^t(S^\mp | R_{\sigma\alpha}^\pm = \rho \wedge S_{\sigma\alpha}^t \wedge p^\mp)$, and the independence in C_α^t of $R_{\sigma\alpha}^\pm = \rho$, $S_{\sigma\alpha}^t$ and p (communication independence again), we get:

$$C_\alpha^t(R_{\sigma\alpha}^\pm = \rho | S_{\sigma\alpha}^t \wedge S^\pm) = C_\alpha^t(R_{\sigma\alpha}^\pm = \rho) C_\alpha^t(S_{\sigma\alpha}^t) \frac{\rho C_\alpha^t(p^\pm) + (1-\rho) C_\alpha^t(p^\mp)}{C_\alpha^t(S_{\sigma\alpha}^t \wedge S^\pm)}$$

$$= C_\alpha^t(R_{\sigma\alpha}^\pm = \rho) \frac{\rho C_\alpha^t(p^\pm) + (1-\rho) C_\alpha^t(p^\mp)}{C_\alpha^t(S^\pm | S_{\sigma\alpha}^t)}$$

Substituted back into (10) this gives:

$$\tau_{\sigma\alpha}^{t+1(\pm)}(\rho) = \{C_\alpha^t(R_{\sigma\alpha}^\pm = \rho | S_{\sigma\alpha}^t \wedge S^\pm) : \rho \in [0,1]\} \tag{11}$$

$$= \left\{ C_\alpha^t(R_{\sigma\alpha}^\pm = \rho) \frac{\rho C_\alpha^t(p^\pm) + (1-\rho) C_\alpha^t(p^\mp)}{C_\alpha^t(S^\pm | S_{\sigma\alpha}^t)} : \rho \in [0,1] \right\} \tag{12}$$

$$= \tau_{\sigma\alpha}^{t+}(\rho) \frac{\rho C_\alpha^t(p^\pm) + (1-\rho) C_\alpha^t(p^\mp)}{C_\alpha^t(S^\pm | S_{\sigma\alpha}^t)} \tag{13}$$

What is the denominator? By applying the continuous version of the conditional expansion theorem, followed by the discrete conditional expansion theorem we have:

$$C_\alpha^t(S^\pm | S_{\sigma\alpha}^t) = \int_0^1 C_\alpha^t(S^\pm | R_{\sigma\alpha}^\pm = \rho \wedge S_{\sigma\alpha}^t \wedge p^\pm) C_\alpha^t(R_{\sigma\alpha}^\pm = \rho \wedge p^\pm | S_{\sigma\alpha}^t)$$
$$+ C_\alpha^t(S^\pm | R_{\sigma\alpha}^\pm = \rho \wedge S_{\sigma\alpha}^t \wedge p^\mp) C_\alpha^t(R_{\sigma\alpha}^\pm = \rho \wedge p^\mp | S_{\sigma\alpha}^t) d\rho$$

By (4), (5) and aforementioned independence assumptions this gives:

$$C_\alpha^t(S^\pm | S_{\sigma\alpha}^t) = \int_0^1 \rho C_\alpha^t(R_{\sigma\alpha}^\pm = \rho) C_\alpha^t(p^\pm) + (1-\rho) C_\alpha^t(R_{\sigma\alpha}^\pm = \rho) C_\alpha^t(p^\mp) d\rho$$

$$= C_\alpha^t(p^\pm) \int_0^1 \rho \tau_{\sigma\alpha}^{t+}(\rho) d\rho + C_\alpha^t(p^\mp) \int_0^1 (1-\rho) \tau_{\sigma\alpha}^{t+}(\rho) d\rho$$

$$= C_\alpha^t(p^\pm) \int_0^1 \rho \tau_{\sigma\alpha}^{t+}(\rho) d\rho + C_\alpha^t(p^\mp) \left(\int_0^1 \tau_{\sigma\alpha}^{t+}(\rho) d\rho - \int_0^1 \rho \tau_{\sigma\alpha}^{t+}(\rho) d\rho \right)$$

$$= C_\alpha^t(p^\pm) \langle \tau_{\sigma\alpha}^{t+} \rangle + C_\alpha^t(p^\mp) \left(1 - \langle \tau_{\sigma\alpha}^{t+} \rangle \right)$$

$$= C_\alpha^t(p^\pm) \langle \tau_{\sigma\alpha}^{t+} \rangle + C_\alpha^t(p^\mp) \langle \tau_{\sigma\alpha}^{t-} \rangle$$

Substituting back into (13) gives the trust update function:

$$\tau_{\sigma\alpha}^{t+1(\pm)}(\rho) = \tau_{\sigma\alpha}^{t+}(\rho) \frac{\rho C_\alpha^t(p^\pm) + (1-\rho) C_\alpha^t(p^\mp)}{C_\alpha^t(p^\pm)\langle \tau_{\sigma\alpha}^{t+}\rangle + C_\alpha^t(p^\mp)\langle \tau_{\sigma\alpha}^{t-}\rangle} \quad (14)$$

Using equations (9) and (14), Laputa can calculate the credence function for each debate participant at $t+1$ from their immediately preceding credence functions in response to novel arguments/reasons received in the period t to $t+1$. By repeating this process for the specified number of rounds it can determine the credence distributions that results in the group from engagement in an exchange of arguments. As the update functions are complex, it helps to derive some qualitative rules for updating against which to check Laputa's performance. The qualitative update rules for credence in p are given in table 1. A '+' means that the current belief is reinforced (i.e. $C_\alpha^{t+1}(p) > C_\alpha^t(p)$ if $C_\alpha^t(p) > 0.5$, and $C_\alpha^{t+1}(p) < C_\alpha^t(p)$ if $C_\alpha^t(p) < 0.5$.), a '−' that the strength of the belief is weakened, and '0' that her credence is unchanged. A source is trusted by an inquirer if its expected reliability is greater than 0.5, and a message is surprising/expected if it contains an argument/reason for something that is disbelieved/believed to some degree by the receiver. See Olsson and Vallinder (Olsson and Vallinder) for derivations.

Source trusted?	Is message surprising?		
	No	Neither	Yes
Yes	+	+	−
Neither	0	0	0
No	−	−	+

Table 1 Qualitative rules for updating credence.

The qualitative rules for updating trust are:

Source trusted?	Is message expected?		
	Yes	Neither	No
Yes	+	0	−
Neither	+	0	−
No	+	0	−

Table 2 Qualitative rules for updating trust.

4 Debates in Laputa

A debate in Laputa is defined by the following parameters:

A duration for the debate: The number N of time steps over which the debate occurs.

The set of participants/inquirers: $K = \{\alpha_1, \alpha_2, \ldots, \alpha_m\}$.

The set of sources for each participant: $\Sigma_\alpha = \{\sigma_i, \sigma_1, \ldots, \sigma_n\}$, where σ_i is their own inquiry source and $\sigma_1 \ldots \sigma_n \in K$ are all the participants in the debate from whom they can receive arguments.

The listen chance for each of α's sources for every α in K: $P(S_{\sigma\alpha})$ is the probability that α receives a novel argument/reason from their source σ in any time step.

An assertion threshold for every α in K: T_α is to be understood as the credence each participant in the debate must have in the conclusion of an argument before they are willing to make that argument to any of their peers. A value above 0.5 indicates an agent that is only prepared to argue for what they believe, whereas a value of less than 0.5 indicates an agent that only argues against what they believe.

A reliability of personal inquiry for each α: $R^{\pm}_{\sigma_i \alpha}$.

A trust at t function for each of α's sources for each α: $\tau^{t+}_{\sigma\alpha}$

A credence in p at t for each α in K: $C^t_\alpha(p)$.

The dynamical functions constraining the step-wise evolution of the final two types of parameters are the update functions. As currently implemented Laputa assumes all the other parameters to be constant through time, though this is an assumption that could be easily relaxed in future development.

Laputa aids the user in specifying these parameters with a directed graphical interface. In this way a debate can be inputted by specifying a number of nodes, representing debate participants, together with arrows between the nodes, depicting the source/recipient relations among the participants; the participant at the base of an arrow is a peer-source for the participant at its head. In Laputa, such a graph is called a social network. Each node and arrow then has a number of parameters that the user specifies corresponding to the above list. Each participant is defined by five parameters: initial degree of belief in p [$C^t_\alpha(p)$], own-inquiry accuracy [$R^{\pm}_{i\alpha}$], own-inquiry chance [$P(S_{i\alpha})$], own-inquiry trust [$\tau^{t+}_{i\alpha}$] and threshold of assertion [T_α]. Likewise, each arrow from a σ_m to α is defined by two parameters: listen chance [$P(S_{\sigma_m \alpha})$] and listen trust [$\tau^{t+}_{\sigma_m \alpha}$].

Finally, it is of interest to study not only particular debates but the procedures such debates exemplify. In Laputa, a deliberative procedure is specified by constraints on the debate parameters listed above and attendant sampling distributions. In more detail, the topology of the deliberative procedure is specified by a set of social networks with a sampling distribution over this set, while the edge and vertex parameter sampling protocols are specified by density functions over the unit interval. These sampling distributions are supposed to be tuned to the actual frequency of the debate parameter values and topologies exhibited by the deliberative procedure. To evaluate a deliberative procedure, Laputa samples a directed graph according to the specified protocol, and then parameterises the edges and vertices according to the specified parameter sampling protocols. The result of this sampling from the constraints is the initial state of a particular debate. Laputa then simulates this debate for a specified number of steps and records the result. Due to the stochastic nature of Laputa simulations, the same debate should be simulated a statistically significant number of times[3] and the results aggregated. This whole process is repeated a statistically significant number of times, so that a sample of debate simulations conforming to the procedure is attained. The results are then aggregated in order to evaluate the procedure as a whole.

While the sampling of edge and vertex parameters is largely unproblematic, the modeling of deliberative procedures does face significant challenges where it comes to the sampling of social networks; challenges that the Laputa research program has yet to overcome (for an introductory discussion and direction to further reading see Masterton (Masterton)). One problem is that because the edge and vertex parameters are independently and identically distributed in Laputa, isomorphic graphs model the same social network. Hence, structure constraints should be specified by a set of isomorphism classes of directed graphs with an attendent sampling distribution. However, most randomn sampling techniques work by sampling graphs and not isomorphism classes of graphs. While a sampling distribution over graphs implies a sampling distribution over isomorphism classes of graphs, one can only discern the latter distribution from the former if one sorts the graphs into their isomorphism classes. This is a non-trivial task. For instance, Nauty—the best extant graph isomorphism identifying program—takes between a thousandth and a tenth of a second to discern whether two graphs are isomorphic on a standard desktop depending on the number of vertices. This may sound pretty good but the typical social network constraint contains trillions of graphs, so it would take Nauty billions of years to sort the typical social network constraint into its isomorphism classes and compute the distribution over these classes implied by a random sampling of graphs. It follows that if social networks are sampled by randomly sampling directed graphs, then it is practically impossible to ascertain whether such a sampling implies the desired sampling of social networks for the model in question.

There is a way of sampling graphs that effectively samples isomorphism classes: one defines the structure constraint graph statisitically and then runs a Markov Chain Monte Carlo (MCMC) simulation to identify the Exponential Random Graph Model (ERGM) that comes closest to producing the desired distribution of graph statistics when used for sampling. As isomorphic graphs are identical in their graph statistics (degree distribution, compactness, etc), so sampling in this manner is effectively sampling isomorphism classes of graphs. This is a superior method of sampling social networks than random sampling of graphs because there is a one to one corrrespondence between isomorphism classes of graphs and social networks; however, this approach also has its limitations. One such limitation is that structurally very dissimilar graphs can have very similar graph statistics. This leads to *instability* in the MCMC simulations used to find the appropriate sampling protocol: repeated such simulations can identify distinct ERGM's capable of producing the same statistics. This leaves us with the difficult problem of deciding which of these non-equivalent models is optimal for the deliberative procedure at hand. Another

---

[3] The computer program does not do this at present, but it is a priority for future development.

limitation is that for certain distirbutions over graph statistics such simulations may fail to find an appropriate ERGM in a reasonable timeframe.

Despite its shortcomings, the graph statistic ERGM approach sketched above is probably the best way of specifying the topology of deliberative procedures available at the present time. This is not the method of sampling social networks currently employed in Laputa. The present approach is to sample the number of vertices in the graph uniformly from a size constraint (an interval of natural numbers) and then to populate this number of vertices with edges, each of the logically possible edges having the same probability of being included in the graph. The resultant distibution heavily favours smaller networks, samples from all logically possible graphs allowed by the size constraint, and is a distirbution over graphs rather than their isomorphism classes. A future development of Laputa would be to replace this simple binomial random sampling protocol with some variant of graph statistic ERGM approach.

5 Applications and outlook

The probabilistic model we have presented is quite complex making it difficult to prove interesting analytical results. For the purposes of studying the consequences of the model, a simulation environment was created (programmer: Staffan Angere). The Laputa simulation environment allows for effortless exploration of various complex (e.g. statistical) properties of the model (see Figure 2).

We will not describe the simulation environment here since we have done so elsewhere (see, e.g., Olsson (2011)). However it is worth noting that due to the stochastic elements in the characterization of debates—the listen chances, the reliability of individual inquiry, etc—simulations of one and the same debate may vary in outcome. Hence it is useful to distinguish between a particular debate, corresponding to a particular simulation by Laputa, and a debate-type, corresponding to a particular parameterized social network. A *deliberative procedure* can then be viewed as a set of constraints on the debate parameters; different deliberative procedures being characterized by different constraints. For instance, jury deliberation is typically undertaken by groups of between 6 and 15 in size, while the number of participants in parliamentary debates can range from the low tens all the way up toward 1000. Such constraints, together with sampling protocols, are specified in the batch window of Laputa. Laputa then generates debate-types conforming to the argumentative practice in question by sampling within these constraints according to the relevant sampling protocols. In this way Laputa can not only simulate a single debate but many such debates conforming to some deliberative procedure.

We will close this article by giving three examples of how the Laputa simulation environment has been applied in the study of statistical properties of argumentative practices.

In the introduction we encountered the influential argumentation model put forward by Dung. The aim of the model is to study the acceptability of arguments in

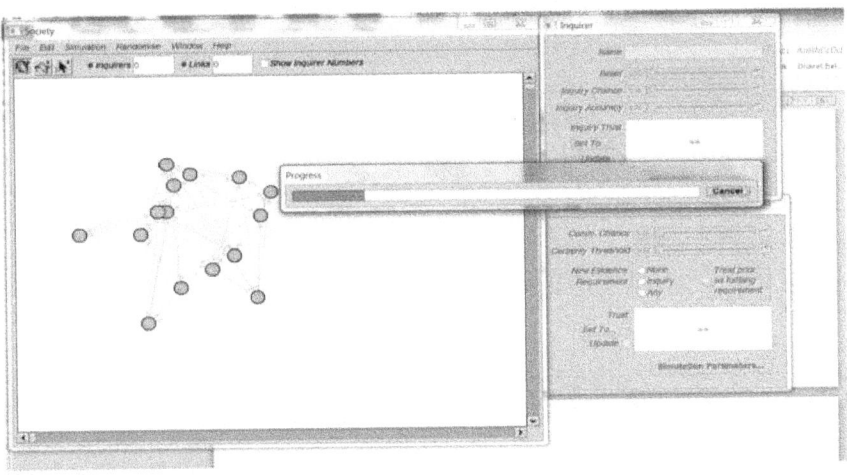

Fig. 1 Laputa calculating the veritisitic value for a random sample of social networks.

a competitive, or adversarial, argumentative context in which each party wants to win over the other. We also noted that Laputa, by contrast, is a model for studying argumentation in a collaborative context where the participants put pro and con considerations on the table as part of a collective inquiry for the sake of the common good. Jury deliberation in court would be the paradigm case. Dung's model can be used, in principle, for automatic detection of the acceptability of an argument. The focus of the Laputa framework is rather on the acceptability of an argumentative practice construed as a social practice of information exchange (Goldman, 1999). More precisely, we wish to know which argumentation practices are conducive to the argumentative goals that we find valuable.

The obvious goal of an argumentation practice is to establish the truth of the proposition under consideration; e.g., the guilt of the defendant. Hence, what we are interested in, first and foremost, is the veritistic value (Goldman, 1999) of an argumentative practice, viz., an assessment of how conducive that practice is to finding the truth. But we may also be interested in the properties of argumentative practices that are not truth-related. We may, for instance, be interested in the conditions under which an argumentative practice leads to polarization whereupon members of a deliberating group predictably move toward a more extreme points in the direction indicated by the members' predeliberation tendencies (Sunstein, 2002, p. 176, italics removed). Polarization can be studied without any consideration of truth. Polarization has been observed in argumentative groups under a variety of different circumstances and is considered an empirically robust phenomenon (Isenberg, 1986).

Olsson (in press) studied the conditions under which inquirers in Laputa polarize. He found that inquirers will polarize under conditions of what he called social calibration: if initially disposed to judge along the same lines, inquirers in Laputa will adopt a more extreme position in the same direction as the effect of group deliberation, just like members of real argumentative bodies. A group is socially calibrated if either everyone correctly believes that everyone else is trustworthy or everyone correctly believes that everyone else is untrustworthy (e.g. systematic liars). Olsson noted that groups that are not socially calibrated tend to diverge in the sense that inquirers will eventually adopt contrary positions. Olsson also studied what happens to mutual trust in the polarization process. He observed that inquirers thereby become increasingly trusting which creates a snowball effect. To the extent that Bayesian reasoning is normatively correct, the bottom line is that polarization and divergence are not necessarily the result of mere irrational group think but that even ideally rational inquirers will predictably polarize or diverge under realistic conditions.

An interesting problem in the context of collaborative inquiry concerns the conditions under which participants are warranted to make an assertion in front of their peers. This problem has a long history in philosophy. Theorists of knowledge can be divided into two camps: those who think that nothing short of certainty or (subjective) probability 1 can warrant assertion and those who disagree with this claim. Vallinder and Olsson (Vallinder and Olsson) addressed this by inquiring into the problem of setting the probability threshold required for assertion in such a way that the social epistemic good is maximized, where the latter is taken to be the veritistic value in the sense of Goldman (1999). Results obtained by means of computer simulation utilising Laputa indicate that the certainty rule is optimal in the (infinite) limit of inquiry and communication but that a lower threshold is preferable in less idealized cases.

Another interesting application is jury deliberation, and in particular the notorious problem of identifying the optimal size of a deliberating jury. Angere and Olsson (forthcoming) studied the effect of jury size and required majority on the quality of group decision making using an extension of the Laputa model. They found that Goldman's measure of veritistic value (Goldman, 1999) is unsuitable for measuring jury competence. Instead, they introduced the idea of J-value (jury value) which takes into account the unique characteristics, asymmetries and principles involved in jury voting. Using the Laputa simulation model, they found that requiring more than a 50% majority should be avoided. Moreover, while it is in principle always better to have a larger jury, given a 50% required majority, the value of having more than 12-15 jurors is likely to be negligible. Finally, they suggested a formula for calculating the optimal jury size given the cost, economic or otherwise, of adding another juror.

Laputa is more a developing framework than an already finished product. In future work, we would like to expand the analysis toolkit with a bandwagon function and other analysis tools that can help explain simulation results. In the statistical (batch) mode, Laputa generates and studies random graphs and their statistical properties. However, it is well-known that realistic social networks are often clustered in various ways. Such networks should be incorporated in future versions of Laputa. Also high on the agenda is the incorporation of division of labor and the extension of the language beyond simple yes-no questions.

References

Alchourrôn, C., P. Gärdenfors, and D. Makinson (1985). On the logic of theory change: Partial meet contraction and revision functions. *Journal of Symbolic Logic 50*, 510–530.

Angere, S. Knowledge in a social network. Forthcoming.

Angere, S. and E. J. Olsson. What is the optimal size of a deliberating jury? Forthcoming.

Bovens, L. and E. J. Olsson (2000). Coherentism, reliability and Bayesian networks. *Mind 109*(436), 685–719.

Doyle, J. (1979). A truth maintenance system. *Artificial Intelligence 12*, 231–272.

Dung, P. M. (1995). On the acceptability of arguments and its fundamental role in nonmonotonic reasoning, logic programming and n-person games. *Artificial Intelligence 77*, 321–357.

Furhmann, A. (1991). Theory contraction through base contraction. *Journal of Philosophical Logic 20*(2), 175–203.

Goldman, A. I. (1999). *Knowledge in a Social World*. Clarendon Press, Oxford.

Hansson, B. (1983). Epistemology and evidence. In P. Gärdenfors, B. Hansson, and N.-E. Sahlin (Eds.), *Evidentiary Value: Philosophical, Judicial and Psychological Aspects of a Theory*, pp. 75–97. Lund: Library of Theoria.

Hansson, S. O. (1991). Belief contraction without recovery. *Studia Logica 50*, 251–260.

Isenberg, D. (1986). Group polarisation: A critical review and meta-analysis. *Journal of Personality and Social Psychology 50*(6), 1141–1151.

Kaplan, M. F. and Miller, C. E. (1977). Judgements and group discussion: Effect of presentation and memory factors on polarization. *Sociometry 40*, 337–343.

Masterton, G. Topological variability of collectives and its import for social epistemology. Forthcoming.

Meacham, C. J. G. (2010). Two mistakes regarding the principal principle. *British Journal for the Philosophy of Science 61*, 407–431.

Olsson, E. J. (2002). Corroborating testimony, probability and surprise. *British Journal for the Philosophy of Science 53*, 273–288.

Olsson, E. J. (2005). *Against Coherence: Truth, Probability and Justification*. Oxford: Oxford University Press.

Olsson, E. J. (2011). A simulation approach to veritistic social epistemology. *Episteme 8*(2), 127–143.

Olsson, E. J. and A. Vallinder. Norms of assertion and communication in social networks. In press.

Pearl, J. (1988). *Probabilisitic Reasoning in Intelligent Systems*. Morgan-Kaufmann.

Sunstein, C. R. (2002). The law of group polarization. *The Journal of Political Philosophy 10*(2), 175–195.

Tennant, N. (2012). *Changes in Mind: An Essay on Rational Belief Revision*. Oxford: Oxford University Press.

Vallinder, A. and E. J. Olsson. Trust and the value of overconfidence: A Bayesian perspective on social network communication. In press.

Vallinder, A. and E. J. Olsson (2012). Does computer simulation support the argument from disagreement? *Synthese*.

Vinokur, A. and E. Burnstein (1978). Novel argumentation and attitude change: The case of polarization following group discussion. *European Journal of Social Psychology*, 335–348.

A Propositional Typicality Logic for Extending Rational Consequence

Richard Booth, Thomas Meyer, and Ivan Varzinczak

Abstract We introduce Propositional Typicality Logic (PTL), a logic for reasoning about typicality. We do so by enriching classical propositional logic with a typicality operator of which the intuition is to capture the most typical (or normal) situations in which a given formula holds. The semantics is in terms of ranked models as studied in KLM-style preferential reasoning. This allows us to show that KLM-style rational consequence relations can be embedded in our logic. Moreover we show that we can define consequence relations on the language of PTL itself, thereby moving beyond the propositional setting. Building on the existing link between propositional rational consequence and belief revision, we show that the same correspondence holds in the case of rational consequence and belief revision defined on the language of PTL. Finally we also investigate different notions of entailment for PTL and propose two appropriate candidates.
The present text is an extended version of a paper which appeared in the Proceedings of the 13th European Conference on Logics in Artificial Intelligence [Booth *et al.*, 2012].

1 Introduction

In artificial intelligence, there has been a great deal of work done on how to introduce nonmonotonic reasoning capabilities in logic-based knowledge representation systems [Brachman and Levesque, 2004; Friedman and Halpern, 2001; Gabbay

---

Richard Booth
Université du Luxembourg, Luxembourg,
e-mail: richard.booth@uni.lu

Thomas Meyer · Ivan Varzinczak
Centre for Artificial Intelligence Research, University of KwaZulu-Natal and CSIR Meraka Institute,South Africa,
e-mail: {tmeyer,ivarzinczak}@csir.co.za

and Schlechta, 2009; Hansson, 1999; Harmelen et al., 2008; Makinson, 2005]. In particular, the approach for *preferential reasoning* introduced by Shoham [1988] and developed by Kraus, Lehmann and Magidor [1990], often called the *KLM approach*, turned out to be one of the most successful. This has been the case due to at least three main reasons. Firstly, their framework is based on semantic constructions that are elegant and neat. Secondly, it provides the foundation for the determination of the important notion of entailment in this context [Lehmann and Magidor, 1992]. Finally, it also offers an alternative perspective on the problem of belief change [Gärdenfors and Makinson, 1994]. Moreover recent work has shown that the KLM approach also provides an appropriate springboard from which to investigate further facets of defeasible reasoning in more expressive logics [Britz et al., 2008; Britz et al., 2011a; Britz et al., 2011b; Britz et al., 2012; Britz et al., 2013a; Britz et al., 2013b; Britz and Varzinczak, 2012; Britz and Varzinczak, 2013; Casini and Straccia, 2010; Giordano et al., 2009b; Lehmann and Magidor, 1990; Moodley et al., 2012], such as modal logic [Chellas, 1980; Blackburn et al., 2006] and description logics (DLs) [Baader et al., 2007].

In their seminal papers [Kraus et al., 1990; Lehmann and Magidor, 1992], Kraus and colleagues enrich classical propositional logic with a defeasible 'implication' $\mid\!\sim$ so that one can write down defeasible implication statements (also called conditional assertions) of the form $\alpha \mid\!\sim \beta$, where α and β are propositional formulae. In this setting, a sentence of the form $\alpha \mid\!\sim \beta$ is given the meaning that "if α is the case, then *usually* (but not necessarily always) β is the case", making it possible to formalize the well-known example that "birds usually fly" (b $\mid\!\sim$ f), "penguins are birds" (p $\rightarrow$ b), but "penguins usually do not fly" (p $\mid\!\sim \neg$f).

A curious aspect of the KLM approach (and of the corresponding belief revision constructions) is that it is crucially, albeit tacitly, based on a notion of *normality* [Boutilier, 1994] or *typicality* [Lehmann, 1998]. More formally, the semantics of a statement of the form $\alpha \mid\!\sim \beta$ says that "all most preferred (i.e., most normal) α-worlds are β-worlds" (leaving it open for α-worlds that are less preferred — or exceptional — not to satisfy β). In other words, the statement $\alpha \mid\!\sim \beta$ captures the intuition that "typical α-cases are β cases", which in our example gives us "typical birds fly" and "typical penguins do not fly". Given this, it seems quite natural to be able to state, for instance, that "penguins are non-typical birds", or that "penguins and ostriches are the only birds that typically do not fly".

It turns out that in the corresponding underlying language it is not possible to refer directly to such a notion of typicality and, importantly, use it in the scope of other logical constructs. According to Britz and Varzinczak [2012],

> This has to do partly with the syntactic restrictions imposed on $\mid\!\sim$, namely no nesting of conditionals, but, more fundamentally, it relates to where and how the notion of normality is used in such statements. ... [I]n a KLM defeasible statement $\alpha \mid\!\sim \beta$, the normality spotlight is somewhat put on α, as though normality was a property of the premise and not of the conclusion. Whether the situations in which β holds are normal or not plays no role in the reasoning that is carried out. In the original KLM framework, normality is linked to the premise as a whole, rather than its constituents. Technically this meant one could not refer directly to normality of a sentence in the scope of logical operators.

In this chapter, we fill this gap with the introduction of an explicit operator to talk about typicality. Intuitively, our new syntactic construction allows us to single out those most typical states of affairs in which a given formula holds. The result is a more expressive language allowing us, for instance, to make statements formalizing the aforementioned examples in a succinct way.

In the rest of this section we set up the notation and conventions that shall be followed in the upcoming sections. The remainder of the present chapter is then structured as follows: In Section 2 we provide the required background on the KLM approach to defeasible reasoning. We then define and investigate PTL, a propositional typicality logic extending propositional logic (Section 3). The semantics of PTL is in terms of ranked models as studied in the literature on preferential reasoning and summarized in Section 2. This allows us to embed propositional KLM-style consequence relations in our new language. In Section 4 we show that, although the addition of the typicality operator increases the expressivity of the logic, the nesting of the typicality does not add anything beyond the inclusion of a non-nested typicality operator. In Section 5 we investigate the link between AGM belief revision and PTL. We show that propositional AGM belief revision can be expressed in terms of typicality, and also that it can be lifted to a version of revision on PTL. We then move to an investigation of rational consequence relations in terms of PTL (Section 6). We show that propositional rational consequence can be expressed in PTL, that it can be extended to PTL in terms of PTL itself, and that the propositional connection between rational consequence and revision carries over to PTL. In Section 7 we raise the question of what an appropriate notion of entailment for PTL is. We propose and investigate different definitions of entailment and identify two appropriate candidates. After a discussion of and comparison with related work (Section 8), we conclude with a summary of the contributions and directions for further investigation.

1.1 Logical Preliminaries

We work in a propositional language over a finite set of propositional *variables* (alias *atoms*) $\mathscr{P}$. (In later sections we shall adopt a richer language.) We shall use $p, q, \ldots$ as meta-variables for the atomic propositions. Propositional formulae (and in later sections, formulae of the richer language) are denoted by $\alpha, \beta, \ldots$, and are recursively defined in the usual way: $\alpha ::= p \mid \neg \alpha \mid \alpha \wedge \alpha$. All the other Boolean truth-functional connectives ($\vee, \rightarrow, \leftrightarrow, \ldots$) are defined in terms of $\neg$ and $\wedge$ in the standard way. We use $\top$ as an abbreviation for $p \vee \neg p$, and $\bot$ for $p \wedge \neg p$, for some atom $p \in \mathscr{P}$. With $\mathscr{L}$ we denote the set of all propositional formulae.

We denote by $\mathscr{U}$ the set of all *valuations* $v : \mathscr{P} \longrightarrow \{0, 1\}$. Sometimes we shall represent the valuations of the logic under consideration as sequences of 0s and 1s, and with the obvious implicit ordering of atoms. Thus, for the logic generated from $\mathscr{P} = \{p, q\}$, the valuation in which p is true and q is false will be represented as 10.

Satisfaction of a formula $\alpha \in \mathscr{L}$ by $v \in \mathscr{U}$ is defined in the usual truth-functional way and is denoted by $v \Vdash \alpha$. With $Mod(\alpha)$ we denote the set of all valuations satisfying α. Logical consequence and logical equivalence are denoted by $\models$ and $\equiv$ respectively. Given sentences α and β, $\alpha \models \beta$ (α entails β) means $Mod(\alpha) \subseteq Mod(\beta)$. $\alpha \equiv \beta$ is an abbreviation of $\alpha \models \beta$ and $\beta \models \alpha$.

A *knowledge base* $\mathscr{K}$ is a finite set of formulae $\mathscr{K} \subseteq \mathscr{L}$. We extend the notions of $Mod(\cdot)$, entailment and logical equivalence to knowledge bases in the usual way: for a finite $\mathscr{K} \subseteq \mathscr{L}$, $Mod(\mathscr{K})$ is the set of all valuations satisfying every formula in $\mathscr{K}$, and $\mathscr{K} \models \alpha$ if and only if $Mod(\mathscr{K}) \subseteq Mod(\alpha)$. With $\models \alpha$ (α is a tautology) we understand $\emptyset \models \alpha$.

2 Defeasible Consequence Relations

In the present section, we provide a brief outline of propositional preferential and rational consequence relations as studied by Lehmann and colleagues in the early 90's with some minor modifications to their initial formulation. (For more details, the reader is referred to the original work of Kraus et al. [1990] and Lehmann and Magidor [1992].)

A *defeasible consequence relation* $\mathrel{\vert\!\sim}$ is defined as a binary relation on formulae of our underlying propositional logic, i.e., $\mathrel{\vert\!\sim} \subseteq \mathscr{L} \times \mathscr{L}$. We say that $\mathrel{\vert\!\sim}$ is a *preferential* consequence relation [Kraus et al., 1990] if it satisfies the following set of properties, alias postulates or Gentzen-style rules, as they are sometimes also referred to (below, $\models$ denotes validity in classical propositional logic):

$$\text{(Ref)}\ \alpha \mathrel{\vert\!\sim} \alpha \qquad \text{(LLE)}\ \frac{\models \alpha \leftrightarrow \beta,\ \alpha \mathrel{\vert\!\sim} \gamma}{\beta \mathrel{\vert\!\sim} \gamma} \qquad \text{(And)}\ \frac{\alpha \mathrel{\vert\!\sim} \beta,\ \alpha \mathrel{\vert\!\sim} \gamma}{\alpha \mathrel{\vert\!\sim} \beta \wedge \gamma}$$

$$\text{(Or)}\ \frac{\alpha \mathrel{\vert\!\sim} \gamma,\ \beta \mathrel{\vert\!\sim} \gamma}{\alpha \vee \beta \mathrel{\vert\!\sim} \gamma} \qquad \text{(RW)}\ \frac{\alpha \mathrel{\vert\!\sim} \beta,\ \models \beta \to \gamma}{\alpha \mathrel{\vert\!\sim} \gamma} \qquad \text{(CM)}\ \frac{\alpha \mathrel{\vert\!\sim} \beta,\ \alpha \mathrel{\vert\!\sim} \gamma}{\alpha \wedge \beta \mathrel{\vert\!\sim} \gamma}$$

The semantics of preferential consequence relations is in terms of *preferential models*; these are partially ordered structures with states labeled by propositional valuations. We make this terminology more precise below.

Let S be a set and $\prec \subseteq S \times S$ be a strict partial order on S, i.e., $\prec$ is *irreflexive* and *transitive*. Given $S' \subseteq S$, we say that $s \in S'$ is *minimal* in S' if there is no $s' \in S'$ such that $s' \prec s$. With $\min_\prec S'$ we denote the minimal elements of $S' \subseteq S$ with respect to $\prec$. We say that $S' \subseteq S$ is *smooth* [Kraus et al., 1990] if for every $s \in S'$ either s is minimal in S' or there is $s' \in S'$ such that s' is minimal in S' and $s' \prec s$.

Definition 1. A preferential model is a tuple $\mathscr{P} = \langle S, \ell, \prec \rangle$ where S is a set of states; $\ell : S \longrightarrow \mathscr{U}$ is a labeling function; $\prec \subseteq S \times S$ is a strict partial order on S satisfying the smoothness condition.[1]

---

[1] That is, for every $\alpha \in \mathscr{L}$, the set $[\![\alpha]\!]^{\mathscr{P}}$ (cf. Definition 2) is smooth.

Propositional Typicality Logic for Extending Rational Consequence

Given a preferential model $\mathscr{P} = \langle S, \ell, \prec \rangle$ and $\alpha \in \mathscr{L}$, with $[\![\alpha]\!]^{\mathscr{P}}$ we denote the set of states satisfying α (α-*states* for short) in $\mathscr{P}$ according to the following definition:

Definition 2. Let $\mathscr{P} = \langle S, \ell, \prec \rangle$ be a preferential model and let $\alpha \in \mathscr{L}$. Then $[\![\alpha]\!]^{\mathscr{P}} := \{s \in S \mid \ell(s) \Vdash \alpha\}$.

In the KLM approach, states lower down in the order are seen as being *more preferred* (or *more normal*) than those higher up.

As an example, let $\mathscr{P} = \{b, f, p\}$, where b stands for the proposition "Tweety is a bird", f for "Tweety flies" and p for the proposition "Tweety is a penguin". Figure 1 below depicts the preferential model $\mathscr{P} = \langle S, \ell, \prec \rangle$ where $S = \{s_i \mid 1 \leq i \leq 6\}$, ℓ is such that $\ell(s_1) = 000$, $\ell(s_2) = 010$, $\ell(s_3) = 110$, $\ell(s_4) = 100$, $\ell(s_5) = 101$ and $\ell(s_6) = 111$, and $\prec$ is the transitive closure of $\{(s_1, s_4), (s_1, s_5), (s_2, s_4), (s_2, s_5), (s_3, s_4), (s_3, s_5), (s_4, s_6), (s_5, s_6)\}$.

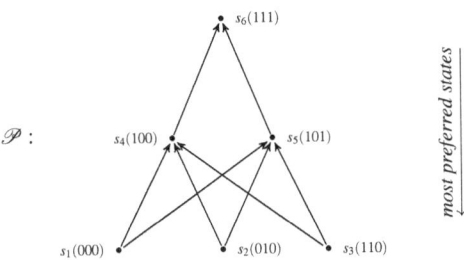

Fig. 1 Example of a preferential model.

Given a preferential model $\mathscr{P} = \langle S, \ell, \prec \rangle$ and a sentence $\alpha \in \mathscr{L}$, we say that α is *satisfiable* in $\mathscr{P}$ if $[\![\alpha]\!]^{\mathscr{P}} \neq \emptyset$, otherwise α is *unsatisfiable* in $\mathscr{P}$. We say that α is *true* in $\mathscr{P}$ (denoted $\mathscr{P} \Vdash \alpha$) if $[\![\alpha]\!]^{\mathscr{P}} = S$.

Given $\mathscr{P} = \langle S, \ell, \prec \rangle$ and $\alpha, \beta \in \mathscr{L}$, the defeasible statement $\alpha \mathrel{\mid\!\sim} \beta$ holds in $\mathscr{P}$ (noted $\alpha \mathrel{\mid\!\sim}_{\mathscr{P}} \beta$) if and only if $\min_\prec [\![\alpha]\!]^{\mathscr{P}} \subseteq [\![\beta]\!]^{\mathscr{P}}$, i.e., every $\prec$-minimal α-state is a β-state. As an example, in the model $\mathscr{P}$ of Figure 1, we have $b \mathrel{\mid\!\sim}_{\mathscr{P}} f$ (since $\min_\prec [\![b]\!]^{\mathscr{P}} = \{s_3\} \subseteq [\![f]\!]^{\mathscr{P}} = \{s_2, s_3, s_6\}$), and $p \mathrel{\mid\!\sim}_{\mathscr{P}} \neg f$ (since $\min_\prec [\![p]\!]^{\mathscr{P}} = \{s_5\} \subseteq [\![\neg f]\!]^{\mathscr{P}} = \{s_1, s_4, s_5\}$).

The representation theorem for preferential consequence relations then states:

Theorem 1 (Kraus et al. [1990]). *A defeasible consequence relation is a preferential consequence relation if and only if it is defined by some preferential model, i.e., $\mathrel{\mid\!\sim}$ is preferential if and only if there exists $\mathscr{P}$ such that $\mathrel{\mid\!\sim}_{\mathscr{P}} := \{(\alpha, \beta) \mid \alpha \mathrel{\mid\!\sim}_{\mathscr{P}} \beta\}$ is such that $\mathrel{\mid\!\sim} \,=\, \mathrel{\mid\!\sim}_{\mathscr{P}}$.*

If, in addition to the preferential properties, the defeasible consequence relation $\mathrel{\mid\!\sim}$ also satisfies the following Rational Monotonicity property [Lehmann and Magidor, 1992], it is said to be a *rational consequence relation*:

$$\text{(RM)} \quad \frac{\alpha \mathrel{\mid\!\sim} \beta,\ \alpha \mathrel{\mid\!\not\sim} \neg\gamma}{\alpha \wedge \gamma \mathrel{\mid\!\sim} \beta}$$

The semantics of rational consequence relations is in terms of *ranked* models, i.e., preferential models in which the preference order is *modular*:

Definition 3. Given a set S, $\prec\ \subseteq S \times S$ is modular if and only if there is a ranking function $rk : S \longrightarrow \mathbb{N}$ such that for every $s, s' \in S$, $s \prec s'$ if and only if $rk(s) < rk(s')$.

Definition 4. A ranked model $\mathscr{R} = \langle S, \ell, \prec \rangle$ is a preferential model such that $\prec$ is modular.

The preferential model in Figure 1 is also an example of a ranked model.

The representation theorem for rational consequence relations then states:

Theorem 2 (Lehmann & Magidor [1992]). *A defeasible consequence relation is a rational consequence relation if and only if it is defined by some ranked model, i.e., $\mathrel{\mid\!\sim}$ is rational if and only if there exists $\mathscr{R}$ such that $\mathrel{\mid\!\sim}_{\mathscr{R}} := \{(\alpha, \beta) \mid \alpha \mathrel{\mid\!\sim}_{\mathscr{R}} \beta\}$ is such that $\mathrel{\mid\!\sim}\ =\ \mathrel{\mid\!\sim}_{\mathscr{R}}$.*

There seems to be an agreement in the nonmonotonic reasoning community that rational consequence constitutes the 'right' type of entailment for nonmonotonic logics; one of the reasons stemming from its confluence with the AGM paradigm for belief revision [Alchourrón et al., 1985; Hansson, 1999] (see also Section 5). In the remainder of this chapter we shall therefore assume that $\mathrel{\mid\!\sim}$ is at least rational.[2]

From a technical point of view, the main advantage of assuming rationality is that we can do away with states and work with a much simpler semantics in which the preferential ordering is placed directly on valuations [Gärdenfors and Makinson, 1994]. Therefore, from now on we shall adopt the following definition of ranked models:

Definition 5. A ranked model $\mathscr{R}$ is a pair $\langle \mathscr{V}, \prec \rangle$, where $\mathscr{V} \subseteq \mathscr{U}$ and $\prec\ \subseteq \mathscr{V} \times \mathscr{V}$ is a modular order over $\mathscr{V}$.

The definition above is of course not Lehmann and Magidor's [1992] original definition of ranked models, but as alluded to above, a characterization of rational consequence à la Theorem 2 can be given in terms of ranked models as we present them here [Gärdenfors and Makinson, 1994].

Definition 6. Let $\alpha \in \mathscr{L}$ and let $\mathscr{R} = \langle \mathscr{V}, \prec \rangle$ be a ranked model. With $[\![\alpha]\!]$ we denote the set of valuations satisfying α in $\mathscr{R}$, defined as follows:

$$[\![p]\!] := \{v \in \mathscr{V} \mid v(p) = 1\}, \quad [\![\neg\alpha]\!] := \mathscr{V} \setminus [\![\alpha]\!], \quad [\![\alpha \wedge \beta]\!] := [\![\alpha]\!] \cap [\![\beta]\!]$$

Given a (simplified) ranked model $\mathscr{R}$, as in the case with states, the intuition is that valuations lower down in the ordering are more preferred than those higher up. Hence, a pair (α, β) is in the consequence relation defined by $\mathscr{R}$ (denoted as $\alpha \mathrel{\mid\!\sim}_{\mathscr{R}} \beta$) if and only if $\min_{\prec} [\![\alpha]\!] \subseteq [\![\beta]\!]$, i.e., the most preferred (with respect to $\prec$) α-valuations are also β-valuations.

---

[2] Even in the context of rational consequence relations, in what follows we shall use preferential semantics and preferential reasoning when referring to the KLM approach.

3 A Logic to Talk about Typicality

As alluded to in the Introduction, there is a need for a formalism in which we can make statements such as "the typical bird-situations", "ostriches are non-typical birds", and "penguins and ostriches are the only typical non-flying birds". In some cases, it may be possible to express these by means of several $\mid\!\sim$-statements, in a sort of '$\mid\!\sim$-normal form'. However, in order for us to do so in a succinct way, we should be able to shift the focus of typicality from the premise of a KLM-style statement and drop the interdiction to nest $\mid\!\sim$'s.

Boutilier's [1994] conditional $\Rightarrow$ as well as Britz et al.'s [2009] internalized $\prec$ as a modality are good candidates for the type of extension that we have in mind here. Nevertheless, these approaches are too expressive for our purposes in that there the preference relation $\prec$ becomes explicit to the user (cf. Section 8). Here we argue for a way to express typicality in which the complexity of the underlying semantics, here expressed as the preference relation $\prec$, is somehow hidden from the users, who, from a knowledge representation perspective, want a formalism which is precise but at the same time concise. (We shall come back to this point at the end of this section.)

The remainder of the present section is devoted to the introduction of the main focus of this chapter, namely a propositional typicality logic, called PTL, which extends classical propositional logic with a *typicality operator* $\bullet$, the semantics of which *implicitly* refers to the preference ordering.

The language of PTL, denoted by $\mathscr{L}^\bullet$, is recursively defined as follows:

$$\alpha ::= p \mid \neg\alpha \mid \alpha \wedge \alpha \mid \bullet\alpha$$

where, as before, p denotes an atom and all the other connectives are defined in terms of $\neg$ and $\wedge$, and $\top$ and $\bot$ are seen as abbreviations. Assuming $\mathscr{P} = \{b, f, p, o\}$, where b, f and p are as before and o stands for "is an ostrich", the following are examples of $\mathscr{L}^\bullet$-sentences: $\bullet b$, $o \rightarrow \neg \bullet b$, $p \vee o \leftrightarrow b \wedge \bullet \neg f$.

Intuitively, a sentence of the form $\bullet\alpha$ is understood to refer to the typical situations in which α holds. (Note that α can itself be a $\bullet$-sentence — more on that in Section 4.) The semantics of our enriched language is in terms of (simplified) ranked models (cf. Definition 5) and we extend the notion of satisfaction from Definition 6 as follows:

Definition 7. Let $\alpha \in \mathscr{L}^\bullet$ and let $\mathscr{R} = \langle \mathscr{V}, \prec \rangle$. Then $[\![\bullet\alpha]\!] := \min_\prec [\![\alpha]\!]$.

Given $\alpha \in \mathscr{L}^\bullet$ and $\mathscr{R}$ a ranked model, we say that α is *satisfiable* in $\mathscr{R}$ if $[\![\alpha]\!] \neq \emptyset$, otherwise α is *unsatisfiable* in $\mathscr{R}$. We say that α is *true* in $\mathscr{R}$ (denoted as $\mathscr{R} \Vdash \alpha$) if $[\![\alpha]\!] = \mathscr{V}$. For $\mathscr{K} \subseteq \mathscr{L}^\bullet$, $\mathscr{R} \Vdash \mathscr{K}$ if $\mathscr{R} \Vdash \alpha$ for every $\alpha \in \mathscr{K}$. We say that α is *valid*, denoted as $\models \alpha$, if $\mathscr{R} \Vdash \alpha$ for every ranked model $\mathscr{R}$.[3]

---

[3] The observant reader would have noticed that this sounds like modal logic [Chellas, 1980]. We shall defer a discussion on our typicality operator as a modality until the appropriate point (Section 8).

As an example, let $\mathscr{P} = \{\mathsf{b},\mathsf{f},\mathsf{p}\}$ and consider the (simplified) ranked model $\mathscr{R}$ depicted in Figure 2. Then we have $[\![\bullet\mathsf{b}]\!] = \{110\}$, $[\![\bullet\mathsf{p}]\!] = \{101\}$ and $[\![\bullet(\mathsf{b}\wedge\neg\mathsf{f})]\!] = \{100, 101\}$.

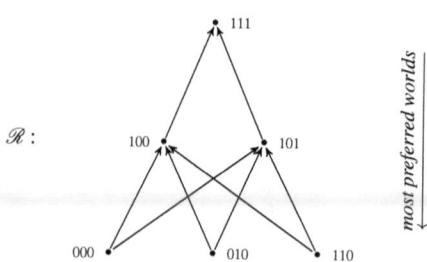

Fig. 2 An $\mathscr{L}^\bullet$-model for $\mathscr{P} = \{\mathsf{b},\mathsf{f},\mathsf{p}\}$.

It is worth noting that for every ranked model $\mathscr{R}$ and every $\alpha \in \mathscr{L}^\bullet$, there is a $\beta \in \mathscr{L}$ (i.e., a propositional formula) such that $\mathscr{R} \Vdash \alpha \leftrightarrow \beta$. That is to say, given $\mathscr{R}$, every α can be expressed as a propositional formula (β) in $\mathscr{R}$. Of course, this does not mean that propositional logic is as expressive as PTL, since the formula β used to express α in the ranked model $\mathscr{R}$ depends on the specific $\mathscr{R}$. Rather, the relationship between PTL and propositional logic is similar to the relationship between modal logic and propositional logic in the sense that both modal logic and PTL add to propositional logic an operator that is not truth-functional. (In Section 8 we discuss in more detail the relationship between PTL and modal approaches to preferential reasoning.)

Next is a result showing that if a ranked model $\mathscr{R}$ is such that $\mathscr{R} \Vdash \bullet\alpha$ for some $\alpha \in \mathscr{L}^\bullet$, then $\mathscr{R}$ consists of only α-worlds in which all worlds are incomparable (alias equally preferred) according to the preference relation $\prec$. In other words, typicality allows a syntactic way of expressing the preference relation is empty.

Proposition 1. Let $\mathscr{R} = \langle \mathscr{V}, \prec \rangle$. Then

1. $\prec = \emptyset$ if and only if there exists $\alpha \in \mathscr{L}^\bullet$ such that $\mathscr{R} \Vdash \bullet\alpha$;
2. For every $\alpha \in \mathscr{L}^\bullet$, $\mathscr{R} \Vdash \bullet\alpha$ if and only if for every $\beta \in \mathscr{L}$ such that $\mathscr{R} \Vdash \alpha \rightarrow \beta$, $\mathscr{R} \Vdash \bullet\beta$.

Proof. Proving Part 1: Suppose that $\prec = \emptyset$. Then it is not hard to verify that $\mathscr{R} \Vdash \bullet\top$. Conversely, suppose that $\mathscr{R} \Vdash \bullet\alpha$ for some $\alpha \in \mathscr{L}^\bullet$. That is, $\min_\prec[\![\alpha]\!] = \mathscr{V}$. From this it follows that $\prec = \emptyset$ since otherwise there would be a valuation $v \in \mathscr{V}$ such that $v \notin \min_\prec[\![\alpha]\!]$.

Proving Part 2: Pick an $\alpha \in \mathscr{L}^\bullet$ and suppose $\mathscr{R} \Vdash \bullet\alpha$. Therefore $\mathscr{R} \Vdash \alpha$ and since $\mathscr{R} \Vdash \alpha \rightarrow \beta$, it follows that $\mathscr{R} \Vdash \beta$. Now, from $\mathscr{R} \Vdash \bullet\alpha$ it also follows that $\prec = \emptyset$, and from $\mathscr{R} \Vdash \beta$ we then have that $\mathscr{R} \Vdash \bullet\beta$. Conversely, suppose that for every $\beta \in \mathscr{L}$ such that $\mathscr{R} \Vdash \alpha \rightarrow \beta$, $\mathscr{R} \Vdash \bullet\beta$. Let β be α itself, from which it then follows that $\mathscr{R} \Vdash \bullet\alpha$.

One of the consequences of this result is that if $\bullet\alpha$ is true in a ranked model, then so is α (but the converse, of course, does not hold).

Another useful property of the typicality operator $\bullet$ is that it allows us to express (propositional) rational consequence, as defined in Section 2.

Proposition 2. *Let $\mathscr{R} = \langle \mathscr{V}, \prec \rangle$. Then for every $\alpha, \beta \in \mathscr{L}$ (i.e., α and β are propositional formulae), $\alpha \mathrel|\!\sim_\mathscr{R} \beta$ if and only if $\mathscr{R} \Vdash \bullet\alpha \to \beta$.*

Proof. Let $\mathscr{R} = \langle \mathscr{V}, \prec \rangle$ be a ranked model. $\alpha \mathrel|\!\sim_\mathscr{R} \beta$ if and only if $\min_\prec [\![\alpha]\!] \subseteq [\![\beta]\!]$ if and only if $\min_\prec [\![\alpha]\!] \cap [\![\neg\beta]\!] = \emptyset$ if and only if $[\![\bullet\alpha \wedge \neg\beta]\!] = \emptyset$ if and only if $[\![\bullet\alpha \to \beta]\!] = \mathscr{V}$ if and only if $\mathscr{R} \Vdash \bullet\alpha \to \beta$.

Proposition 2 shows that the introduction of a typicality operator into the object language allows us to express KLM-style rational consequence as defined in Section 2.[4] This forms part of our argument to show that our semantics for typicality is the correct one, but it does not provide a justification for introducing all the additional expressivity obtained from typicality. The next two results provide such a justification.

Proposition 3. *There is an $\mathscr{L}^\bullet$-sentence that cannot be expressed as a single KLM-style $\mathrel|\!\sim$-statement. That is, there is $\alpha \in \mathscr{L}^\bullet$ such that there exists $\mathscr{R}$ such that $\mathscr{R} \Vdash \alpha$ and for every $\beta, \gamma \in \mathscr{L}$, $\beta \mathrel|\!\not\sim_\mathscr{R} \gamma$.*

The sentence $\alpha = q \wedge (\bullet q \to \neg p)$ is such that it cannot be expressed as a *single* $\mathrel|\!\sim$-statement. The proof is tedious and we shall omit it here. However, note that α can be expressed as the *set* of defeasible statements $\{\neg q \mathrel|\!\sim \bot, q \mathrel|\!\sim \neg p\}$.

This raises the question whether the $\mathscr{L}^{\mathrel|\!\sim}$-language, i.e., the set of all $\mathrel|\!\sim$-statements built up from a propositional language $\mathscr{L}$, is as expressive as $\mathscr{L}^\bullet$. This is so if and only if for every $\alpha \in \mathscr{L}^\bullet$ there is a subset X of $\mathscr{L}^{\mathrel|\!\sim}$ such that, for every ranked model $\mathscr{R}$, $\mathscr{R} \Vdash \alpha$ if and only if $\mathscr{R} \Vdash X$. The answer is 'no', as witnessed by the following result.

Proposition 4. *There are $\mathscr{L}^\bullet$-sentences that cannot be expressed as a set of KLM-style $\mathrel|\!\sim$-statements.*

Proof. Assume $\mathscr{P} = \{p, q\}$ and let α be the sentence $\bullet p$. This sentence is true in exactly four ranked models (all of which have empty orderings): one with $\mathscr{V} = \{11, 10\}$, one with $\mathscr{V} = \{11\}$, one with $\mathscr{V} = \{10\}$ and one with $\mathscr{V} = \emptyset$. Assume that there is a set X of $\mathrel|\!\sim$-statements that are all true in exactly these ranked models. Then for each ranked model built up from $\mathscr{P}$ other than these four, there exists at least one $\mathrel|\!\sim$-statement contained in X that is not true in it. In particular there exists such a $\mathrel|\!\sim$-statement, call it $\beta \mathrel|\!\sim \gamma$, for the ranked model in which $\mathscr{V} = \{11, 10\}$ and 11 is preferred to 10. Since $\beta \mathrel|\!\sim \gamma$ is not true in this model, there is a world in $\min_\prec [\![\beta]\!]$ which is not in $[\![\gamma]\!]$. If 11 is such a world, then $\beta \mathrel|\!\sim \gamma$ cannot be true

---

[4] Observe that Proposition 2 shows that rational consequence for *propositional* logic can be expressed in PTL. In Section 6 we shall see that it is also possible to express, in PTL itself, the extended notion of rational consequence for the more expressive language of $\mathscr{L}^\bullet$.

in the ranked model $\langle\{11\},\emptyset\rangle$ mentioned above. (Remember that β and γ are both propositional and therefore their truth values are entirely determined by the propositional valuations.) Alternatively, if 10 is such a world, then $\beta \mathrel{\vert\!\sim} \gamma$ cannot be true in $\langle\{10\},\emptyset\rangle$ either. Hence there cannot be such a world falsifying the statement $\beta \mathrel{\vert\!\sim} \gamma$, and therefore X does not rule out the ranked models with a non-empty ordering, from which we derive a contradiction. Hence there cannot be such a set X.

A corollary of this result is that PTL does indeed add to the expressivity of the KLM approach. In the next section we assess how much expressivity is actually added by our typicality operator.

4 Unnesting the Birds

In the previous section we have argued for the need to include typicality explicitly in the object language. The observant reader would have noticed that the language of PTL allows for any arbitrary (finite) nesting of the typicality operator. An important point to consider is whether this much expressivity is needed, and whether it is not perhaps sufficient to restrict the language to non-nested applications of typicality.

In this section we show that once typicality is added to the language, nesting does not increase the expressivity any further, provided that we are allowed to add new propositional atoms. We shall thus be working with languages in which the set of propositional atoms $\mathscr{P}$ may vary, and more specifically, with languages with respect to a given knowledge base. So, given a knowledge base $\mathscr{K}$, we denote by $\mathscr{P}^{\mathscr{K}}$ the set of propositional atoms occurring in $\mathscr{K}$. Furthermore, by a *ranked model on* $\mathscr{P}^{\mathscr{K}}$ we mean a ranked model built up using only the propositional atoms occurring in $\mathscr{P}^{\mathscr{K}}$.

Now, given any finite $\mathscr{K} \subseteq \mathscr{L}^{\bullet}$ we: (*i*) Show how to transform every sentence $\beta \in \mathscr{L}^{\bullet}$ into a new sentence $\widehat{\beta}$ containing no nested instances of the $\bullet$ operator (and therefore also how to transform $\mathscr{K}$ into a knowledge base $\widehat{\mathscr{K}}$, containing no nested instances of the $\bullet$ operator); (*ii*) Show how to construct an auxiliary set of formulae $\widehat{E}$, containing no nested instances of the $\bullet$ operator, regulating the behavior of the newly introduced propositional atoms, and (*iii*) Show how to transform every ranked model $\mathscr{R}$ on $\mathscr{P}^{\mathscr{K}}$ into its 'appropriate representative' $\widehat{\mathscr{R}}$ on the set of atoms $\mathscr{P}^{\widehat{\mathscr{K}}}$ such that, for every $\beta \in \mathscr{L}^{\bullet}$, β is true in $\mathscr{R}$ if and only if $\widehat{\beta}$ is true in $\widehat{\mathscr{R}}$. Using these constructions we show that $\widehat{\mathscr{K}} \cup \widehat{E}$ is the non-nested version of the original knowledge base $\mathscr{K}$ in the sense that the ranked models in which $\widehat{\mathscr{K}} \cup \widehat{E}$ are true are precisely the 'appropriate representatives' of the ranked models in which $\mathscr{K}$ is true.

To be more precise, let $\mathscr{K} \subseteq \mathscr{L}^{\bullet}$ be a knowledge base, let $S^{\mathscr{K}}$ denote all the subformulae of $\mathscr{K}$, and let $B^{\mathscr{K}} := \{\bullet\alpha \in S^{\mathscr{K}} \mid \alpha \in \mathscr{L}\}$. So $B^{\mathscr{K}}$ contains all occurrences of subformulae in $\mathscr{K}$ containing a single $\bullet$ operator. Informally, the idea is to substitute (all occurrences of) every element $\bullet\alpha$ of $B^{\mathscr{K}}$ with a newly introduced atom $p^{\bullet\alpha}$, and to require that $p^{\bullet\alpha}$ be equivalent to $\bullet\alpha$. In doing so we reduce the level of nesting in $\mathcal{K}$ by a factor of 1. Now, let $E^{\mathcal{K}} := \{p^{\bullet\alpha} \leftrightarrow \bullet\alpha \mid \bullet\alpha \in B^{\mathcal{K}}\}$, and for every $\beta \in \mathcal{L}^{\bullet}$, let $\beta^{\mathcal{K}}$ be obtained from β by the simultaneous substitution in β of (every occurrence of) every $\bullet\alpha \in B^{\mathcal{K}}$ by $p^{\bullet\alpha}$ (observe that $\beta^{\mathcal{K}} = \beta$ if β is a propositional formula). We refer to $\beta^{\mathcal{K}}$ as the $\mathcal{K}$-transform of β. Also, let $\mathcal{K}^{\bullet} := \{\beta^{\mathcal{K}} \mid \beta \in \mathcal{K}\}$. The idea is that $\mathcal{K}^{\bullet} \cup E^{\mathcal{K}}$ is a version of $\mathcal{K}$ with one fewer level of nesting.

Example 1. Let $\mathcal{K} = \{\bullet(p \wedge q) \to r, \bullet(\bullet p \vee r), \bullet(p \wedge \bullet(q \vee \bullet r))\}$. Then we have $S^{\mathcal{K}} = \mathcal{K} \cup \{\bullet(p \wedge q), p, q, r, \bullet p, \bullet(q \vee \bullet r), \bullet r\}$. Then $B^{\mathcal{K}} = \{\bullet(p \wedge q), \bullet p, \bullet r\}$ and $E^{\mathcal{K}} = \{p^{\bullet(p \wedge q)} \leftrightarrow \bullet(p \wedge q), p^{\bullet p} \leftrightarrow \bullet p, p^{\bullet r} \leftrightarrow \bullet r\}$. Now $(\bullet(p \wedge q) \to r)^{\mathcal{K}} = p^{\bullet(p \wedge q)} \to r$, $(\bullet(\bullet p \vee r))^{\mathcal{K}} = \bullet(p^{\bullet p} \vee r)$, $(\bullet(p \wedge \bullet(q \vee \bullet r)))^{\mathcal{K}} = \bullet(p \wedge \bullet(q \vee p^{\bullet r}))$. Hence $\mathcal{K}^{\bullet} = \{p^{\bullet(p \wedge q)} \to r, \bullet(p^{\bullet p} \vee r), \bullet(p \wedge \bullet q \vee p^{\bullet r})\}$. Observe that $\mathcal{K}$ has a nesting level of 3, while $\mathcal{K}^{\bullet}$ has a nesting level of 2.

Let $\mathcal{R} = \langle \mathcal{V}, \prec \rangle$ be a ranked model on $\mathcal{P}^{\mathcal{K}}$. We define $\mathcal{R}^{\bullet} = \langle \mathcal{V}^{\bullet}, \prec^{\bullet} \rangle$ on $\mathcal{P}^{\mathcal{K}^{\bullet}}$ as follows: for all $v \in \mathcal{V}$, let $v^{\bullet}$ be a valuation on $\mathcal{P}^{\mathcal{K}^{\bullet}}$ such that (i) $v^{\bullet}(p) = v(p)$ for every $p \in \mathcal{P}^{\mathcal{K}}$, and (ii) for every $p^{\bullet\alpha} \in \mathcal{P}^{\mathcal{K}^{\bullet}} \setminus \mathcal{P}^{\mathcal{K}}$, $v^{\bullet}(p^{\bullet\alpha}) = 1$ if and only if $v \in [\![\bullet\alpha]\!]$ in the ranked model $\mathcal{R}$. And for all $v^{\bullet}, v^{\bullet\prime} \in \mathcal{V}^{\bullet}$, $v^{\bullet} \prec^{\bullet} v^{\bullet\prime}$ if and only if $v \prec v'$. So $\mathcal{R}^{\bullet}$ is an extended version of $\mathcal{R}$ with every valuation v in $\mathcal{R}$ replaced with an extended valuation $v^{\bullet}$ in which the truth values of the atoms occurring in v remain unchanged, and the truth values of the new atoms are constrained by the requirement that every $p^{\bullet\alpha}$ be equivalent to $\bullet\alpha$ (for $\bullet\alpha \in B^{\mathcal{K}}$). We refer to $\mathcal{R}^{\bullet}$ as the $\mathcal{K}$-extended version of $\mathcal{R}$. From this we obtain the following result.

Proposition 5. *Let $\mathcal{K}$ be an $\mathcal{L}^{\bullet}$-knowledge base. Then*

1. *For every ranked model $\mathcal{R}$ on $\mathcal{P}^{\mathcal{K}}$, $\mathcal{R}^{\bullet} \Vdash E^{\mathcal{K}}$;*
2. *A ranked model $\mathcal{R}'$ on $\mathcal{P}^{\mathcal{K}^{\bullet}}$ is such that $\mathcal{R}' \Vdash E^{\mathcal{K}}$ if and only if there is a ranked model $\mathcal{R}$ on $\mathcal{P}^{\mathcal{K}}$ such that $\mathcal{R}^{\bullet} = \mathcal{R}'$;*
3. *For all $\beta \in \mathcal{L}^{\bullet}$, $\mathcal{R} \Vdash \beta$ if and only if $\mathcal{R}^{\bullet} \Vdash \beta^{\mathcal{K}}$.*

Proof. Let $\mathcal{K} \subseteq \mathcal{L}^{\bullet}$ be finite.

Proving 1: Let $\mathcal{R}$ be a ranked model on $\mathcal{P}^{\mathcal{K}}$ and consider $\mathcal{R}^{\bullet} = \langle \mathcal{V}^{\bullet}, \prec^{\bullet} \rangle$. Pick any $p^{\bullet\alpha} \leftrightarrow \bullet\alpha \in E^{\mathcal{K}}$ and pick any $v \in \mathcal{V}^{\bullet}$. By construction, $v(p^{\bullet\alpha}) = 1$ if and only if $v \in [\![\bullet\alpha]\!]$, and so $v \in [\![p^{\bullet\alpha} \leftrightarrow \bullet\alpha]\!]$. From this it follows that $\mathcal{R} \Vdash E^{\mathcal{K}}$.

Proving 2: Pick a ranked model $\mathcal{R}'$ on $\mathcal{P}^{\mathcal{K}^{\bullet}}$ and suppose that $\mathcal{R}' \Vdash E^{\mathcal{K}}$ is the case. Let $\mathcal{R} = \langle \mathcal{V}, \prec \rangle$ be the ranked model on $\mathcal{P}^{\mathcal{K}}$ obtained from $\mathcal{R}'$ by restricting $\mathcal{V}$ (and therefore $\prec$ as well) to $\mathcal{P}^{\mathcal{K}}$. It follows immediately that the ranked model $\mathcal{R}^{\bullet}$ on $\mathcal{P}^{\mathcal{K}^{\bullet}}$ obtained from $\mathcal{R}$ is therefore equal to $\mathcal{R}'$. Conversely, suppose there is a ranked model $\mathcal{R}$ on $\mathcal{P}^{\mathcal{K}}$ such that $\mathcal{R}^{\bullet} = \mathcal{R}'$. By construction, $v^{\bullet}(p^{\bullet\alpha}) = 1$ if and only if $v^{\bullet} \in [\![\bullet\alpha]\!]$ for every $p^{\bullet\alpha} \in \mathcal{P}^{\mathcal{K}^{\bullet}} \setminus \mathcal{P}^{\mathcal{K}}$, from which it follows that $\mathcal{R}^{\bullet} \Vdash p^{\bullet\alpha} \leftrightarrow \bullet\alpha$ for every $p^{\bullet\alpha} \leftrightarrow \bullet\alpha \in E^{\mathcal{K}}$. So $\mathcal{R}^{\bullet}$, and therefore $\mathcal{R}'$, is a model of $E^{\mathcal{K}}$.

Proving 3: Pick any ranked model $\mathcal{R} = \langle \mathcal{V}, \prec \rangle$ on $\mathcal{P}^{\mathcal{K}}$. Suppose that $\mathcal{R} \Vdash \mathcal{K}$ and pick any $\beta \in \mathcal{L}$. Pick a $v \in \mathcal{V}$ and suppose that $v \in [\![\beta]\!]$. Now consider $\mathcal{R}^{\bullet} = \langle \mathcal{V}^{\bullet}, \prec^{\bullet} \rangle$ and, in particular, $v^{\bullet} \in \mathcal{V}^{\bullet}$. From the construction of $v^{\bullet}$ and $\beta^{\mathcal{K}}$ it follows immediately that $v^{\bullet} \in [\![\beta^{\mathcal{K}}]\!]$ in $\mathcal{R}^{\bullet}$. Conversely, suppose that $\mathcal{R}^{\bullet} \Vdash \mathcal{K}^{\bullet}$ and pick any $\beta \in \mathscr{L}^\bullet$. Pick a $v^\bullet \in \mathscr{V}^\bullet$ and suppose that $v^\bullet \in [\![\beta^{\mathscr{K}}]\!]$. Now consider $\mathscr{R} = \langle \mathscr{V}, \prec \rangle$ and, in particular, $v \in \mathscr{V}$. From the construction of $v^\bullet$ and $\bullet\beta$ it follows immediately that $v \in [\![\beta^{\mathscr{K}}]\!]$.

Proposition 5 above shows that the $\mathscr{K}$-extended version of a ranked model $\mathscr{R}$ is the only 'appropriate representative' of $\mathscr{R}$ in the class of ranked models based on the extended language of $\mathscr{K}^\bullet$. In addition, the $\mathscr{K}$-extended versions of the ranked models based on the language of $\mathscr{K}$ are the only ones satisfying $E^{\mathscr{K}}$.

From Proposition 5 it also follows that all sentences β of $\mathscr{L}^\bullet$ and their $\mathscr{K}$-transforms behave exactly the same with respect to, respectively, any ranked model $\mathscr{R}$ and its $\mathscr{K}$-extended version $\mathscr{R}^\bullet$. This applies, in particular, to the elements of $\mathscr{K}$, as the next corollary shows.

Corollary 1. *Let $\mathscr{K}$ be an $\mathscr{L}^\bullet$-knowledge base and let $\mathscr{R}$ be a ranked model on $\mathscr{P}^{\mathscr{K}}$. Then $\mathscr{R} \Vdash \mathscr{K}$ if and only if $\mathscr{R}^\bullet \Vdash \mathscr{K}^\bullet$.*

Proof. Follows from Part 3 of Proposition 5.

These results show that the move from a knowledge base $\mathscr{K}$ to $\mathscr{K}^\bullet$ ensures that we can reduce the level of nesting of $\bullet$ operators by a factor of 1. To arrive at a set $\widehat{\mathscr{K}}$ not containing any nested occurrences of $\bullet$ we just need to iterate the transform process a sufficient number of times. So, we define $\widehat{\mathscr{K}}$ as follows: Let $\mathscr{K}_0 := \mathscr{K}$, and for every $i > 0$, let $B_i := B^{\mathscr{K}_{i-1}}$, $\mathscr{K}_i := \mathscr{K}_{i-1}^\bullet$, and let $n := \min_<\{i \mid B_{i+1} = \emptyset\}$. We then let $\widehat{\mathscr{K}} := \mathscr{K}_n$. So for every $i = 1, \ldots, n$, $\mathscr{K}_i$ has one fewer level of nesting of $\bullet$ than $\mathscr{K}_{i-1}$ until we get to $\mathscr{K}_n = \widehat{\mathscr{K}}$, which has no nested occurrences of $\bullet$. Similarly, for every $\beta \in \mathscr{L}^\bullet$, we define $\widehat{\beta}$ as follows: Let $\beta_0 := \beta$, for every $i = 1, \ldots, n$, let $\beta_i := \beta^{\mathscr{K}_{i-1}}$, and let $\widehat{\beta} := \beta_n$. We refer to $\widehat{\beta}$ as the *full $\mathscr{K}$-transform of β*. In a similar vein, we let $\widehat{E} := \bigcup_{i=0}^{i=n-1} E^{\mathscr{K}_i}$.

Example 2. Continuing Example 1, let $\mathscr{K}_0 = \mathscr{K}$. Then $B_1 = B^{\mathscr{K}_0} = B^{\mathscr{K}}$, and $\mathscr{K}_1 = \mathscr{K}^\bullet$ with $E_0 = E^{\mathscr{K}}$. Then

$$S^{\mathscr{K}_1} = S^{\mathscr{K}^\bullet} = \mathscr{K}^\bullet \cup \{p^{\bullet(p \wedge q)}, r, p^{\bullet p}, p, \bullet(q \vee p^{\bullet r}), q, p^{\bullet r}\}$$

and then $B_2 = B^{\mathscr{K}_1} = \{\bullet(q \vee p^{\bullet r})\}$, and $E_1 = \{p^{\bullet(q \vee p^{\bullet r})} \leftrightarrow \bullet(q \vee p^{\bullet r})\}$. Now we have $\mathscr{K}_2 = \mathscr{K}_1^\bullet = \mathscr{K}^{\bullet\bullet} = \{p^{\bullet(p \wedge q)} \to r, \bullet(p^{\bullet p} \vee r), \bullet(p \wedge p^{\bullet(q \vee p^{\bullet r})})\}$. Then, in the second iteration, we get

$$S^{\mathscr{K}_2} = \mathscr{K}_2 \cup \{p^{\bullet(p \wedge q)}, r, p^{\bullet r}, p, p^{\bullet(q \vee p^{\bullet r})}\}$$

and then $B_3 = B^{\mathscr{K}_2} = \{\bullet(p^{\bullet p} \vee r), \bullet(p \wedge p^{\bullet(q \vee p^{\bullet r})})\}$ with $E_2 = \{p^{\bullet(p^{\bullet p} \vee r)} \leftrightarrow \bullet(p^{\bullet p} \vee r), p^{\bullet(p \wedge p^{\bullet(q \vee p^{\bullet r})})} \leftrightarrow \bullet(p \wedge p^{\bullet(q \vee p^{\bullet r})})\}$. Then we get $\mathscr{K}_3 = \mathscr{K}_2^\bullet = \{p^{\bullet(p \wedge q)} \to r, p^{\bullet(p^{\bullet p} \vee r)}, p^{\bullet(p \wedge p^{\bullet(q \vee p^{\bullet r})})}\}$. In the next iteration, we have

$$S^{\mathscr{K}_3} = \mathscr{K}_3 \cup \{p^{\bullet(p \wedge q)}, r, p^{\bullet(p^{\bullet p} \vee r)}, p^{\bullet(p \wedge p^{\bullet(\bullet q \vee p^{\bullet r})})}\}$$

and therefore $B_4 = \emptyset$. Hence $n = 3$, and then

$$\widehat{\mathcal{K}} = \left\{ p^{\bullet(p\wedge q)} \to r,\ p^{\bullet(p^{\bullet p}\vee r)},\ p^{\bullet(p\wedge p^{\bullet(q\vee p^{\bullet r})})} \right\}$$

$$\widehat{E} = \left\{ \begin{array}{l} p^{\bullet(p\wedge q)} \leftrightarrow \bullet(p\wedge q),\ p^{\bullet p} \leftrightarrow \bullet p,\ p^{\bullet r} \leftrightarrow \bullet r,\ p^{\bullet(q\vee p^{\bullet r})} \leftrightarrow \bullet(q\vee p^{\bullet r}), \\ p^{\bullet(p^{\bullet p}\vee r)} \leftrightarrow \bullet(p^{\bullet p}\vee r),\ p^{\bullet(p\wedge p^{\bullet(q\vee p^{\bullet r})})} \leftrightarrow \bullet(p \wedge p^{\bullet(q\vee p^{\bullet r})}) \end{array} \right\}$$

Finally, for any ranked model $\mathscr{R}$ on the set of propositional variables $\mathscr{P}^{\mathscr{K}}$, we define its *full $\mathscr{K}$-extended version* $\widehat{\mathscr{R}}$ as follows: Let $\mathscr{R}_0 := \mathscr{R}$, and for all $i = 1,\ldots,n$, $\mathscr{R}_i := \mathscr{R}_{i-1}^{\bullet}$. Then we let $\widehat{\mathscr{R}} := \mathscr{R}_n$.

Using Proposition 5 and Corollary 1 we then obtain the result we require.

Theorem 3. *Let $\mathscr{K}$ be an $\mathscr{L}^{\bullet}$-knowledge base. Then*

1. *For every $\mathscr{R}$ on $\mathscr{P}^{\mathscr{K}}$, its full $\mathscr{K}$-extended version $\widehat{\mathscr{R}}$ is such that $\widehat{\mathscr{R}} \Vdash \widehat{E}$;*
2. *A ranked model $\mathscr{R}'$ on $\mathscr{P}^{\mathscr{K}}$ is a model of $\widehat{E}$ if and only if there is a ranked model $\mathscr{R}$ on $\mathscr{P}^{\mathscr{K}}$ such that $\mathscr{R}' = \widehat{\mathscr{R}}$;*
3. *For all $\beta \in \mathscr{L}^{\bullet}$, $\mathscr{R} \Vdash \beta$ if and only if $\widehat{\mathscr{R}} \Vdash \widehat{\beta}$;*
4. *Let $\mathscr{R}$ be a ranked model $\mathscr{R}$ on $\mathscr{P}^{\mathscr{K}}$. Then $\mathscr{R} \Vdash \mathscr{K}$ if and only if $\widehat{\mathscr{R}} \Vdash \widehat{\mathscr{K}}$.*

Proof. The proofs of Parts 1–3 follow by induction from the proofs of the corresponding Parts 1–3 of Proposition 5. The proof of Part 4 follows by induction from Corollary 1.

5 Belief Revision and Typicality

Given the well-known connection between propositional rational consequence relations and AGM-style belief revision [Alchourrón et al., 1985], as developed by Gärdenfors and Makinson [1994], it is perhaps not surprising that propositional AGM belief revision can be expressed using the typicality operator. In this section we make this claim more precise. The formal representation of propositional AGM revision that we provide below is based on that given by Katsuno and Mendelzon [1991].

The starting point here is to fix a non-empty subset $\mathscr{V}$ of $\mathscr{U}$ (as done by Kraus et al. [1990]), and to assume that everything is done within the context of $\mathscr{V}$. In that sense, $\mathscr{V}$ becomes the set of *all* valuations that are available to us. This is slightly more general than the Katsuno-Mendelzon framework, which assumes $\mathscr{V}$ to be equal to $\mathscr{U}$, but is a special case of the original AGM approach. To reflect this restriction, we use $Mod_{\mathscr{V}}(\alpha)$ to denote the set $Mod(\alpha) \cap \mathscr{V}$. In the same vein, in the rationality postulates below, validity is understood to be modulo $\mathscr{V}$. That is, for $\alpha \in \mathscr{L}$ (i.e., α is a propositional sentence), we let $\models \alpha$ if and only if $Mod_{\mathscr{V}}(\alpha) = \mathscr{V}$.

Next, we fix a knowledge base $\kappa \in \mathscr{L}$ (i.e., represented as a propositional formula) such that $Mod_{\mathscr{V}}(\kappa) \neq \emptyset$. A revision operator $\circ$ on $\mathscr{L}$ for κ is a function from $\mathscr{L}$ to $\mathscr{L}$. Intuitively, $\kappa \circ \alpha$ is the result of revising κ by α (clearly the models of $\kappa \circ \alpha$ should be in $\mathscr{V}$). An AGM revision operator $\circ$ on $\mathscr{L}$ for κ is a revision operator on $\mathscr{L}$ for κ which satisfies the following six properties:

(R1) $\models (\kappa \circ \alpha) \to \alpha$
(R2) If $\not\models \neg(\kappa \wedge \alpha)$, then $\models (\kappa \circ \alpha) \leftrightarrow (\kappa \wedge \alpha)$
(R3) If $\not\models \neg\alpha$, then $\not\models \neg(\kappa \circ \alpha)$
(R4) If $\models \kappa_1 \leftrightarrow \kappa_2$ and $\models \alpha_1 \leftrightarrow \alpha_2$, then $\models (\kappa_1 \circ \alpha_1) \leftrightarrow (\kappa_2 \circ \alpha_2)$
(R5) $\models ((\kappa \circ \alpha) \wedge \beta) \to (\kappa \circ (\alpha \wedge \beta))$
(R6) If $\not\models \neg((\kappa \circ \alpha) \wedge \beta)$, then $\models (\kappa \circ (\alpha \wedge \beta)) \to ((\kappa \circ \alpha) \wedge \beta)$

Given $\mathscr{V}$, a ranked model $\mathscr{R} = \langle \mathscr{V}, \prec \rangle$ is defined as κ-*faithful* if and only if $\min_\prec \mathscr{V} = [\![\kappa]\!]$. We say that a revision operator $\circ_\mathscr{R}$ (on $\mathscr{L}$) is *defined by* a κ-faithful ranked model $\mathscr{R}$ if and only if $Mod_\mathscr{V}(\kappa \circ_\mathscr{R} \alpha) = \min_\prec [\![\alpha]\!]$. Katsuno and Mendelzon [1991] proved that, for $\mathscr{V} = \mathscr{U}$, (i) every revision operator $\circ_\mathscr{R}$ defined by a κ-faithful ranked model $\mathscr{R}$ is an AGM revision operator (on $\mathscr{L}$), and (ii) for every AGM revision operator $\circ$ (on $\mathscr{L}$) for κ, there is a κ-faithful ranked model $\mathscr{R}$ such that $Mod_\mathscr{V}(\kappa \circ \alpha) = Mod_\mathscr{V}(\kappa \circ_\mathscr{R} \alpha)$.

In what follows, we show that the revision operator $\circ$ can be expressed in $\mathscr{L}^\bullet$ using typicality. The key insight is to identify the knowledge base κ to be revised with the formula $\bullet\top$, while $\kappa \circ \alpha$ is identified with $\bullet\alpha$.

Proposition 6. *Let $\kappa \in \mathscr{L}$ such that $Mod_\mathscr{V}(\kappa) \neq \emptyset$ and let $\mathscr{R} = \langle \mathscr{V}, \prec \rangle$ be a κ-faithful ranked model.*

1. *For every $\alpha \in \mathscr{L}$, $[\![\kappa \circ_\mathscr{R} \alpha]\!] = [\![\bullet\alpha]\!]$;*
2. *Let $\circ$ be any AGM revision operator (on $\mathscr{L}$) for κ. Then there is a κ-faithful ranked model $\mathscr{R} = \langle \mathscr{V}, \prec \rangle$ such that $Mod_\mathscr{V}(\kappa \circ \alpha) = [\![\bullet\alpha]\!]$.*

Proof. Let $\kappa \in \mathscr{L}$ and let $\mathscr{R} = \langle \mathscr{V}, \prec \rangle$ be a κ-faithful ranked model.

Proving 1: Pick any $\alpha \in \mathscr{L}$. By definition, $[\![\kappa \circ_\mathscr{R} \alpha]\!] = \min_\prec [\![\alpha]\!]$ and $[\![\bullet\alpha]\!] = \min_\prec [\![\alpha]\!]$, so the result holds.

Proving 2: Pick any AGM belief revision operator $\circ$ (on $\mathscr{L}$) for κ. First we need to prove that there is a κ-faithful ranked model $\mathscr{R} = \langle \mathscr{V}, \prec \rangle$ such that $Mod_\mathscr{V}(\kappa \circ \alpha) = \min_\prec [\![\alpha]\!]$. The proof is exactly the same as the proof of Theorem 3.3 provided by Katsuno and Mendelzon [1991], but with $\mathscr{U}$ replaced by $\mathscr{V}$. The result then follows from the fact that $[\![\bullet\alpha]\!] = \min_\prec [\![\alpha]\!]$.

This result shows that propositional AGM belief revision can be embedded in PTL. But we can take this a step further and extend revision to apply to the language of PTL, i.e., to $\mathscr{L}^\bullet$, as well. So, with a given (non-empty) $\mathscr{V}$ still fixed, and for $\alpha \in \mathscr{L}^\bullet$, we let $\mathscr{R}^\alpha_\mathscr{V} := \{\mathscr{R} \mid \mathscr{R} = \langle \mathscr{V}, \prec \rangle$ and $\min_\prec [\![\top]\!] = [\![\alpha]\!]\}$. Then we fix a $\kappa \in \mathscr{L}^\bullet$ such that $\mathscr{R}^\kappa_\mathscr{V} \neq \emptyset$. The definition of a revision operator $\circ$ is then the same as above, except that it is now with respect to $\mathscr{L}^\bullet$. In order to define an AGM belief revision operator on $\mathscr{L}^\bullet$ we need to rephrase the six revision postulates with respect to κ and the elements $\mathscr{R}$ of $\mathscr{R}^\kappa_\mathscr{V}$:

(R1$^\bullet$) $\mathscr{R} \Vdash (\kappa \circ \alpha) \to \alpha$
(R2$^\bullet$) If $\mathscr{R} \nVdash \neg(\kappa \wedge \alpha)$, then $\mathscr{R} \Vdash (\kappa \circ \alpha) \leftrightarrow (\kappa \wedge \alpha)$
(R3$^\bullet$) If $\mathscr{R} \nVdash \neg\alpha$, then $\mathscr{R} \nVdash \neg(\kappa \circ \alpha)$
(R4$^\bullet$) If $\mathscr{R} \Vdash \alpha_1 \leftrightarrow \alpha_2$, then $\mathscr{R} \Vdash (\kappa \circ \alpha_1) \leftrightarrow (\kappa \circ \alpha_2)$

(R5$^\bullet$) $\mathscr{R} \Vdash ((\kappa \circ \alpha) \wedge \beta) \to (\kappa \circ (\alpha \wedge \beta))$
(R6$^\bullet$)   If $\mathscr{R} \nVdash \neg((\kappa \circ \alpha) \wedge \beta)$, then $\mathscr{R} \Vdash (\kappa \circ (\alpha \wedge \beta)) \to ((\kappa \circ \alpha) \wedge \beta)$

This gives us a representation result similar to that of Katsuno and Mendelzon in the propositional case, but with the revision operator now defined on the more expressive language $\mathscr{L}^\bullet$.

Theorem 4. *Let $\kappa \in \mathscr{L}^\bullet$ such that $\mathscr{R}_\mathscr{V}^\kappa \neq \emptyset$ and let $\mathscr{R} \in \mathscr{R}_\mathscr{V}^\kappa$. Then $\circ_\mathscr{R}$ satisfies the postulates (R1$^\bullet$)–(R6$^\bullet$). Conversely, let $\circ$ be a revision operator (for a fixed $\mathscr{V}$ and κ) satisfying the postulates (R1$^\bullet$)–(R6$^\bullet$). Then there is an $\mathscr{R} \in \mathscr{R}_\mathscr{V}^\kappa$ such that $Mod_\mathscr{V}(\kappa \circ \alpha) = Mod_\mathscr{V}(\kappa \circ_\mathscr{R} \alpha)$.*

Proof. Let $\mathscr{R} = \langle \mathscr{V}, \prec \rangle \in \mathscr{R}_\mathscr{V}^\kappa$ be a κ-faithful ranked model. First we prove that $\circ_\mathscr{R}$ satisfies Postulates (R1$^\bullet$)–(R6$^\bullet$). For (R1$^\bullet$), observe that $[\![\kappa \circ_\mathscr{R} \alpha]\!] = \min_\prec [\![\alpha]\!] \subseteq [\![\alpha]\!]$. For (R2$^\bullet$), suppose there is a $v \in \mathscr{V}$ such that $v \in [\![\kappa \wedge \alpha]\!]$. Since $\mathscr{R}$ is κ-faithful, it follows that for every $w \in [\![\kappa \wedge \alpha]\!]$, $w \in \min_\prec [\![\top]\!]$. Therefore $[\![\kappa \circ_\mathscr{R} \alpha]\!] = \min_\prec [\![\top]\!] = [\![\kappa \wedge \alpha]\!]$ from which it follows that $\mathscr{R} \Vdash (\kappa \circ_\mathscr{R} \alpha) \leftrightarrow (\kappa \wedge \alpha)$. For (R3$^\bullet$), suppose there is a $v \in \mathscr{V}$ such that $v \in [\![\alpha]\!]$. Then $[\![\kappa \circ_\mathscr{R} \alpha]\!] = \min_\prec [\![\alpha]\!] \neq \emptyset$, and so $\mathscr{R} \nVdash \neg(\kappa \circ_\mathscr{R} \alpha)$. For (R4$^\bullet$), suppose that $\mathscr{R} \Vdash \alpha_1 \leftrightarrow \alpha_2$. Then $\min_\prec [\![\alpha_1]\!] = \min_\prec [\![\alpha_2]\!]$ from which it follows that $\mathscr{R} \Vdash (\kappa \circ_\mathscr{R} \alpha_1) \leftrightarrow (\kappa \circ_\mathscr{R} \alpha_2)$. For (R5$^\bullet$), pick any $v \in \mathscr{V}$ and suppose that $v \in [\![(\kappa \circ_\mathscr{R} \alpha) \wedge \beta]\!]$. That is, $v \in \min_\prec [\![\alpha]\!] \cap [\![\beta]\!]$. We need to show that $v \in \min_\prec [\![\alpha \wedge \beta]\!]$. If this is not the case, there is a $w \in \min_\prec [\![\alpha \wedge \beta]\!]$ such that $w \prec v$. But this cannot be since $v \in \min_\prec [\![\alpha]\!]$. So (R5$^\bullet$) is satisfied. For (R6$^\bullet$), suppose that $\mathscr{R} \nVdash \neg(\kappa \circ_\mathscr{R} \alpha) \wedge \beta$. That is, there is a $w \in \min_\prec [\![\alpha]\!] \cap [\![\beta]\!]$. Now pick a $v \in \mathscr{V}$ and suppose that $v \in [\![\kappa \circ_\mathscr{R} (\alpha \wedge \beta)]\!]$. That is, $v \in \min_\prec [\![\alpha \wedge \beta]\!]$. We need to show that $v \in \min_\prec ([\![\alpha]\!] \cap [\![\beta]\!])$. Since $v \in \min_\prec [\![\beta]\!]$, if this is not the case, there is an $x \in \min_\prec ([\![\alpha]\!] \cap [\![\neg \beta]\!])$ such that $x \prec v$. But then $w \prec v$ as well, which is impossible (because $v \in \min_\prec [\![\alpha \wedge \beta]\!]$). So (R6$^\bullet$) is satisfied.

Conversely, let $\circ$ be a revision operator for κ satisfying Postulates (R1$^\bullet$)–(R6$^\bullet$). We construct a ranked model $\mathscr{R} \in \mathscr{R}_\mathscr{V}^\kappa$ such that $Mod_\mathscr{V}(\kappa \circ \alpha) = Mod_\mathscr{V}(\kappa \circ_\mathscr{R} \alpha)$. The construction is essentially the same as that of Katsuno and Mendelzon [1991, Theorem 3.3]. We let $\mathscr{R} = \langle \mathscr{V}, \prec \rangle$, where $\prec$ is obtained as follows. Define a binary relation $\preceq$ on $\mathscr{V}$ such that for every $v, w \in \mathscr{V}$, $v \preceq w$ if and only if either $v \in [\![\kappa]\!]$ or $v \in [\![\kappa \circ f(v,w)]\!]$ where $f(v,w)$ is any element of $\mathscr{L}$ (i.e., a propositional formula) such that $[\![f(v,w)]\!] = \{v,w\}$. Katsuno and Mendelzon show that $\preceq$ is a total preorder (a binary relation that is reflexive, transitive, and connected). We let $\prec$ be the strict version of $\preceq$: $v \prec w$ if and only if $v \preceq w$ and $w \npreceq v$. It follows immediately that $\prec$ is a modular ordering and that $\mathscr{R}$ is therefore a ranked model. To show that $\mathscr{R} \in \mathscr{R}_\mathscr{V}^\kappa$ we need to show that $\min_\prec [\![\top]\!] = [\![\kappa]\!]$. We can split this into two parts: (*i*) If $v, w \in [\![\kappa]\!]$, then $v \nprec w$; (*ii*) If $v \in [\![\kappa]\!]$ and $w \notin [\![\kappa]\!]$, then $v \prec w$. Part (*i*) follows immediately from the definition of the relation $\preceq$. For Part (*ii*), suppose that $v \in [\![\kappa]\!]$ and $w \notin [\![\kappa]\!]$. From Postulate (R2$^\bullet$) it follows that $[\![\kappa \circ f(v,w)]\!] = \{v\}$. By the definition of $\preceq$, it then follows that $v \prec w$. It remains to show that $[\![\kappa \circ \alpha]\!] = [\![\kappa \circ_\mathscr{R} \alpha]\!] = \min_\prec [\![\alpha]\!]$. The proof is the same as that of Katsuno and Mendelzon [1991, Theorem 3.3].

6 Rational Consequence on $\mathscr{L}^\bullet$

We have seen in Section 5 that typicality can be used to express propositional AGM belief revision, as well as AGM belief revision defined for PTL. From Proposition 2 we know that rational consequence for propositional logic can be expressed in PTL itself. In this section we shall complete the picture by showing that (*i*) the expected connection between rational consequence relations and AGM belief revision for PTL holds, and that (*ii*) rational consequence for PTL can be expressed in PTL itself, a result analogous to Theorem 4.

As in Section 5, we start by fixing a set of valuations $\mathscr{V} \subseteq \mathscr{U}$. In this case, however, $\mathscr{V}$ is allowed to be empty as well. Let $\mathscr{R}_\mathscr{V} := \{\mathscr{R} \mid \mathscr{R} = \langle \mathscr{V}, \prec \rangle\}$. Then we let $\mathrel{\mid\!\sim}$ be a binary relation on $\mathscr{L}^\bullet$. We say that $\mathrel{\mid\!\sim}$ is a *rational consequence relation on* $\mathscr{L}^\bullet$ (with respect to $\mathscr{R} \in \mathscr{R}_\mathscr{V}$) if and only if $\mathrel{\mid\!\sim}$, viewed as a *binary connective* on $\mathscr{L}^\bullet$, satisfies the following seven properties, adopted from the rationality properties from Section 2:

(Ref) $\mathscr{R} \Vdash \alpha \mathrel{\mid\!\sim} \alpha$

(LLE) $\dfrac{\mathscr{R} \Vdash \alpha \leftrightarrow \beta,\ \mathscr{R} \Vdash \alpha \mathrel{\mid\!\sim} \gamma}{\mathscr{R} \Vdash \beta \mathrel{\mid\!\sim} \gamma}$

(And) $\dfrac{\mathscr{R} \Vdash \alpha \mathrel{\mid\!\sim} \beta,\ \mathscr{R} \Vdash \alpha \mathrel{\mid\!\sim} \gamma}{\mathscr{R} \Vdash \alpha \mathrel{\mid\!\sim} \beta \wedge \gamma}$

(Or) $\dfrac{\mathscr{R} \Vdash \alpha \mathrel{\mid\!\sim} \gamma,\ \mathscr{R} \Vdash \beta \mathrel{\mid\!\sim} \gamma}{\mathscr{R} \Vdash \alpha \vee \beta \mathrel{\mid\!\sim} \gamma}$

(RW) $\dfrac{\mathscr{R} \Vdash \alpha \mathrel{\mid\!\sim} \beta,\ \mathscr{R} \Vdash \beta \to \gamma}{\mathscr{R} \Vdash \alpha \mathrel{\mid\!\sim} \gamma}$

(CM) $\dfrac{\mathscr{R} \Vdash \alpha \mathrel{\mid\!\sim} \beta,\ \mathscr{R} \Vdash \alpha \mathrel{\mid\!\sim} \gamma}{\mathscr{R} \Vdash \alpha \wedge \beta \mathrel{\mid\!\sim} \gamma}$

(RM) $\dfrac{\mathscr{R} \Vdash \alpha \mathrel{\mid\!\sim} \beta,\ \mathscr{R} \not\Vdash \alpha \mathrel{\mid\!\sim} \neg \gamma}{\mathscr{R} \Vdash \alpha \wedge \gamma \mathrel{\mid\!\sim} \beta}$

As in Section 2, given a ranked model $\mathscr{R}$, a pair (α, β) is in the consequence relation defined by $\mathscr{R}$ (denoted as $\alpha \mathrel{\mid\!\sim}_\mathscr{R} \beta$) if and only if $\min_\prec \llbracket \alpha \rrbracket \subseteq \llbracket \beta \rrbracket$. In this case, however, α and β are taken to be elements of $\mathscr{L}^\bullet$ and not just of $\mathscr{L}$.

In order for us to describe the connection between rational consequence and AGM revision for PTL, we first consider the following additional property on defeasible consequence relations:

(Cons) $\mathscr{R} \not\Vdash \top \mathrel{\mid\!\sim} \bot$

It is easy to see that for a ranked model $\mathscr{R} = \langle \mathscr{V}, \prec \rangle$, $\top \mathrel{\mid\!\sim}_\mathscr{R} \bot$ holds if and only if $\mathscr{V} = \emptyset$. By insisting that Property (Cons) holds, we are restricting ourselves to ranked models in which $\mathscr{V} \neq \emptyset$, a restriction that is necessary to comply with Postulate (R3) for AGM belief revision (cf. Section 5). So, we consider only the case where the (fixed) set $\mathscr{V}$ is non-empty. (It is not hard to see that if $\mathscr{V} = \emptyset$, then $\mathscr{L}^\bullet \times \mathscr{L}^\bullet$ is the only rational consequence relation satisfying all the seven rationality properties above, that $\mathscr{R} = \langle \emptyset, \emptyset \rangle$ is the only ranked model, and that $\mathrel{\mid\!\sim}_\mathscr{R} = \mathscr{L}^\bullet \times \mathscr{L}^\bullet$.)

Intuitively, given a rational consequence relation $\mathrel{\mid\!\sim}$ and a belief revision operator $\circ$ for a given knowledge base κ, the idea is to (*i*) associate κ with all the βs such that

Propositional Typicality Logic for Extending Rational Consequence

⊤ |∼ β holds and *(ii)* to associate the consequences of κ ∘ α with all the βs such that α |∼ β holds. Such is the approach adopted by Gärdenfors and Makinson [1994] in the propositional case.

For a rational relation |∼ on $\mathscr{L}^\bullet$, let $C^{|\sim} := \{\alpha \in \mathscr{L}^\bullet \mid \top \mid\sim \alpha\}$ and let $\mathscr{K}^{|\sim}$ be the set of all logically strongest formulae (modulo $\mathscr{R}_\mathscr{V}$) to be defeasibly concluded from ⊤. That is,

$$\mathscr{K}^{|\sim} := \{\alpha \in C^{|\sim} \mid \text{for all } \beta \in C^{|\sim}, \text{ if } \models \beta \to \alpha, \text{ then } \models \alpha \to \beta\},$$

where $\models$ is understood to mean validity modulo $\mathscr{R}_\mathscr{V}$.

The following result establishes the connection between AGM revision and rational consequence for $\mathscr{L}^\bullet$. The result is closely related to that by Gärdenfors and Makinson [1994]. (In fact, parts of the proof of Proposition 7 rely heavily on results first obtained by Gärdenfors and Makinson.)

Proposition 7.

1. *Let |∼ be a rational consequence relation on $\mathscr{L}^\bullet$ also satisfying (Cons), and let $\kappa \in \mathscr{K}^{|\sim}$. Then there is a κ-faithful ranked model $\mathscr{R} \in \mathscr{R}_\mathscr{V}$ for which the AGM revision operator $\circ_\mathscr{R}$ on $\mathscr{L}^\bullet$ for κ is such that $\mathscr{R} \Vdash \alpha \mid\sim \beta$ if and only if $\mathscr{R} \Vdash (\kappa \circ_\mathscr{R} \alpha) \to \beta$.*
2. *Let κ be any element of $\mathscr{L}^\bullet$ such that $\not\models \neg \kappa$, and let ∘ be an AGM revision operator on $\mathscr{L}^\bullet$ for κ. Let $\mathscr{R} \in \mathscr{R}_\mathscr{V}^\kappa$ be the (κ-faithful) ranked model $\mathscr{R} \in \mathscr{R}_\mathscr{V}^\kappa$ such that $\circ = \circ_\mathscr{R}$.[5] Let |∼ be the defeasible consequence relation on $\mathscr{L}^\bullet$ such that $\alpha \mid\sim \beta$ if and only if $\mathscr{R} \Vdash (\kappa \circ \alpha) \to \beta$. Then |∼ is a rational consequence relation (with respect to $\mathscr{R}$) also satisfying (Cons).*

Proof. Proving 1: Consider the following translated versions of the seven rationality properties and the (Cons) property (making use of $\mathscr{R} \Vdash \alpha \mid\sim \beta$ if and only if $\mathscr{R} \Vdash (\kappa \circ_\mathscr{R} \alpha) \to \beta$).

(Ref) $\mathscr{R} \Vdash (\kappa \circ \alpha) \to \alpha$

(LLE) $\dfrac{\mathscr{R} \Vdash \alpha \leftrightarrow \beta, \ \mathscr{R} \Vdash (\kappa \circ \alpha) \to \gamma}{\mathscr{R} \Vdash (\kappa \circ \beta) \to \gamma}$

(And) $\dfrac{\mathscr{R} \Vdash (\kappa \circ \alpha) \to \beta, \ \mathscr{R} \Vdash (\kappa \circ \alpha) \to \gamma}{\mathscr{R} \Vdash (\kappa \circ \alpha) \to \beta \wedge \gamma}$

(Or) $\dfrac{\mathscr{R} \Vdash (\kappa \circ \alpha) \to \gamma, \ \mathscr{R} \Vdash (\kappa \circ \beta) \to \gamma}{\mathscr{R} \Vdash (\kappa \circ (\alpha \vee \beta)) \to \gamma}$

(RW) $\dfrac{\mathscr{R} \Vdash (\kappa \circ \alpha) \to \beta, \ \mathscr{R} \Vdash \beta \to \gamma}{\mathscr{R} \Vdash (\kappa \circ \alpha) \to \gamma}$

---

[5] Remember that $\mathscr{R}$ exists by Theorem 4.

$$\text{(CM)} \quad \frac{\mathscr{R} \Vdash (\kappa \circ \alpha) \to \beta, \ \mathscr{R} \Vdash (\kappa \circ \alpha) \to \gamma}{\mathscr{R} \Vdash (\kappa \circ (\alpha \wedge \beta)) \to \gamma}$$

$$\text{(RM)} \quad \frac{\mathscr{R} \Vdash (\kappa \circ \alpha) \to \beta, \ \mathscr{R} \nVdash (\kappa \circ \alpha) \to \neg\gamma}{\mathscr{R} \Vdash (\kappa \circ (\alpha \wedge \gamma)) \to \beta}$$

$$\text{(Cons)} \quad \mathscr{R} \nVdash (\kappa \circ \top) \to \bot$$

We show below that any $\mathscr{R} \in \mathscr{R}_\gamma^\kappa$ satisfying these properties, also satisfies the properties (R1•)–(R6•). From Theorem 4 the result then follows.

(R1•) follows directly from (Ref), and (R4•) follows from (LLE). For (R5•), observe that it is equivalent to the following property

$$\text{(Con)} \quad \frac{\mathscr{R} \Vdash (\kappa \circ (\alpha \wedge \beta)) \to \gamma}{\mathscr{R} \Vdash (\kappa \circ \alpha) \wedge \beta \to \gamma}$$

which, in turn, is equivalent to (Or). For Postulate (R6•), observe that it is equivalent to (RM).

For (R3•), it can be shown that $\mathrel{\mid\!\sim}$ satisfies the following property:

$$\text{(WRM)} \quad \frac{\mathscr{R} \nVdash (\kappa \circ \top) \to \neg\alpha, \ \mathscr{R} \Vdash (\kappa \circ \top) \to (\alpha \to \beta)}{\mathscr{R} \Vdash (\kappa \circ \alpha) \to \beta}$$

and that (R3•) follows from (WRM).

For (R2•), it can be shown that $\mathrel{\mid\!\sim}$ satisfies the following property:

$$\text{(WC)} \quad \frac{\mathscr{R} \Vdash (\kappa \circ \alpha) \to \beta}{\mathscr{R} \Vdash (\kappa \circ \top) \to (\alpha \to \beta)}$$

and that (R2•) follows from (WRC).

Proving 2: We need to show that $\mathrel{\mid\!\sim}$ satisfies the seven rationality properties, plus (Cons). This can be done by replacing each of these properties with the translated versions in Part 1 of this proof (involving $\circ_\mathscr{R}$), and checking whether the revision operator $\circ_\mathscr{R}$ satisfies the translated properties.

Showing that (Ref), (LLE) and (RW) hold is easy. To show that (Cons) holds observe that, since $\mathscr{R}$ is a κ-faithful ranked model, it follows that $\min_\prec [\![\top]\!] \neq \emptyset$. For (And), observe that from the fact that $\min_\prec [\![\alpha]\!] \subseteq [\![\beta]\!]$ and $\min_\prec [\![\alpha]\!] \subseteq [\![\gamma]\!]$ it follows that $\min_\prec [\![\alpha]\!] \subseteq [\![\beta]\!] \cap [\![\gamma]\!]$. For (Or), observe that from the fact that $\min_\prec [\![\alpha]\!] \subseteq [\![\gamma]\!]$ and $\min_\prec [\![\beta]\!] \subseteq [\![\gamma]\!]$ it follows that $\min_\prec [\![\alpha \vee \beta]\!] \subseteq [\![\gamma]\!]$ (since $\min_\prec [\![\alpha \vee \beta]\!] \subseteq \min_\prec [\![\alpha]\!] \cup \min_\prec [\![\beta]\!]$). For (CM), suppose that $\min_\prec [\![\alpha]\!] \subseteq [\![\beta]\!]$ and $\min_\prec [\![\alpha]\!] \subseteq [\![\gamma]\!]$ and pick a $v \in \min_\prec [\![\alpha \wedge \beta]\!]$. Then it must be the case that $v \in \min_\prec [\![\alpha]\!]$ (if there were a $w \in \min_\prec [\![\alpha]\!]$ such that $w \prec v$, then it would be the case that $w \in \min_\prec [\![\alpha \wedge \beta]\!]$), from which it follows that $v \in \min_\prec [\![\gamma]\!]$. For (RM), suppose that $\min_\prec [\![\alpha]\!] \subseteq [\![\beta]\!]$ and $\min_\prec [\![\alpha]\!] \nsubseteq [\![\neg\gamma]\!]$ and pick a $v \in \min_\prec [\![\alpha \wedge \gamma]\!]$. Then it must be the case that

$v \in \min_\prec \llbracket \alpha \rrbracket$ (from $\min_\prec \llbracket \alpha \rrbracket \not\subseteq \llbracket \neg \gamma \rrbracket$ we know there is at least one $w \in \min_\prec \llbracket \alpha \rrbracket$ such that $w \in \llbracket \gamma \rrbracket$), from which it follows that $v \in \min_\prec \llbracket \beta \rrbracket$.

Proposition 7 then allows us to obtain a representation result for rational consequence relations on $\mathscr{L}^\bullet$, as the following corollary shows.

Corollary 2. *For every ranked model $\mathscr{R}$, $\mathrel{\mid\kern-0.4em\sim}_\mathscr{R}$ is a rational consequence relation on $\mathscr{L}^\bullet$. Conversely, for every rational consequence relation $\mathrel{\mid\kern-0.4em\sim}$ on $\mathscr{L}^\bullet$ there exists a ranked model $\mathscr{R}$ such that $\mathrel{\mid\kern-0.4em\sim}_\mathscr{R} = \mathrel{\mid\kern-0.4em\sim}$.*

Proof. Pick a ranked model $\mathscr{R} = \langle \mathscr{V}, \prec \rangle$. If $\mathscr{V} \neq \emptyset$, then $\mathscr{R} \in \mathscr{R}_\mathscr{V}$. Let κ be such that $\min_\prec \llbracket \top \rrbracket = \llbracket \kappa \rrbracket$. So $\mathscr{R} \in \mathscr{R}_\mathscr{V}^\kappa$ and by Part 2 of Proposition 7 it follows that $\mathrel{\mid\kern-0.4em\sim}_\mathscr{R}$ is a rational consequence relation on $\mathscr{L}^\bullet$. If $\mathscr{V} = \emptyset$, then $\mathrel{\mid\kern-0.4em\sim}_\mathscr{R} = \mathscr{L}^\bullet \times \mathscr{L}^\bullet$. And it is easy to see that $\mathrel{\mid\kern-0.4em\sim}$ is then a rational consequence relation.

Conversely, let $\mathrel{\mid\kern-0.4em\sim}$ be a rational consequence relation on $\mathscr{L}^\bullet$ and let $\kappa \in \mathscr{K}^\mathrel{\mid\kern-0.4em\sim}$. If $\top \mathrel{\mid\kern-0.4em\not\sim} \bot$ is the case, then, from Part 1 of Proposition 7 it follows that there is an $\mathscr{R} \in \mathscr{R}_\mathscr{V}^\kappa$ such that $\mathrel{\mid\kern-0.4em\sim}_\mathscr{R} = \mathrel{\mid\kern-0.4em\sim}$. And since $\mathscr{R}_\mathscr{V}^\kappa \subseteq \mathscr{R}_\mathscr{V}$, the result follows. If $\top \mathrel{\mid\kern-0.4em\sim} \bot$ holds, then by (RW), (LLE) and (Or), $\mathrel{\mid\kern-0.4em\sim} = \mathscr{L}^\bullet \times \mathscr{L}^\bullet$. And it is easy to see that for the ranked model $\mathscr{R} = \langle \emptyset, \emptyset \rangle$, $\mathrel{\mid\kern-0.4em\sim}_\mathscr{R} = \mathrel{\mid\kern-0.4em\sim}$.

7 Entailment for PTL

In this section we focus on what is perhaps the central question concerning PTL from the perspective of knowledge representation and reasoning: What does it mean for a PTL formula to be *entailed* by a (finite) knowledge base $\mathscr{K}$?

Formally, we view an entailment relation as a binary relation $\models_*$ from the power set of the language under consideration (in this case $\mathscr{L}^\bullet$) to the language itself. Its associated *consequence* relation is defined as:

$$Cn_*(\mathscr{K}) \equiv_{\mathrm{def}} \{\alpha \mid \mathscr{K} \models_* \alpha\}$$

Before looking at specific candidates, we propose some desired properties for such an entailment relation. The obvious place to start is to consider the properties for Tarskian consequence relations [Tarski, 1941].

(Inclusion) $\mathscr{K} \subseteq Cn_*(\mathscr{K})$
(Idempotency) $Cn_*(\mathscr{K}) = Cn_*(Cn_*(\mathscr{K}))$
(Monotonicity)   If $\mathscr{K}_1 \subseteq \mathscr{K}_2$, then $Cn_*(\mathscr{K}_1) \subseteq Cn_*(\mathscr{K}_2)$

Inclusion and Idempotency are both properties we want to have satisfied, but Monotonicity is not. To see why not, it is enough to refer to the classic example from the Introduction: Let $\mathscr{K}_1 = \{\mathrm{p} \to \mathrm{b}, \bullet\mathrm{b} \to \mathrm{f}\}$ ("penguins are birds" and "birds typically fly"), and let $\mathscr{K}_2 = \mathscr{K}_1 \cup \{\bullet\mathrm{p} \to \neg\mathrm{f}\}$ (add to $\mathscr{K}_1$ that "penguins typically do not fly"). Given this, we want $\bullet\mathrm{p} \to \mathrm{f} \in Cn_*(\mathscr{K}_1)$ ("penguins typically fly" as a consequence of $\mathscr{K}_1$), but we want $\bullet\mathrm{p} \to \mathrm{f} \notin Cn_*(\mathscr{K}_2)$ ("penguins typically fly" *not* as a consequence of $\mathscr{K}_2$), thereby invalidating Monotonicity.

In addition to Inclusion and Idempotency we require $\models_*$ to behave classically when presented with propositional information only (below $\models$ denotes classical entailment):

(Classic) If $\mathcal{K} \subseteq \mathcal{L}$, then for every $\alpha \in \mathcal{L}$, $\mathcal{K} \models_* \alpha$ if and only if $\mathcal{K} \models \alpha$

Therefore, we also require that the classical consequences of a knowledge base expressed in $\mathcal{L}^\bullet$ be classically closed — below the $Cn(\cdot)$ operator refers to classical consequence of the propositional language $\mathcal{L}$:

(Classic Closure) $Cn_*(\mathcal{K}) \cap \mathcal{L} = Cn(Cn_*(\mathcal{K}) \cap \mathcal{L})$

We now consider four obvious candidates for the notion of entailment in PTL:

$\mathcal{K} \models_1 \alpha$ if and only if for all $\mathscr{R}$ such that $\mathscr{R} \Vdash \mathcal{K}$, $\mathscr{R} \Vdash \alpha$('Global') (1)

$\mathcal{K} \models_2 \alpha$ if and only if for all $\mathscr{R}$, $[\![\bigwedge \mathcal{K}]\!] \subseteq [\![\alpha]\!]$('Local') (2)

$\mathcal{K} \models_3 \alpha$ if and only if for all $\mathscr{R}$ such that $\mathscr{R} \Vdash \bullet \bigwedge \mathcal{K}$, $\mathscr{R} \Vdash \alpha$('Classical') (3)

$\mathcal{K} \models_4 \alpha$ if and only if for all $\mathscr{R}$, $[\![\bullet \bigwedge \mathcal{K}]\!] \subseteq [\![\alpha]\!]$('Supra-Local') (4)

In what follows we shall analyze each of these candidates against the aforementioned properties.

The entailment relation $\models_1$ in (1) corresponds to the standard Tarskian notion of entailment [Tarski, 1941] applied to the semantics of PTL. It also resembles the notion of *global entailment* in modal logic [Blackburn et al., 2001]. It is not difficult to see that $\models_1$ satisfies Inclusion, Idempotency, Classic, and Classic Closure, properties that we want. Note, though, that $\models_1$ also satisfies Monotonicity, a property that we do *not* want.

Entailment relation $\models_2$ in (2) is the 'local' version of $\models_1$ in the modal sense and, as such, is stronger than $\models_1$. $\models_1$ does not imply $\models_2$, as shown by the following example: we have $\{\bullet(\alpha \wedge \beta)\} \models_1 \bullet \alpha$ but $\{\bullet(\alpha \wedge \beta)\} \not\models_2 \bullet \alpha$. It is not hard to see that $\models_2$ is also a Tarskian consequence relation and, as such, it satisfies the Monotonicity property as well.

The entailment relation $\models_3$ in (3) above boils down to a version of classical entailment in that only ranked models with an empty preference relation take part in its definition — remember Proposition 1. It is easy to see that $\models_3$ is monotonic and therefore insufficient according to our desiderata.

Finally, note that the option represented by $\models_4$ in (4) is weaker than $\models_2$ in (2), and therefore we have $\models_2 \subseteq \models_4$. Moreover, it is not hard to see that $\models_4$ is *non-monotonic*, which puts it in a good position as an appropriate candidate for PTL-entailment. Unfortunately, there is an argument which eliminates $\models_4$ from contention as a viable form of entailment. We explore this in more detail in what follows.

Definition 8. Let $\mathcal{K} \subseteq \mathcal{L}^\bullet$ and let $\models_n$, $1 \leq n \leq 4$, be one of the entailment relations above. Let

$$\mid\!\sim_n^{\mathcal{K}} := \{(\alpha, \beta) \mid \alpha, \beta \in \mathcal{L},\ \mathcal{K} \models_n \bullet\alpha \to \beta\}$$

Propositional Typicality Logic for Extending Rational Consequence 143

We say $\mid\!\sim_n^{\mathscr{K}}$ is the (propositional) consequence relation generated by $\mathscr{K}$ and $\models_n$.

Proposition 8. *There is $\mathscr{K} \subseteq \mathscr{L}^\bullet$ for which $\mid\!\sim_4^{\mathscr{K}}$ is not a preferential consequence relation.*

Proof. Let $\mathscr{K} = \{\bullet b \to f, \bullet b \to w\}$ ("typical birds fly" and "typical birds have wings"). Let $\mathscr{R} = \langle \mathscr{V}, \prec \rangle$, where $\mathscr{V} = \{110, 100\}$ and $\prec = \{(100, 110)\}$ (100 is more preferred than 110). Then $[\![\bullet((\bullet b \to f) \land (\bullet b \to w))]\!] = 110$, but $110 \notin [\![\bullet(b \land w) \to f]\!]$. So we have both $b \mid\!\sim_4^{\mathscr{K}} f$ and $b \mid\!\sim_4^{\mathscr{K}} w$, but $b \land w \not\mid\!\sim_4^{\mathscr{K}} f$. Hence $\mid\!\sim_4^{\mathscr{K}}$ does not satisfy cautious monotonicity (CM) and is therefore not a preferential consequence relation.

This result rules $\models_4$ out as an appropriate notion of entailment for PTL, as the propositional defeasible consequence relation it induces does not comply with the basic KLM properties from Section 2. The proposition above also raises the obvious question on how the other entailment relations behave under a similar scrutiny.

Proposition 9.

1. *For every $\mathscr{K} \subseteq \mathscr{L}^\bullet$, $\mid\!\sim_1^{\mathscr{K}}$ is a preferential consequence relation;*
2. *There is $\mathscr{K} \subseteq \mathscr{L}^\bullet$ for which $\mid\!\sim_2^{\mathscr{K}}$ is not a preferential consequence relation;*
3. *For every $\mathscr{K} \subseteq \mathscr{L}^\bullet$, $\mid\!\sim_3^{\mathscr{K}}$ satisfies Monotonicity:*

$$\frac{\alpha \mid\!\sim \beta}{\alpha \land \gamma \mid\!\sim \beta}$$

Proof. Proving Part 1: Let $\mathscr{K} \subseteq \mathscr{L}^\bullet$ and let $\mathscr{R}$ be a ranked model. Then, given $\alpha \in \mathscr{L}^\bullet$, we clearly have $\mathscr{R} \Vdash \bullet \alpha \to \alpha$, so (Ref) holds. Let $\alpha, \beta, \gamma \in \mathscr{L}$ and suppose that $\mathscr{R} \Vdash \bullet \alpha \to \beta$ and $\models \beta \to \gamma$ ($\models$ here refers to classical validity). Then $\mathscr{R} \Vdash \bullet \alpha \to \gamma$ and then (RW) holds. Now suppose that $\mathscr{R} \Vdash \bullet \alpha \to \gamma$ and $\models \alpha \leftrightarrow \beta$. Then $\bullet \beta \to \gamma$ so (LLE) also holds. Suppose $\mathscr{R} \Vdash \bullet \alpha \to \beta$ and $\mathscr{R} \Vdash \bullet \alpha \to \gamma$. Then clearly $\mathscr{R} \Vdash \bullet \alpha \to \beta \land \gamma$ and then (And) holds. Suppose $\mathscr{R} \Vdash \bullet \alpha \to \gamma$ and $\mathscr{R} \Vdash \bullet \beta \to \gamma$. It is not hard to check that $\mathscr{R} \Vdash \bullet(\alpha \lor \beta) \to \gamma$, so (Or) also holds. Now suppose $\mathscr{R} \Vdash \bullet \alpha \to \beta$ and $\mathscr{R} \Vdash \bullet \alpha \to \gamma$. Pick any $w \in \min_\prec [\![\alpha \land \gamma]\!]$, i.e., w is a "best $\alpha \land \gamma$-world". If $w \in \min_\prec [\![\alpha]\!]$, then $w \in [\![\beta]\!]$ (since $\mathscr{R} \Vdash \bullet \alpha \to \beta$) and so $w \in [\![\bullet(\alpha \land \gamma) \to \beta]\!]$. If $w \notin \min_\prec [\![\alpha]\!]$, then there is w' such that $w' \prec w$ and $w' \in \min_\prec [\![\alpha]\!]$. But then $w' \in [\![\neg \gamma]\!]$, since $w \in \min_\prec [\![\alpha \land \gamma]\!]$. So $\mathscr{R} \not\Vdash \bullet \alpha \to \gamma$, which is a contradiction. Hence $\mathscr{R} \Vdash \bullet(\alpha \land \gamma) \to \beta$, and therefore (CM) holds.

Proving Part 2: Let $\mathscr{K}$ and $\mathscr{R}$ be as in the proof of Proposition 8. Then $110 \in [\![\bigwedge \mathscr{K}]\!]$ (since $110 \notin \min_\prec [\![b]\!]$). However, $110 \notin [\![\bullet(b \land w) \to f]\!]$. From this it is easy to see that $\mid\!\sim_2^{\mathscr{K}}$ does not satisfy (CM).

Proving Part 3: Let $\mathscr{K} \subseteq \mathscr{L}^\bullet$ and pick any $\mathscr{R} = \langle \mathscr{V}, \prec \rangle$ such that $\mathscr{R} \Vdash \bullet \bigwedge \mathscr{K}$. Then, by Proposition 1, we have $\min_\prec [\![\bigwedge \mathscr{K}]\!] = [\![\bigwedge \mathscr{K}]\!]$ and $\prec = \emptyset$. Now suppose $\mathscr{R} \Vdash \bullet \alpha \to \beta$. This means that for every $w \in \mathscr{V}$, if $w \in [\![\alpha]\!]$, then $w \in [\![\beta]\!]$. Now pick any $w \in [\![\alpha \land \gamma]\!]$. Since $w \in [\![\alpha]\!]$, we also have $w \in [\![\beta]\!]$. The same argument holds for any $w \in \min_\prec [\![\alpha \land \gamma]\!]$ (since $\prec = \emptyset$), and therefore we have $\mathscr{R} \Vdash \bullet(\alpha \land \gamma) \to \beta$.

From the results above we can conclude that even though none of $\models_i$, $1 \leq i \leq 4$, is an appropriate notion of entailment in the context of PTL, $\models_1$ turns out to be the best of all options in that it delivers a consequence relation that is preferential. There is, however, an additional argument against the use of $\models_1$ as well, one that is based on an adaptation of a result obtained by Lehmann and Magidor [1992] in the propositional case. To make the argument, we first present a result showing that all formulae of $\mathscr{L}^\bullet$ can be rewritten as statements of rational consequence:

Lemma 1. *For every $\mathscr{R}$ and $\alpha \in \mathscr{L}^\bullet$, $\mathscr{R} \Vdash \alpha$ if and only if $\mathscr{R} \Vdash \bullet \neg \alpha \to \bot$ if and only if $\neg \alpha \mathrel{\vert\!\sim}_\mathscr{R} \bot$. Conversely for every $\mathscr{R}$ and $\alpha, \beta \in \mathscr{L}^\bullet$, $\alpha \mathrel{\vert\!\sim}_\mathscr{R} \beta$ if and only if $\mathscr{R} \Vdash \bullet \alpha \to \beta$.*

Proof. Let $\mathscr{R} = \langle \mathscr{V}, \prec \rangle$ and $\alpha \in \mathscr{L}^\bullet$. $\mathscr{R} \Vdash \alpha$ if and only if $[\![\alpha]\!] = \mathscr{V}$ if and only if $[\![\neg\alpha]\!] = \emptyset$ if and only if (i) $[\![\bullet\neg\alpha]\!] = \emptyset$ if and only if $[\![\neg\bullet\neg\alpha]\!] = \mathscr{V}$ if and only if $[\![\bullet\alpha \to \bot]\!] = \mathscr{V}$ if and only if $\mathscr{R} \Vdash \bullet\neg\alpha \to \bot$. But (i) is the case if and only if $\min_\prec [\![\neg\alpha]\!] = \emptyset$ if and only if $\min_\prec [\![\neg\alpha]\!] \subseteq [\![\bot]\!]$ if and only if $\neg\alpha \mathrel{\vert\!\sim}_\mathscr{R} \bot$.

The proof of the converse part is analogous to that of Proposition 2.

We can therefore think of $\mathscr{L}^\bullet$ as a language for expressing defeasible consequence on $\mathscr{L}^\bullet$ with $\mathrel{\vert\!\sim}$ viewed as the only main connective. More precisely, let $\mathscr{L}^\bullet_{\vert\!\sim} := \{\alpha \mathrel{\vert\!\sim} \beta \mid \alpha, \beta \in \mathscr{L}^\bullet\}$, and for any ranked model $\mathscr{R}$, let $\mathscr{R} \Vdash \alpha \mathrel{\vert\!\sim} \beta$ if and only if $\alpha \mathrel{\vert\!\sim}_\mathscr{R} \beta$. The next result shows that the languages $\mathscr{L}^\bullet$ and $\mathscr{L}^\bullet_{\vert\!\sim}$ are equally expressive.

Proposition 10. *For every $\mathscr{R}$ and $\alpha \mathrel{\vert\!\sim} \beta \in \mathscr{L}^\bullet_{\vert\!\sim}$, $\mathscr{R} \Vdash \alpha \mathrel{\vert\!\sim} \beta$ if and only if $\mathscr{R} \Vdash \bullet\alpha \to \beta$. Conversely, for every $\mathscr{R}$ and $\alpha \in \mathscr{L}^\bullet$, $\mathscr{R} \Vdash \alpha$ if and only if $\mathscr{R} \Vdash \neg\alpha \mathrel{\vert\!\sim} \bot$.*

Proof. Straightforward, by applying Lemma 1.

$\mathscr{L}^\bullet_{\vert\!\sim}$ is similar to the language for conditional knowledge bases studied by Lehmann and Magidor [1992], but with the propositional component replaced by $\mathscr{L}^\bullet$ (i.e., $\mathrel{\vert\!\sim} \subseteq \mathscr{L}^\bullet \times \mathscr{L}^\bullet$).

Based on this, we restate entailment in terms of the language $\mathscr{L}^\bullet_{\vert\!\sim}$, and propose an additional property that any appropriate notion of entailment ought to satisfy. Let $\mathscr{K}$ be a (finite) subset of $\mathscr{L}^\bullet_{\vert\!\sim}$, let $\models_*$ be a (potential) entailment relation from $\mathscr{P}(\mathscr{L}^\bullet_{\vert\!\sim})$ to $\mathscr{L}^\bullet_{\vert\!\sim}$, and let $\mathrel{\vert\!\sim}^\mathscr{K}_*$ be a defeasible consequence relation on $\mathscr{L}^\bullet$ obtained from $\models_*$ as follows: $\alpha \mathrel{\vert\!\sim}^\mathscr{K}_* \beta$ if and only if $\mathscr{K} \models_* \alpha \mathrel{\vert\!\sim} \beta$.

(Rationality) For every finite $\mathscr{K} \subseteq \mathscr{L}^\bullet_{\vert\!\sim}$, the consequence relation $\mathrel{\vert\!\sim}^\mathscr{K}_*$ obtained from $\models_*$ should be *rational*.

Rationality is essentially the property for the entailment of propositional conditional knowledge bases proposed by Lehmann and Magidor [1992], but applied to the language of $\mathscr{L}^\bullet_{\vert\!\sim}$ (cf. Section 2). Based on their results, it follows that the consequence relation $\mathrel{\vert\!\sim}^\mathscr{K}_1$ obtained from $\models_1$ does not satisfy Rationality. In fact, analogous to one of their results [Lehmann and Magidor, 1992, Section 4.2], we have the following one.

Propositional Typicality Logic for Extending Rational Consequence

Proposition 11. *For finite $\mathscr{K} \subseteq \mathscr{L}^\bullet_{\mathrel{\mathpalette\raisebox{-1pt}{\sim}}}$, let $\mathrel{\mathpalette\raisebox{-1pt}{\sim}}^{\mathscr{K}} = \{(\alpha, \beta) \mid \alpha \mathrel{\mathpalette\raisebox{-1pt}{\sim}} \beta \in \mathscr{K}\}$, and let $\mathrel{\mathpalette\raisebox{-1pt}{\sim}}^P$ be the intersection of all preferential consequence relations on $\mathscr{L}^\bullet$ containing $\mathrel{\mathpalette\raisebox{-1pt}{\sim}}^{\mathscr{K}}$.[6] For the consequence relation $\mathrel{\mathpalette\raisebox{-1pt}{\sim}}^{\mathscr{K}}_1$ obtained from $\models_1$, it follows that $\mathrel{\mathpalette\raisebox{-1pt}{\sim}}^{\mathscr{K}}_1 = \mathrel{\mathpalette\raisebox{-1pt}{\sim}}^P$ is a preferential consequence relation, but not necessarily a rational consequence relation.*

Proof. The proof is analogous to that given by Lehmann and Magidor [1992, Section 4.2] in the propositional case.

Since $\mathscr{L}^\bullet$ and $\mathscr{L}^\bullet_{\mathrel{\mathpalette\raisebox{-1pt}{\sim}}}$ are equally expressive (cf. Proposition 10), Proposition 11 provides additional evidence that $\models_1$ is not an appropriate form of entailment.

Having shown that none of the obvious notions of entailment above is an appropriate form of entailment for PTL, we now turn our attention to an alternative proposal. It is the notion of the *rational closure* of a conditional knowledge base, proposed by Lehmann and Magidor [1992] for the propositional case, but here applied to the language of $\mathscr{L}^\bullet_{\mathrel{\mathpalette\raisebox{-1pt}{\sim}}}$.

Definition 9. *Let $\mathrel{\mathpalette\raisebox{-1pt}{\sim}}_0$ and $\mathrel{\mathpalette\raisebox{-1pt}{\sim}}_1$ be rational consequence relations. $\mathrel{\mathpalette\raisebox{-1pt}{\sim}}_0$ is preferable to $\mathrel{\mathpalette\raisebox{-1pt}{\sim}}_1$ (written $\mathrel{\mathpalette\raisebox{-1pt}{\sim}}_0 \ll \mathrel{\mathpalette\raisebox{-1pt}{\sim}}_1$) if and only if*

- *there is an $\alpha \mathrel{\mathpalette\raisebox{-1pt}{\sim}} \beta \in \mathrel{\mathpalette\raisebox{-1pt}{\sim}}_1 \setminus \mathrel{\mathpalette\raisebox{-1pt}{\sim}}_0$ such that for all γ such that $\gamma \vee \alpha \mathrel{\mathpalette\raisebox{-1pt}{\sim}}_0 \neg \alpha$ and for all δ such that $\gamma \mathrel{\mathpalette\raisebox{-1pt}{\sim}}_0 \delta$, we also have $\gamma \mathrel{\mathpalette\raisebox{-1pt}{\sim}}_1 \delta$;*
- *for every $\gamma, \delta \in \mathscr{L}$, if $\gamma \mathrel{\mathpalette\raisebox{-1pt}{\sim}} \delta$ is in $\mathrel{\mathpalette\raisebox{-1pt}{\sim}}_0 \setminus \mathrel{\mathpalette\raisebox{-1pt}{\sim}}_1$, then there is an assertion $\rho \mathrel{\mathpalette\raisebox{-1pt}{\sim}} \nu$ in $\mathrel{\mathpalette\raisebox{-1pt}{\sim}}_1 \setminus \mathrel{\mathpalette\raisebox{-1pt}{\sim}}_0$ such that $\rho \vee \gamma \mathrel{\mathpalette\raisebox{-1pt}{\sim}}_1 \neg \gamma$.*

The motivation for $\ll$ here is essentially that for the same ordering for the propositional case provided by Lehmann and Magidor [1992]. Given $\mathscr{K} \subseteq \mathscr{L}^\bullet_{\mathrel{\mathpalette\raisebox{-1pt}{\sim}}}$, the idea is now to define the rational closure as the most preferred (with respect to $\ll$) of all those rational consequence relations which include $\mathscr{K}$.

Lemma 2. *Let $\mathscr{K} \subseteq \mathscr{L}^\bullet_{\mathrel{\mathpalette\raisebox{-1pt}{\sim}}}$ be finite, and let $\mathrel{\mathpalette\raisebox{-1pt}{\sim}}^{\mathscr{K}} := \{(\alpha, \beta) \mid \alpha \mathrel{\mathpalette\raisebox{-1pt}{\sim}} \beta \in \mathscr{K}\}$. Then there is a unique rational consequence relation containing $\mathrel{\mathpalette\raisebox{-1pt}{\sim}}^{\mathscr{K}}$ which is preferable (with respect to $\ll$) to all other rational consequence relations containing $\mathrel{\mathpalette\raisebox{-1pt}{\sim}}^{\mathscr{K}}$.*

This allows us to define the rational closure $\models_{rc}$ of a knowledge base on $\mathscr{L}^\bullet_{\mathrel{\mathpalette\raisebox{-1pt}{\sim}}}$.

Definition 10. *For finite $\mathscr{K} \subseteq \mathscr{L}^\bullet_{\mathrel{\mathpalette\raisebox{-1pt}{\sim}}}$, let $\mathrel{\mathpalette\raisebox{-1pt}{\sim}}^{\mathscr{K}} := \{(\alpha, \beta) \mid \alpha \mathrel{\mathpalette\raisebox{-1pt}{\sim}} \beta \in \mathscr{K}\}$, and let $\mathrel{\mathpalette\raisebox{-1pt}{\sim}}^{\mathscr{K}}_{rc}$ be the (unique) rational consequence relation containing $\mathrel{\mathpalette\raisebox{-1pt}{\sim}}^{\mathscr{K}}$ which is preferable (with respect to $\ll$) to all other rational consequence relations containing $\mathrel{\mathpalette\raisebox{-1pt}{\sim}}^{\mathscr{K}}$. Then $\alpha \mathrel{\mathpalette\raisebox{-1pt}{\sim}} \beta$ is in the rational closure of $\mathscr{K}$ (written as $\mathscr{K} \models_{rc} \alpha \mathrel{\mathpalette\raisebox{-1pt}{\sim}} \beta$) if and only if $\alpha \mathrel{\mathpalette\raisebox{-1pt}{\sim}}^{\mathscr{K}}_{rc} \beta$.*

Definition 10 gives us a notion of rational closure for $\mathscr{L}^\bullet_{\mathrel{\mathpalette\raisebox{-1pt}{\sim}}}$. Since $\mathscr{L}^\bullet$ and $\mathscr{L}^\bullet_{\mathrel{\mathpalette\raisebox{-1pt}{\sim}}}$ are equally expressive (remember Proposition 10), we can use Definition 10 to define rational closure for $\mathscr{L}^\bullet$ as well:

---

[6] Recall that a preferential consequence relation is one satisfying the first six properties discussed in Section 2.

Definition 11. Let $\mathcal{K} \subseteq \mathcal{L}^\bullet$, $\alpha \in \mathcal{L}^\bullet$, and let $\mathcal{K}^\vdash := \{\neg \beta \mathrel{\mid\!\sim} \bot \mid \beta \in \mathcal{K}\}$. Then α is in the rational closure of $\mathcal{K}$ (written as $\mathcal{K} \models_{rc} \alpha$) if and only if $\neg \alpha \mathrel{\mid\!\sim} \bot$ is in the rational closure of $\mathcal{K}^\vdash$.

It is not hard to show that rational closure satisfies Inclusion, Idempotency, Classic, Classic Closure, and Rationality, but not Monotonicity. It is therefore a reasonable candidate for entailment for PTL.

We conclude this section with an outlook on another proposal for entailment for $\mathcal{L}^\bullet$ based on a semantic construction. It is inspired by a proposal by Giordano et al. [2012]. The idea is to define a partial order on a certain subclass of ranked models satisfying a knowledge base $\mathcal{K} \subseteq \mathcal{L}^\bullet$, with models lower down in the ordering being viewed as more 'conservative', in the sense that one can draw fewer conclusions from them, and therefore being more preferred.

For $\mathcal{K} \subseteq \mathcal{L}^\bullet$, let $\mathscr{V}^\mathcal{K}$ be the elements of $\mathscr{U}$ 'permitted' by $\mathcal{K}$:

$$\mathscr{V}^\mathcal{K} := \{v \mid v \in \mathscr{V} \text{ for some } \mathscr{R} = \langle \mathscr{V}, \prec \rangle \text{ such that } \mathscr{R} \Vdash \mathcal{K}\}$$

Moreover, let $\mathscr{R}^\mathcal{K} := \{\mathscr{R} = \langle \mathscr{V}^\mathcal{K}, \prec \rangle \mid \mathscr{R} \Vdash \mathcal{K}\}$. Now, for any $\mathscr{R} = \langle \mathscr{V}^\mathcal{K}, \prec \rangle \in \mathscr{R}^\mathcal{K}$, let $\mathscr{V}_0^\mathscr{R} := \min_\prec \mathscr{V}^\mathcal{K}$, and for $i > 0$ let $\mathscr{V}_i^\mathscr{R} := \min_\prec \left(\mathscr{V}^\mathcal{K} \setminus (\cup_{j=0}^{j=i-1} \mathscr{V}_j^\mathscr{R}) \right)$. So $\mathscr{V}_0^\mathscr{R}$ contains the elements of $\mathscr{V}^\mathcal{K}$ lowest down with respect to $\prec$, $\mathscr{V}_1^\mathscr{R}$ contains the elements of $\mathscr{V}^\mathcal{K}$ just above $\mathscr{V}_0^\mathscr{R}$ with respect to $\prec$, etc. Next, for every $v \in \mathscr{V}^\mathcal{K}$ we define the *height* of v in $\mathscr{R}$ as $h^\mathscr{R}(v) = i$ if and only if $v \in \mathscr{V}_i^\mathscr{R}$. And based on that, we define the partial order $\preceq$ on $\mathscr{R}^\mathcal{K}$ as follows: $\mathscr{R}_1 \preceq \mathscr{R}_2$ if and only if for every $v \in \mathscr{V}^\mathcal{K}$, $h^{\mathscr{R}_1}(v) \leq h^{\mathscr{R}_2}(v)$. From this we get the following result which is a special case of Theorem 2 in the recent paper by Giordano et al. [2012].

Proposition 12 (Giordano et al. [2012]). *For every $\mathcal{K} \subseteq \mathcal{L}^\bullet$, the partial order $\preceq$ on the elements of $\mathscr{R}^\mathcal{K}$ has a unique minimum element.*

This allows us to provide a definition for the notion of *minimum entailment* of a PTL knowledge base.

Definition 12. Let $\mathcal{K} \subseteq \mathcal{L}^\bullet$, $\alpha \in \mathcal{L}^\bullet$, and $\mathscr{R}^\mathcal{K}$ be the (unique) minimum element of $\mathscr{R}^\mathcal{K}$ with respect to the partial order $\preceq$ on $\mathscr{R}^\mathcal{K}$. Then α is in the minimum entailment of $\mathcal{K}$ ($\mathcal{K} \models_{\min} \alpha$) if and only if $\mathscr{R}^\mathcal{K} \Vdash \alpha$.

It can be shown that minimum entailment as defined above satisfies Inclusion, Idempotency, Classic, Classic Closure, and Rationality, but not Monotonicity. As for rational closure, it is a reasonable candidate for entailment for $\mathcal{L}^\bullet$. In fact, the connection between rational closure and minimal entailment may even be closer than that. There is strong evidence to support the conjecture that they actually coincide, but this remains to be investigated.

8 Discussion and Related Work

To the best of our knowledge, the first attempt to formalize a notion of typicality in the context of defeasible reasoning was that by Delgrande [1987]. Given the relationship between our constructions and those by Kraus and colleagues, most of the remarks in the comparison made by Lehmann and Magidor [1992, Section 3.7] are applicable in comparing Delgrande's approach to ours and therefore we do not repeat them here.

Crocco and Lamarre [1992] as well as Boutilier [1994] have explored the links between defeasible consequence relations and notions of normality similar to the one we investigate here. In particular, Boutilier defines a family of conditional logics of normality in which a statement of the form "if α, then normally β" is formalized via a binary modality $\Rightarrow$ as a conditional $\alpha \Rightarrow \beta$. Here we achieve the same with a unary operator.

Roughly speaking, Boutilier's semantic intuition is the same as that of KLM (and therefore same as ours). The main difference is that Boutilier defines a conditional connective $\Rightarrow$ in the language, whereas Kraus et al. define $\mid\!\sim$ at a meta-level to the language. In this respect, Boutilier's approach is more general in that it allows for nested conditionals. If these are omitted, i.e., if one works in the 'flat' conditional logic in which $\Rightarrow$ is the main connective and no nesting is allowed, then one gets the same results for both preferential and rational entailment with both systems. So Boutilier achieves with modalities (he works in a bi-modal language) what Kraus and colleagues achieve with a (meta-level) preference order.

It turns out that in Boutilier's approach one cannot always capture the notion of "most typical α's" (but for a different reason than that given in Proposition 4). In Boutilier's modal logic, such a set (of *most* normal α-worlds) need not exist in general. This is because Boutilier drops the smoothness condition [Boutilier, 1994, p. 103] and therefore at any point in a ranked model one can have infinitely descending chains of more and more normal α-worlds. If one imposes smoothness in Boutilier's approach, which can be done by e.g. requiring the ordering determined by Boutilier's $\Box$ also to be *Noetherian*[7], one could then define his conditional $\Rightarrow$ more elegantly as follows:

$$\alpha \Rightarrow \beta \equiv_{\text{def}} \bullet\alpha \to \beta \tag{5}$$

where, in Boutilier's notation, $\bullet\alpha$ would be given by

$$\bullet\alpha \equiv_{\text{def}} \alpha \wedge \Box\neg\alpha \tag{6}$$

(Of course negated conditionals of the form $\alpha \not\Rightarrow \beta$ can then be expressed as $\neg(\bullet\alpha \to \beta)$.) In adopting smoothness and defining conditionals in this way one would expect both approaches to become equivalent modulo the underlying language — ours is propositional, whereas Boutilier's is modal. However, our statement $\bullet\alpha \to \beta$ differs from Boutilier's $\alpha \Rightarrow \beta$ in a significant way. In Boutilier's approach, a statement of

---
[7] By doing so Boutilier's framework becomes very close to Britz et al.'s [2009].

the form $\alpha \Rightarrow \beta$ is true at some world (in a ranked model) if and only if it is true at *all* worlds in that ranked model [Boutilier, 1994, p. 114]. On the other hand, it is not hard to find a ranked model in which $\bullet\alpha \to \beta$ holds at a world without being true in the whole model. This establishes Boutilier's conditional as a 'global' statement, while ours has the (more general) 'local flavor'. We can easily simulate Boutilier's notion of *acceptance* [Boutilier, 1994, p. 115] by stating $\top \to (\bullet\alpha \to \beta)$.

It is also worth mentioning that our interpretation of the conditional $\Rightarrow$ in (5) above and Boutilier's differ in another subtle way, which also relates to whether one adopts smoothness or not. In (5), $\alpha \Rightarrow \beta$ is defined as "the normal α's are β's", whereas, strictly speaking, Boutilier's definition of $\alpha \Rightarrow \beta$ reads as "there is a point from which $\alpha \to \beta$ is not violated". Such a 'frontier' for normality, implicitly referred to in Boutilier's definition of $\alpha \Rightarrow \beta$, is not as crisp as ours in the sense that the point where one draws the normality line might be too 'far away' (in the ordering) from the more and more normal α-worlds. One can definitely make a case for dropping the smoothness condition, but requiring it is a small price to pay compared to the much simpler account of typicality one gets and that we have investigated in this chapter.

In a description logic setting, Giordano et al. [2009b] also study notions of typicality. Semantically, they do so by placing an (absolute) ordering on *objects* in first-order domains in order to define versions of defeasible subsumption relations in the description logic $\mathcal{ALC}$. The authors moreover extend the language of $\mathcal{ALC}$ with an explicit typicality operator **T** of which the intended meaning is to single out instances of a concept that are deemed as 'typical'. That is, given an $\mathcal{ALC}$ concept C, $\mathbf{T}(C)$ denotes the most typical individuals having the property of being C in a particular DL interpretation.

Giordano et al.'s approach defines rational versions of the DL subsumption relation $\sqsubseteq$ satisfying the corresponding rationality properties stated in DL terms. Nevertheless, they do not provide representation results *à la* KLM and do not address the links with belief revision either. Recently Britz et al. [2011b; 2013b] have provided such representation results in the DL case. Even though here we have investigated typicality in a propositional setting, we expect that our representation result and constructions for the rational closure (as well as the links with belief revision) can be lifted to the DL case, thereby filling the mentioned gaps in Giordano et al.'s approach and shedding some light on the issues related to typicality and defeasible reasoning in more expressive logics.

Britz et al. [2009] investigate another embedding of propositional preferential reasoning in modal logic. In their setting, the modular ordering is an accessibility relation on possible worlds, axiomatized via a modal operator $\square$. Without getting into the technical details of the axiomatization of their underlying modal logic, other than mentioning that their accessibility relation is a modular ordering, it is worth noting that our typicality operator can be defined in terms of their modality as $\bullet\alpha \equiv_{\text{def}} \square\neg\alpha \wedge \alpha$ (just as the alternative formulation of Boutilier's approach in (6) above). The modal sentence $\square\neg\alpha \wedge \alpha$ says that the worlds satisfying it are α-worlds and whatever world is more preferable than these is a $\neg\alpha$-world. In other words, these are the minimal α-worlds. The general case of defining Britz et al.'s modality in terms of our typicality operator is not possible, but in a finitely generated language as we consider here, the logics become identical.

In our enriched language the preference relation is not explicit in the syntax. The meaning of the typicality operator is *informed* by the preference relation, but the latter remains nevertheless tacit. This stands in contrast to the approaches of Baltag and Smets [2008], Boutilier [1994], Britz et al. [2009] and Giordano et al. [2009a], which cast the preference relation as an (explicit) extra modality in the language. From a knowledge representation perspective, our approach has the advantage of hiding some complex aspects of the semantics from the user (e.g. a knowledge engineer who will write down sentences in an agents knowledge base).

Finally, Britz and Varzinczak [2012; 2013] investigate another, complementary aspect of defeasibility by introducing (non-standard) modal operators allowing us to talk about relative normality in accessible worlds. With their defeasible versions of modalities, namely $\boxdot$ and $\diamondsuit$, formalizing respectively the notions of *defeasible necessity* and *distinct possibility*, it becomes possible to make statements of the form "α holds in all of the normal (typical) accessible worlds", thereby capturing defeasibility of what is 'expected' in target worlds. (Note that this is different from stating something like $\Box \bullet \alpha$, which says that all accessible worlds are typical α-worlds.) Such preferential versions of modalities allow for the definition of a family of modal logics in which defeasible modes of inference such as defeasible actions, knowledge and obligations can be expressed. These can be integrated either with existing $\mid\!\sim$-based modal logics [Britz *et al.*, 2011a; Britz *et al.*, 2012] or with a modal extension of our typicality operator in striving towards a coherent theory of defeasible reasoning in more expressive languages.

9 Concluding Remarks

The main contributions of the work reported in the present chapter can be summarized as follows:

- We present the logic PTL which provides a formal account of typicality in a propositional language allowing us to refer directly and concisely to the most typical situations in which a given formula holds;
- We show that we can embed the (propositional) KLM framework within the more expressive language of PTL, and we also define rational consequence relations on the language of PTL itself;
- We establish a connection between rational consequence and belief revision, both on PTL, and
- We investigate appropriate notions of entailment for PTL and propose two candidates.

For future work we are interested in algorithms for computing the appropriate forms of entailment for PTL, specifically algorithms that can be reduced to validity checking for PTL. It follows indirectly from results by Lehmann and Magidor [1992] that this type of entailment has the same worst-case complexity of validity checking for PTL. Given the aforementioned links with modal logic, we know that this is at least a PSPACE-complete problem.

As briefly alluded to above, we also plan to extend PTL to more expressive logics such as description logics and modal logics. With the introduction of a typicality operator in these languages, it becomes possible to extend the propositional properties for rational consequence and obtain a characterization that reflects the additional structure of these languages, in the syntax and, more importantly, in the semantics as well.

Finally, from a knowledge representation and reasoning perspective, when dealing with knowledge bases, issues related to modularization [Cuenca Grau et al., 2006; Herzig and Varzinczak, 2005b; Herzig and Varzinczak, 2006], consistency checking [Herzig and Varzinczak, 2004; Herzig and Varzinczak, 2005a; Lang et al., 2003; Zhang et al., 2002], knowledge base integration [Meyer et al., 2005] and maintenance [Herzig et al., 2006; Varzinczak, 2008; Varzinczak, 2010] as well as versioning [Franconi et al., 2010; Noy and Musen, 2002] show up. These are tasks acknowledged as important by the community in the classical case [Herzig and Varzinczak, 2007; Konev et al., 2008; Moodley, 2011; Thielscher, 2011; Varzinczak, 2006] and that also make sense in a nonmonotonic setting. When moving to a defeasible approach, though, such tasks have to be reassessed and specific methods and techniques redesigned. This constitutes an avenue worthy of exploration.

Acknowledgements

The authors are grateful to the anonymous referees for their constructive and useful remarks on an earlier version of this text.

This work is based upon research supported in part by the South African National Research Foundation (UID 81225, IFR1202160021). Any opinion, findings and conclusions or recommendations expressed in this material are those of the author(s) and therefore the NRF do not accept any liability in regard thereto. This work was partially funded by Project number 247601, Net2: Network for Enabling Networked Knowledge, from the FP7-PEOPLE-2009-IRSES call.

Richard Booth is supported by the National Research Fund (FNR), Luxembourg (DYNGBaT project).

References

[Alchourrón et al., 1985] C. Alchourrón, P. Gärdenfors, and D. Makinson. On the logic of theory change: Partial meet contraction and revision functions. *Journal of Symbolic Logic*, 50:510–530, 1985.

[Baader et al., 2007] F. Baader, D. Calvanese, D. McGuinness, D. Nardi, and P. Patel-Schneider, editors. *The Description Logic Handbook: Theory, Implementation and Applications*. Cambridge University Press, 2 edition, 2007.

[Baltag and Smets, 2008] A. Baltag and S. Smets. A qualitative theory of dynamic interactive belief revision. In G. Bonanno, W. van der Hoek, and M. Wooldridge, editors, *Logic and the Foundations of Game and Decision Theory (LOFT7)*, number 3 in Texts in Logic and Games, pages 13–60. Amsterdam University Press, 2008.

[Blackburn et al., 2001] P. Blackburn, M. de Rijke, and Y. Venema. *Modal Logic*. Cambridge Tracts in Theoretical Computer Science. Cambridge University Press, 2001.

[Blackburn et al., 2006] P. Blackburn, J. van Benthem, and F. Wolter. *Handbook of Modal Logic*. Elsevier North-Holland, 2006.

[Booth et al., 2012] R. Booth, T. Meyer, and I. Varzinczak. PTL: A propositional typicality logic. In L. Fariñas del Cerro, A. Herzig, and J. Mengin, editors, *Proceedings of the 13th European Conference on Logics in Artificial Intelligence (JELIA)*, number 7519 in LNCS, pages 107–119. Springer, 2012.

[Boutilier, 1994] C. Boutilier. Conditional logics of normality: A modal approach. *Artificial Intelligence*, 68(1):87–154, 1994.

[Brachman and Levesque, 2004] R. Brachman and H. Levesque. *Knowledge Representation and Reasoning*. Morgan Kaufmann, 2004.

[Britz and Varzinczak, 2012] K. Britz and I. Varzinczak. Defeasible modes of inference: A preferential perspective. In *Proceedings of the 14th International Workshop on Nonmonotonic Reasoning (NMR)*, 2012.

[Britz and Varzinczak, 2013] K. Britz and I. Varzinczak. Defeasible modalities. In *Proceedings of the 14th Conference on Theoretical Aspects of Rationality and Knowledge (TARK)*, pages 49–60, 2013.

[Britz et al., 2008] K. Britz, J. Heidema, and T. Meyer. Semantic preferential subsumption. In J. Lang and G. Brewka, editors, *Proceedings of the 11th International Conference on Principles of Knowledge Representation and Reasoning (KR)*, pages 476–484. AAAI Press/MIT Press, 2008.

[Britz et al., 2009] K. Britz, J. Heidema, and W. Labuschagne. Semantics for dual preferential entailment. *Journal of Philosophical Logic*, 38:433–446, 2009.

[Britz et al., 2011a] K. Britz, T. Meyer, and I. Varzinczak. Preferential reasoning for modal logics. *Electronic Notes in Theoretical Computer Science*, 278:55–69, 2011. Proceedings of the 7th Workshop on Methods for Modalities (M4M'2011).

[Britz et al., 2011b] K. Britz, T. Meyer, and I. Varzinczak. Semantic foundation for preferential description logics. In D. Wang and M. Reynolds, editors, *Proceedings of the 24th Australasian Joint Conference on Artificial Intelligence*, number 7106 in LNAI, pages 491–500. Springer, 2011.

[Britz et al., 2012] K. Britz, T. Meyer, and I. Varzinczak. Normal modal preferential consequence. In M. Thielscher and D. Zhang, editors, *Proceedings of the 25th Australasian Joint Conference on Artificial Intelligence*, number 7691 in LNAI, pages 505–516. Springer, 2012.

[Britz et al., 2013a] K. Britz, G. Casini, T. Meyer, K. Moodley, and I. Varzinczak. Ordered interpretations and entailment for defeasible description logics. Technical report, CAIR, CSIR Meraka and UKZN, South Africa, 2013.

[Britz et al., 2013b] K. Britz, G. Casini, T. Meyer, and I. Varzinczak. Preferential role restrictions. In *Proceedings of the 26th International Workshop on Description Logics*, pages 93–106, 2013.

[Casini and Straccia, 2010] G. Casini and U. Straccia. Rational closure for defeasible description logics. In T. Janhunen and I. Niemelä, editors, *Proceedings of the 12th European Conference on Logics in Artificial Intelligence (JELIA)*, number 6341 in LNCS, pages 77–90. Springer-Verlag, 2010.

[Chellas, 1980] B. Chellas. *Modal logic: An introduction*. Cambridge University Press, 1980.

[Crocco and Lamarre, 1992] G. Crocco and P. Lamarre. On the connections between nonmonotonic inference systems and conditional logics. In R. Nebel, C. Rich, and W. Swartout, editors, *Proceedings of the 3rd International Conference on Principles of Knowledge Representation and Reasoning (KR)*, pages 565–571. Morgan Kaufmann Publishers, 1992.

[Cuenca Grau et al., 2006] B. Cuenca Grau, B. Parsia, E. Sirin, and A. Kalyanpur. Modularity and web ontologies. In P. Doherty, J. Mylopoulos, and C. Welty, editors, *Proceedings of the 10th International Conference on Principles of Knowledge Representation and Reasoning (KR)*, pages 198–208. Morgan Kaufmann, 2006.

[Delgrande, 1987] J. Delgrande. A first-order logic for prototypical properties. *Artificial Intelligence*, 33:105–130, 1987.

[Franconi et al., 2010] E. Franconi, T. Meyer, and I. Varzinczak. Semantic diff as the basis for knowledge base versioning. In *Proceedings of the 13th International Workshop on Nonmonotonic Reasoning (NMR)*, 2010.

[Friedman and Halpern, 2001] N. Friedman and J.Y. Halpern. Plausibility measures and default reasoning. *Journal of the ACM*, 48(4):648–685, 2001.

[Gabbay and Schlechta, 2009] D.M. Gabbay and K. Schlechta. Roadmap for preferential logics. *Journal of Applied Non-Classical Logic*, 19(1):43–95, 2009.

[Gärdenfors and Makinson, 1994] P. Gärdenfors and D. Makinson. Nonmonotonic inference based on expectations. *Artificial Intelligence*, 65(2):197–245, 1994.

[Giordano et al., 2009a] L. Giordano, V. Gliozzi, N. Olivetti, and G.L. Pozzato. Analytic tableaux calculi for KLM logics of nonmonotonic reasoning. *ACM Transactions on Computational Logic*, 10(3):18:1–18:47, 2009.

[Giordano et al., 2009b] L. Giordano, N. Olivetti, V. Gliozzi, and G.L. Pozzato. $\mathscr{ALC} + T$: a preferential extension of description logics. *Fundamenta Informaticae*, 96(3):341–372, 2009.

[Giordano et al., 2012] L. Giordano, N. Olivetti, V. Gliozzi, and G.L. Pozzato. A minimal model semantics for nonmonotonic reasoning. In L. Fariñas del Cerro, A. Herzig, and J. Mengin, editors, *Proceedings of the 13th European Conference on Logics in Artificial Intelligence (JELIA)*, number 7519 in LNCS, pages 228–241. Springer, 2012.

[Hansson, 1999] S.O. Hansson. *A Textbook of Belief Dynamics: Theory Change and Database Updating*. Kluwer Academic Publishers, 1999.

[Harmelen et al., 2008] F. van Harmelen, V. Lifschitz, and B. Porter, editors. *Handbook of Knowledge Representation*. Elsevier Science, 2008.

[Herzig and Varzinczak, 2004] A. Herzig and I. Varzinczak. Domain descriptions should be modular. In R. López de Mántaras and L. Saitta, editors, *Proceedings of the 16th European Conference on Artificial Intelligence (ECAI)*, pages 348–352. IOS Press, 2004.

[Herzig and Varzinczak, 2005a] A. Herzig and I. Varzinczak. Cohesion, coupling and the metatheory of actions. In L. Kaelbling and A. Saffiotti, editors, *Proceedings of the 19th International Joint Conference on Artificial Intelligence (IJCAI)*, pages 442–447. Morgan Kaufmann Publishers, 2005.

[Herzig and Varzinczak, 2005b] A. Herzig and I. Varzinczak. On the modularity of theories. In R. Schmidt, I. Pratt-Hartmann, M. Reynolds, and H. Wansing, editors, *Advances in Modal Logic*, 5, pages 93–109. King's College Publications, 2005.

[Herzig and Varzinczak, 2006] A. Herzig and I. Varzinczak. A modularity approach for a fragment of $\mathscr{ALC}$. In M. Fisher, W. van der Hoek, B. Konev, and A. Lisitsa, editors, *Proceedings of the 10th European Conference on Logics in Artificial Intelligence (JELIA)*, number 4160 in LNAI, pages 216–228. Springer-Verlag, 2006.

[Herzig and Varzinczak, 2007] A. Herzig and I. Varzinczak. Metatheory of actions: beyond consistency. *Artificial Intelligence*, 171:951–984, 2007.

[Herzig et al., 2006] A. Herzig, L. Perrussel, and I. Varzinczak. Elaborating domain descriptions. In G. Brewka, S. Coradeschi, A. Perini, and P. Traverso, editors, *Proceedings of the 17th European Conference on Artificial Intelligence (ECAI)*, pages 397–401. IOS Press, 2006.

[Katsuno and Mendelzon, 1991] H. Katsuno and A. Mendelzon. Propositional knowledge base revision and minimal change. *Artificial Intelligence*, 3(52):263–294, 1991.

[Konev et al., 2008] B. Konev, D. Walther, and F. Wolter. The logical difference problem for description logic terminologies. In A. Armando, P. Baumgartner, and G. Dowek, editors, *Proceedings of the 4th International Joint Conference on Automated Reasoning (IJCAR)*, number 5195 in LNAI, pages 259–274. Springer-Verlag, 2008.

[Kraus et al., 1990] S. Kraus, D. Lehmann, and M. Magidor. Nonmonotonic reasoning, preferential models and cumulative logics. *Artificial Intelligence*, 44:167–207, 1990.

[Lang et al., 2003] J. Lang, F. Lin, and P. Marquis. Causal theories of action – a computational core. In V. Sorge, S. Colton, M. Fisher, and J. Gow, editors, *Proceedings of the 18th International Joint Conference on Artificial Intelligence (IJCAI)*, pages 1073–1078. Morgan Kaufmann, 2003.

[Lehmann and Magidor, 1990] D. Lehmann and M. Magidor. Preferential logics: the predicate calculus case. In *Proceedings of TARK*, pages 57–72, 1990.

[Lehmann and Magidor, 1992] D. Lehmann and M. Magidor. What does a conditional knowledge base entail? *Artificial Intelligence*, 55:1–60, 1992.

[Lehmann, 1998] D. Lehmann. Stereotypical reasoning: logical properties. *Logic Journal of IGPL*, 6(1):49–58, 1998.

[Makinson, 2005] D. Makinson. How to go nonmonotonic. In *Handbook of Philosophical Logic*, volume 12, pages 175–278. Springer, 2 edition, 2005.

[Meyer et al., 2005] T. Meyer, K. Lee, and R. Booth. Knowledge integration for description logics. In *Proceedings of the 21st National Conference on Artificial Intelligence (AAAI)*, pages 645–650. AAAI Press/MIT Press, 2005.

[Moodley et al., 2012] K. Moodley, T. Meyer, and I. Varzinczak. A protégé plug-in for defeasible reasoning. In *Proceedings of the 25th International Workshop on Description Logics*, 2012.

[Moodley, 2011] K. Moodley. Debugging and repair of description logic ontologies. M.Sc. thesis, University of KwaZulu-Natal, Durban, South Africa, 2011.

[Noy and Musen, 2002] N. Noy and M. Musen. PromptDiff: A fixed-point algorithm for comparing ontology versions. In R. Dechter, M. Kearns, and R. Sutton, editors, *Proceedings of the 18th National Conference on Artificial Intelligence (AAAI)*, pages 744–750. AAAI Press/MIT Press, 2002.

[Shoham, 1988] Y. Shoham. *Reasoning about Change: Time and Causation from the Standpoint of Artificial Intelligence*. MIT Press, 1988.

[Tarski, 1941] A. Tarski. *Introduction to logic*. Oxford University Press, 1941.

[Thielscher, 2011] M. Thielscher. A unifying action calculus. *Artificial Intelligence*, 175(1):120–141, 2011.

[Varzinczak, 2006] I.J. Varzinczak. *What is a good domain description? Evaluating and revising action theories in dynamic logic*. PhD thesis, Université Paul Sabatier, Toulouse, 2006.

[Varzinczak, 2008] I.J. Varzinczak. Action theory contraction and minimal change. In J. Lang and G. Brewka, editors, *Proceedings of the 11th International Conference on Principles of Knowledge Representation and Reasoning (KR)*, pages 651–661. AAAI Press/MIT Press, 2008.

[Varzinczak, 2010] I.J. Varzinczak. On action theory change. *Journal of Artificial Intelligence Research*, 37:189–246, 2010.

[Zhang et al., 2002] D. Zhang, S. Chopra, and N. Foo. Consistency of action descriptions. In M. Ishizuka and A. Sattar, editors, *Proceedings of the 7th Pacific Rim International Conference on Artificial Intelligence (PRICAI): Trends in Artificial Intelligence*, number 2417 in LNCS, pages 70–79. Springer, 2002.

Credibility-based Selective Revision by Deductive Argumentation

Luciano H. Tamargo, Matthias Thimm, Patrick Krümpelmann, Alejandro J. García, Marcelo A. Falappa, Guillermo R. Simari, and Gabriele Kern-Isberner

Abstract In this chapter we describe recent approaches in which argumentation is applied to the process of revising an agent's beliefs. We first present an approach to selective revision with a preprocessing step based on deductive argumentation. In this approach, a non-prioritized revision operator is proposed that only accepts new information if the information is justifiable with respect to an argumentative evaluation. We integrate the developed argumentative approach into selective revision in a multi-agent scenario with information stemming from different agents with different degrees of credibility. In this context an agent has to choose carefully which information is to be accepted for revision in order to avoid believing faulty or untrustworthy information. We extend our approach of selective revision by deductive argumentation for this setting by including credibility information in the argumentative process. New information is evaluated based on the credibility of the source in combination with all arguments favoring and opposing the new information. The evaluation process determines which part of the new information is to be accepted for revision and thereupon incorporated into the belief base by an appropriate revision operator.

Luciano H. Tamargo · Alejandro J. García · Marcelo A. Falappa · Guillermo R. Simari
Artificial Intelligence Research and Development Laboratory, Department of Computer Science and Engineering, Universidad Nacional del Sur, Bahia Blanca, Argentina,
e-mail: {lt,ajg,mfalappa,grs}@cs.uns.edu.ar

Luciano H. Tamargo · Alejandro J. García · Marcelo A. Falappa
Consejo Nacional de Investigaciones Cientficas y Tcnicas (CONICET), Buenos Aires, Argentina.

Matthias Thimm
Institute for Web Science and Technologies, University of Koblenz, Germany,
e-mail: thimm@uni-koblenz.de

Patrick Krümpelmann · Gabriele Kern-Isberner
Department of Computer Science, TU Dortmund, Germany,
e-mail: patrick.kruempelmann@udo.edu, gabriele.kern-isberner@cs.uni-dortmund.de

1 Introduction

In this chapter we consider the issue of non-prioritized revision in a multi-agent setting in which the agent that is revising its belief base can receive information from multiple informants. This chapter joins two recent works on non-prioritized operators combining selected revision and deductive argumentation [Krümpelmann et al., 2011] and on argumentative credibility-based revision in multi-agent systems [Tamargo et al., 2012b] in order to come up with a unified view.

Belief revision [Alchourron et al., 1985; Hansson, 2001] is concerned with changing beliefs in light of new information. Usually, the beliefs of an agent are not static but change when new information is available. In order to be able to act reasonably in a changing environment the agent has to integrate new information and give up outdated beliefs. In particular, if the agent learns that some beliefs have been misleadingly assumed to be true its beliefs have to be *revised*. The research field of belief revision distinguishes between several different types of operations which can differ in the representation of beliefs, the type on input and the properties of the operation itself. The beliefs of the agent are represented as a set of sentences which can be closed under logical consequence, and thus be an infinite one, called a *belief set*, or as a finite, non closed se, called a *belief base*. We consider the latter case of belief base revision, mainly developed by Hansson [Hansson, 2001] in this work. It comes with the advantage of computational realizability and greater cognitive realism. It moreover allows to distinguish foundational from inferred beliefs which is closer to cognitive realism, and also to argumentation theory. The input of a revision operator can be a single sentence of a logical language or a set of such. Here we consider the more general latter case in which the operator is called a *multiple operator*. They were first considered independently by Fuhrmann, Hansson, Nieder, and Rott; for an overview see [Fuhrmann and Hansson, 1994]. A prominent property to distinguish types of revision operations is the *success* property which leads to *prioritized* revision if satisfied and to *non-prioritized* belief revision if not. In prioritized belief revision [Alchourron et al., 1985; Hansson, 2001] new information is always assumed to represent the most reliable and correct information available and revising the agent's beliefs by the new information is expected to result in believing the new information. This has become known as the *success* postulate which demands that new information is believed after revision. However, this postulate has been questioned as it seems to imply that the new information should be blindly accepted instead of being weighted against current beliefs. The field of non-prioritized belief revision [Hansson, 1999] investigates change operations where revising some beliefs by new information may not result in believing the new information. Imagine a multi-agent system where agents exchange information. In general, agents may be cooperative or competitive. Information that is passed from one agent to another may be intentionally wrong, mistakenly wrong, or correct. It is up to the receiver of the information to evaluate whether it should be integrated into the beliefs or not. In particular, in non-prioritized belief revision the satisfaction of the *success* postulate is not desirable. In this context, a specific class of non-prioritized belief revision operators called selective revision

[Fermé and Hansson, 1999] is particularly interesting. A *selective revision* is a two-step revision that consists of 1.) filtering new information using a *transformation function* and 2.) revising the beliefs with the result of the filtering in a prioritized way. In [Fermé and Hansson, 1999], no concrete implementations of the transformation function are given although several results are proven that show how specific properties for the transformation function and the *inner* prioritized revision translate to specific properties for the *outer* non-prioritized revision.

In this chapter we present a specific implementation of a transformation function for selective revision that makes use of *deductive argumentation* [Besnard and Hunter, 2001]. A deductive argumentation theory is a set of propositional sentences, and an argument consists of some sentence ϕ and a minimal proof for ϕ. If the theory is inconsistent there may also be proofs for the complement of a sentence $\neg\phi$ and in order to decide whether ϕ or $\neg\phi$ is to be believed, an argumentative evaluation is performed that compares arguments with counterarguments. We use the framework of [Besnard and Hunter, 2001] to implement a transformation function for selective revision that decides for each individual piece of information whether to accept it for revision or not, based on its argumentative evaluation. In particular, we consider the case that revision is to be performed based on a set of pieces of information instead of just a single piece of information. By doing so, we allow new information to contain arguments. A non-prioritized revision operator is then proposed that only accepts new information if the information is justifiable with respect to an argumentative evaluation. The proposed operator can accept all, part, or none of the received information. As a result, an agent decides whether to accept some new information on the basis of its own evaluation of the information and the arguments that may be contained in this information. This allows us to implement a decision procedure to find out if the *success* postulate is adequate or not.

We motivate the main ideas of this approach by a (running) example. Consider the following scenario where an agent (*Anna*) has to decide where to spend her holidays and receives new information from her mother that has to be considered.

Example 1. Anna is a surf fanatic (s) and believes that a surf fanatic should travel to Hawaii ($s \Rightarrow h$). Anna has taken a loan (l), and taking a loan means having money available ($l \Rightarrow m$). Having money implies she should travel to Hawaii ($m \Rightarrow h$), and having money also implies she does not have financial problems ($m \Rightarrow \neg f$). Below we show the belief base $\mathcal{K}_1$ with Anna's beliefs. Observe that $\mathcal{K}_1 \vdash h$, i. e., from $\mathcal{K}_1$ Anna concludes she should go to Hawaii.

$$\mathcal{K}_1 = \{s, \ s \Rightarrow h, \ l, \ l \Rightarrow m, \ m \Rightarrow h, \ m \Rightarrow \neg f\}.$$

Now consider the new information $\Phi_1 = \{f, \ f \Rightarrow \neg h, \ v, \ v \Rightarrow \neg h\}$ that Anna has received from her mother, in order to tell her not to travel to Hawaii. In particular, Φ_1 states that Anna has financial problems (f), that having financial problems Anna should not travel to Hawaii ($f \Rightarrow \neg h$), that there is also volcano activity on Hawaii (v), and that given volcano activity Anna should not travel to Hawaii ($v \Rightarrow \neg h$).

Observe that in the example above, $\mathcal{K}_1 \cup \Phi_1$ is inconsistent, that is, Anna can not accept all this information without withdrawing some sentences from $\mathcal{K}_1$. In this situation, Anna could reject all the new information; she could accept all the information but withdraw some of her current beliefs; or some combination. For instance, she could reject f (financial problems) because she has taken a loan, but she could accept $\{f \Rightarrow \neg h,\ v,\ v \Rightarrow \neg h\}$ provided she eliminates part of her beliefs.

In order to consider a situation such as the one described in Example 1, in Section 4 we will introduce a selective revision approach based on deductive argumentation. Deductive argumentation will be used for deciding whether to accept the whole received set, to reject all, or to accept part of the set.

Then in Section 5.2 we will show how to integrate the developed argumentative approach into selective revision in a multi-agent scenario with information stemming from different agents with different degrees of credibility. Thus, an agent can choose which information is to be accepted based on the credibility of its informants. For instance, consider the following example in which an agent (*Sam*) has to revise his knowledge and can receive information from other agents he does not consider equally credible.

Example 2. Consider an agent *Sam* (A_S) interacts with three other agents at his work: his boss *Bob* (A_B), his assigned client *Carl* (A_C) and his colleague *Paul* (A_P). Regarding beliefs related to his job, Sam considers that these agents are not equally credible: for Sam the most credible is A_C, then A_B, then A_P, and finally himself (A_S).

Now consider that Sam believes that he has no work to do ($\neg w$), and that if he has no assigned work then he can go on vacation ($\neg w \to v$). His colleague A_P has also told him that he can replace Sam at his work (r), and in such a case he can go on vacation ($r \to v$). Finally, his client has also informed Sam, that there is no work to do $\neg w$. Hence, Sam has three arguments supporting v (to go on vacation). One supported by $\neg w$ and $\neg w \to v$; the second supported by the information received from A_P: r and $r \to v$; and the third one supported by the information received from his client that he was no work to do and can therefore go on vacation.

Consider now that his boss (A_B) informs Sam that he has work to do (w), that Paul has informed him that he is ill (i) and that if Paul is ill then he has no replacement ($i \to \neg r$). Thus Sam has to revise his beliefs in order to incorporate part or all this new information, and it may be possible that the arguments that support v will not longer exist upon revision.

In order to consider a situation such as the one described in Example 2, we will show in Section 5.2 an approach of selective revision by deductive argumentation for this setting by including credibility information in the argumentative process. New information will be evaluated based on the credibility of the source in combination with all arguments favoring and opposing the new information. The evaluation process determines which part of the new information is to be accepted for revision and thereupon incorporated into the belief base by an appropriate revision operator.

The rest of this chapter is organized as follows. In Section 2 we introduce some necessary technical preliminaries; and we provide an overview on the notions of belief revision and extending the approach of selective revision to selective multiple base revision. We continue in Section 3 with presenting the framework of deductive argumentation. In Section 4 we propose our implementation of selective multiple base revision via deductive argumentation and investigate its properties. Section 5 presents an epistemic model based on credibility for an agent situated in a multi-agent environment. Afterwards we present our approach to argumentative credibility-based revision of epistemic models and go on with an analysis of this approach. In Section 6 we review some related work and in Section 7 we conclude.

2 Background

In this section we fix notations and recall approaches to selective revision. We go into more details on selective multiple revision on belief bases as this will provide most important postulates and techniques for the approaches to be presented here.

2.1 Formal preliminaries

We suppose that the beliefs of an agent are given in the form of propositional sentences. Let At be a propositional signature, i. e., a set of propositional atoms. Let $\mathscr{L}(\mathsf{At})$ be the corresponding propositional language generated by the atoms in At and the connectives $\wedge$ (*and*), $\vee$ (*or*), $\Rightarrow$ (*implication*), and $\neg$ (*negation*). As a notational convenience we assume some arbitrary total order $\ll$ on the elements of $\mathscr{L}(\mathsf{At})$ which is used to enumerate elements of each finite $\Phi \subseteq \mathscr{L}(\mathsf{At})$ in a unique way, cf. [Besnard and Hunter, 2001]. For a finite subset $\Phi \subseteq \mathscr{L}(\mathsf{At})$ the *canonical enumeration* of Φ is the vector $\langle \phi_1, \ldots, \phi_n \rangle$ such that $\{\phi_1, \ldots, \phi_n\} = \Phi$ and $\phi_i \ll \phi_j$ for every $i < j$ with $i, j = 1, \ldots, n$. As $\ll$ is total the canonical enumeration of every finite subset $\Phi \subseteq \mathscr{L}(\mathsf{At})$ is uniquely defined.

We use the operator $\vdash$ to denote classical entailment, i. e., for sets of propositional sentences $\Phi_1, \Phi_2 \subseteq \mathscr{L}(\mathsf{At})$ we say that Φ_2 *follows* from Φ_1, denoted by $\Phi_1 \vdash \Phi_2$, if and only if every $\phi \in \Phi_2$ is entailed by Φ_1 in the classical logical sense. For sentences $\phi, \phi' \in \mathscr{L}(\mathsf{At})$ we write $\phi \vdash \phi'$ instead of $\{\phi\} \vdash \{\phi'\}$. We define the deductive closure $Cn(\cdot)$ of a set of sentences Φ as $Cn(\Phi) = \{\phi \in \mathscr{L}(\mathsf{At}) \mid \Phi \vdash \phi\}$. Two sets of sentences $\Phi, \Phi' \subseteq \mathscr{L}(\mathsf{At})$ are *equivalent*, denoted by $\Phi \equiv \Phi'$, if and only if it holds that $\Phi \vdash \Phi'$ and $\Phi' \vdash \Phi$. We also use the equivalence relation $\cong$ which is defined as $\Phi \cong \Phi'$ if and only if there is a bijection $\sigma : \Phi \to \Phi'$ such that for every $\phi \in \Phi$ it holds that $\phi \equiv \sigma(\phi)$. This means that $\Phi \cong \Phi'$ if Φ and Φ' are *element-wise* equivalent. Note that $\Phi \cong \Phi'$ implies $\Phi \equiv \Phi'$ but not vice versa. In particular, it holds that e. g $\{a \wedge b\} \equiv \{a, b\}$ but $\{a \wedge b\} \not\cong \{a, b\}$. For sentences $\phi, \phi' \in \mathscr{L}(\mathsf{At})$ we write $\phi \sim \phi'$ instead of $\{\phi\} \equiv \{\phi'\}$ if $\sim$ is one of $\{\equiv, \cong\}$. If $\Phi \vdash \bot$ we say that Φ is *inconsistent*.

For a set S let $\mathfrak{P}(S)$ denote the power set of S, i. e., the set of all subsets of S. For a set S let $\mathfrak{PP}(S)$ denote the set of multi-sets of S, i. e., the set of all subsets of S where an element may occur more than once. To distinguish sets from multi-sets we use brackets "$\langle$" and "$\rangle$" for the latter.

2.2 Selective Revision

The field of belief revision is concerned with the change of beliefs when more recent or more reliable information is at hand. The most important description of properties of *prioritized* belief change operators are given by Gärdenfors [Gärdenfors, 1978; 1982], and then by Alchourrón, Gärdenfors and Makinson in their seminal paper [Alchourrón et al., 1985]. There, the authors consider a model of change in which the epistemic state is represented by a *belief set*, that is, a set of sentences closed under logical consequence, and the epistemic input (the new information) is represented by a single sentence. A *belief set* $\mathbf{K}$ is a subset of $\mathscr{L}(\mathsf{At})$ that is deductively closed, *i. e.*, $\mathbf{K} = Cn(\mathbf{K})$.

Selective Revision [Fermé and Hansson, 1999] is a kind of revision categorized as *non-prioritized*: that is, a revision operator in which the new information is not always accepted. There, the problem of revising a belief set $\mathbf{K}$ with a single sentence α is realized by applying a *transformation function* f to α, obtaining a new sentence α', and then revising $\mathbf{K}$ by α' in a prioritized way. The transformation function f is supposed to determine whether α should be accepted as a whole or whether it should be somewhat weakened. So, a main feature of a selective revision operator is that the new information can be partially accepted.

The selective revision operator on belief sets takes a belief set $\mathbf{K}$ and a sentence α, and produces a new belief set in which 'a selective part' of α can be accepted. More formally, a selective revision operator $\circ$ is defined by the equality $\mathbf{K} \circ \alpha = \mathbf{K} * f(\alpha)$, where $*$ is an AGM revision operator [Alchourron et al., 1985] and f is a function, typically with the property $\vdash \alpha \to f(\alpha)$. Fermé and Hansson provide a set of postulates and different constructions for the operator $\circ$, some properties for the function f, and they present representation theorems for three different kinds of selective revision operators. The representation theorems indicate that these constructions provide a fairly faithful extension of the AGM framework to allow for less than total acceptance of new information [Fermé and Hansson, 1999].

Selective revision on belief bases is a generalization of selective revision proposed in [Fermé and Hansson, 1999] and base revision proposed, among others, in [Dalal, 1988; Fuhrmann, 1991; Hansson, 1994b; 1993; Katsuno and Mendelzon, 1991]. Let $\mathscr{K} \subseteq \mathscr{L}(\mathsf{At})$ be a belief base, and α be a sentence of the language. Then, a selective revision $\circ$ of the belief base $\mathscr{K}$ with respect to α, noted by $\mathscr{K} \circ \alpha$, is a new belief base such that the *success postulate* ($\mathscr{K} \circ \alpha \vdash \alpha$), in general, does not hold.

2.3 Selective Multiple Revision on Belief Bases

In this chapter, we consider the problem of *multiple belief base revision* [Krümpelmann et al., 2011], i.e., revising a finite set of sentences by another such set; cf. also the notions of *multiple change* [Fuhrmann and Hansson, 1994; Hansson, 2001] and *parallel belief revision* [Delgrande and Jin, 2012; Kern-Isberner and Krümpelmann, 2011]. Let $\mathcal{K} \subseteq \mathscr{L}(\mathrm{At})$ be a belief base, $\Phi \subseteq \mathscr{L}(\mathrm{At})$ be some set of sentences, and consider the problem of changing $\mathcal{K}$ in order to entail Φ. If $\mathcal{K} \cup \Phi$ is consistent then there is no need for contracting the existing beliefs and the problem can be solved via *expansion* $\mathcal{K} + \Phi$ which is characterized via $\mathcal{K} \cup \Phi$. If $\mathcal{K} \cup \Phi$ is inconsistent, conflicts arising from the addition of Φ to $\mathcal{K}$ have to be resolved. In general, this means that some of the current beliefs have to be given up in order to come up with a consistent belief base. The AGM framework [Alchourron et al., 1985] proposes several basic postulates a revision operator should obey. As we consider belief bases for knowledge representation we start with the corresponding postulates for belief base revision [Hansson, 2001] adapted to revision by sets of sentences [Falappa et al., 2002]. Let $*$ be a multiple base revision operator—i.e., if $\mathcal{K}$ and Φ are sets of sentences so is $\mathcal{K} * \Phi$—and consider the following postulates:

Success. $\quad \mathcal{K} * \Phi \vdash \Phi$.
Inclusion. $\quad \mathcal{K} * \Phi \subseteq \mathcal{K} + \Phi$.
Vacuity. $\quad$ If $\mathcal{K} \cup \Phi \not\vdash \bot$ then $\mathcal{K} + \Phi \subseteq \mathcal{K} * \Phi$.
Consistency. $\quad$ If Φ is consistent then $\mathcal{K} * \Phi$ is consistent.
Core Retainment. $\quad$ If $\alpha \in (\mathcal{K} \cup \Phi) \setminus (K * \Phi)$ then there is a set H such that $H \subseteq K \cup \Phi$ and H is consistent but $H \cup \{\alpha\}$ is inconsistent.
Relevance. $\quad$ If $\alpha \in (\mathcal{K} \cup \Phi) \setminus (K * \Phi)$ then there is a set H such that $\mathcal{K} * \Phi \subseteq H \subseteq K \cup \Phi$ and H is consistent but $H \cup \{\alpha\}$ is inconsistent.

Success states that the new beliefs in Φ have primacy over beliefs in $\mathcal{K}$. Inclusion determines that the belief base revised by Φ is included in the belief base expanded by Φ. Vacuity establishes that if Φ is consistent with the original belief base, then nothing is removed in the revised belief base. Consistency determines that if Φ is consistent then so is the revised belief base. Core Retainment and Relevance establish the intuition that nothing is removed from the original belief base unless its removal in some way contributes to making the new belief base consistent. It is clear that relevance implies core retainment; the difference among these postulates arises from the construction of contraction operators. If contraction is defined by a kernel contraction operator [Hansson, 1994a] then core retainment is satisfied; if contraction is defined by a partial meet contraction operator [Alchourron et al., 1985] then relevance is satisfied. Since we are presenting a revision operator, it is important to remark that revision operators can be defined from the Levi Identity [Levi, 1977; Gärdenfors, 1988].

Another important property for the framework of [Alchourron et al., 1985] is extensionality which can be phrased for multiple base revision as follows:

Extensionality. $\quad$ If $\Phi \equiv \Psi$, then $\mathcal{K} * \Phi \equiv \mathcal{K} * \Psi$.

The above property is not usually considered for the problem of base revision as base revision is motivated by observing explicitly given beliefs and not (only) semantic contents. In particular, for the problem of multiple base revision, satisfaction of *extensionality* imposes that $\mathcal{K} * \{a,b\} \equiv \mathcal{K} * \{a \wedge b\}$ as $\{a,b\} \equiv \{a \wedge b\}$. Identifying the "comma"-operator with the logical "AND"-operator is not always a reasonable thing to do, see e.g. [Delgrande and Jin, 2012; Kern-Isberner and Krümpelmann,

2011] for a discussion. However, we consider the following weakened form of *extensionality*.

Weak Extensionality.   If $\Phi \cong \Phi'$ then $\mathcal{K} * \Phi \equiv \mathcal{K} * \Phi'$.

The *weak extensionality* property only states that the outcomes of the revisions $\mathcal{K} * \Phi$ and $\mathcal{K} * \Phi'$ are equivalent if Φ and Φ' are element-wise equivalent.

Definition 1. A revision operator $*$ is called a *prioritized multiple base revision operator* if $*$ satisfies *success, inclusion, vacuity, consistency, relevance,* and *weak extensionality*.

For non-prioritized multiple base revision the properties *inclusion, vacuity, consistency, relevance,* and *weak extensionality* can also be regarded as desirable. This is not the case for *success* in general but we can replace *success* by weakened versions, cf. [Hansson, 1999]. We denote with ∘ a non-prioritized belief revision operator, i.e., $\mathcal{K} \circ \Phi$ is the non-prioritized revision of $\mathcal{K}$ by Φ. Then consider the following properties for ∘, cf. [Fermé and Hansson, 1999].

Weak Success.   If $\mathcal{K} \cup \Phi \not\vdash \bot$ then $\mathcal{K} \circ \Phi \vdash \Phi$.
Consistent Expansion.   If $\mathcal{K} \not\subseteq \mathcal{K} \circ \Phi$ then $\mathcal{K} \cup (\mathcal{K} \circ \Phi) \vdash \bot$.

Note that *weak success* follows from *vacuity*, and *consistent expansion* follows from *vacuity* and *success*, cf. [Fermé and Hansson, 1999].

Definition 2. A revision operator ∘ is called *non-prioritized multiple base revision operator* if ∘ satisfies *inclusion, consistency, weak extensionality, weak success,* and *consistent expansion*.

We do not require *relevance* to be satisfied by non-prioritized multiple base revisions as it is hardly achievable in the context of selective revision, see below. For the following, bear in mind that the main difference between a prioritized multiple base revision operator $*$ and a non-prioritized multiple base revision operator ∘ is that $\mathcal{K} * \Phi \vdash \Phi$ is required but $\mathcal{K} \circ \Phi \vdash \Phi$ is not.

We adopt the notions of [Fermé and Hansson, 1999] for the problem of *selective multiple belief base revision* and still consider the problem of revising a belief base $\mathcal{K}$ by some set Φ of sentences. Following the ideas of [Fermé and Hansson, 1999] we define the selective multiple base revision.

Definition 3. Let $\mathcal{K}$ be a belief base, $f_\mathcal{K} : \mathfrak{P}(\mathcal{L}(\mathsf{At})) \to \mathfrak{P}(\mathcal{L}(\mathsf{At}))$ be a transformation function, $*$ be some prioritized multiple base revision, and Φ be a set of beliefs. Then, the *selective multiple base revision* of $\mathcal{K}$ by Φ, noted by $\mathcal{K} \circ \Phi$, is defined as follows:

$$\mathcal{K} \circ \Phi = \mathcal{K} * f_\mathcal{K}(\Phi)$$

In [Fermé and Hansson, 1999] several properties for transformation functions in the context of belief set revision are discussed which often corresponds to properties of revision operators. We rephrase some of them here slightly to fit the framework of multiple base revision. Let $\mathcal{K} \subseteq \mathcal{L}(\mathsf{At})$ be consistent and let $\Phi, \Phi' \subseteq \mathcal{L}(\mathsf{At})$.

Inclusion. $f_\mathcal{K}(\Phi) \subseteq \Phi$.
Weak Inclusion. If $\mathcal{K} \cup \Phi$ is consistent then $f_\mathcal{K}(\Phi) \subseteq \Phi$.
Extensionality. If $\Phi \equiv \Phi'$ then $f_\mathcal{K}(\Phi) \equiv f_\mathcal{K}(\Phi')$.
Consistency Preservation. If Φ is consistent then $f_\mathcal{K}(\Phi)$ is consistent.
Consistency. $f_\mathcal{K}(\Phi)$ is consistent.
Maximality. $f_\mathcal{K}(\Phi) = \Phi$.
Weak Maximality. If $\mathcal{K} \cup \Phi$ is consistent then $f_\mathcal{K}(\Phi) = \Phi$.

We also consider the following novel property which corresponds directly to *Weak Extensionality* for multiple base revision operators introduced above.

Weak Extensionality. If $\Phi \cong \Phi'$ then $f_\mathcal{K}(\Phi) \cong f_\mathcal{K}(\Phi')$.

Not all of the above properties may be desirable for a transformation function that is to be used for selective revision. For example, the property *maximality* states that $f_\mathcal{K}$ should not modify the set Φ. Satisfaction of this property makes Definition 3 equivalent to $\mathcal{K} * \Phi$. As $*$ is meant to be a prioritized revision function we lose the possibility for non-prioritized revision.

Note that for *weak extensionality* we demand $f_\mathcal{K}(\Phi)$ and $f_\mathcal{K}(\Phi')$ to be element-wise equivalent instead of just equivalent (in contrast to the property *weak extensionality* for revision). We do this because $f_\mathcal{K}$ is supposed to be applied in the context of base revision which is sensitive to syntactic variants. We introduce the postulate *weak extensionality* for transformation functions with the same motivation as we do for multiple base revision. However, for the case of transformation functions the problem with satisfaction of *extensionality* is more apparent. Consider again $\Phi = \{a,b\}$ and $\Phi' = \{a \wedge b\}$. It follows that $\Phi \equiv \Phi'$ and if $f_\mathcal{K}$ satisfies *extensionality* this results in $f_\mathcal{K}(\{a,b\}) \equiv f_\mathcal{K}(\{a \wedge b\})$. If $f_\mathcal{K}$ also satisfies *inclusion* it follows that $f_\mathcal{K}(\{a \wedge b\}) \in \{\emptyset, \{a \wedge b\}\}$ and therefore $f_\mathcal{K}(\{a,b\}) \in \{\emptyset, \{a,b\}\}$. In general, if $f_\mathcal{K}$ satisfies both *inclusion* and *extensionality* it follows that either $f_\mathcal{K}(\Phi) = \emptyset$ or $f_\mathcal{K}(\Phi) = \Phi$ for every $\Phi \subseteq \mathcal{L}(\text{At})$ (as Φ is equivalent to a Φ' that consists of a single formula that is the conjunction of the formulas in Φ and $f_\mathcal{K}(\Phi') = \emptyset$ or $f_\mathcal{K}(\Phi') = \Phi'$ due to *inclusion*). As we are interested in a more graded approach to belief revision we want to be able to accept or reject specific pieces of Φ and not just Φ as a whole. Consequently, we consider *weak extensionality* as a desirable property instead of *extensionality*. Note that *extensionality* implies *weak extensionality* as $\Phi \cong \Phi'$ implies $\Phi \equiv \Phi'$.

In [Fermé and Hansson, 1999] several representation theorems are given that characterize non-prioritized belief revision by selective revision via (see Def. 3) and specific properties of $*$ and $f_\mathcal{K}$. In particular, it is shown that a reasonable non-prioritized belief revision operator $\circ$ can be characterized by an AGM revision $*$ and a transformation function $f_\mathcal{K}$ that satisfies *extensionality*, *consistency preservation*, and *weak maximality*. Note, however, that [Fermé and Hansson, 1999] deals with the problem of revising a belief set by a single sentence. Nonetheless, we can carry over the results of [Fermé and Hansson, 1999] to the problem of multiple base revision and obtain the following result; the proof can be found in [Krümpelmann et al., 2011]).

Proposition 1. *Let $*$ be a prioritized multiple base revision operator and let $f_\mathcal{K}$ satisfy* inclusion, weak extensionality, consistency preservation, *and* weak maximality. *Then $\circ$ defined according to Def. (3) is a non-prioritized multiple base revision operator.*

Note that *relevance* does not hold for $\mathcal{K} \circ \Phi$ defined via Definition 3 in general. Consider for example the transformation function $f_\mathcal{K}^0$ defined via $f_\mathcal{K}^0(\Phi) = \Phi$ if $\mathcal{K} \cup \Phi$ is consistent and $f_\mathcal{K}^0(\Phi) = \emptyset$ otherwise. Then $f_\mathcal{K}^0$ satisfies all properties for transformation functions except *maximality*. But it is easy to see that $\mathcal{K} \circ \Phi$ defined via Definition 3 using $f_\mathcal{K}^0$ and a prioritized multiple base revision operator $*$ fails to satisfy *relevance*. We leave it to future work to investigate further properties for transformation functions that may enable *relevance* to hold in general.

In the following we aim at implementing a selective multiple base revision using deductive argumentation and go on with introducing the latter.

3 Deductive Argumentation

Argumentation frameworks [Bench-Capon and Dunne, 2007] allow for reasoning with inconsistent information based on the notions of arguments, counterarguments and their relationships. Since the seminal paper [Dung, 1995] interest has grown in research in computational models for argumentation that allow for a coherent procedure for consistent reasoning in the presence of inconsistency. In this work we use the framework of *deductive argumentation* as proposed by Besnard and Hunter [Besnard and Hunter, 2001]. This framework is based on classical propositional logic and is therefore apt for our aim to use argumentation to realize a transformation function f. The central notion of the framework of deductive argumentation is that of an *argument*.

Definition 4 (Argument). *Let $\Phi \subseteq \mathcal{L}(\text{At})$ be a set of sentences. An* argument $\mathscr{A}$ *for a sentence $\alpha \in \mathcal{L}(\text{At})$ in Φ is a tuple $\mathscr{A} = \langle \Psi, \alpha \rangle$ with $\Psi \subseteq \Phi$ that satisfies*

1. $\Psi \not\vdash \bot$,
2. $\Psi \vdash \alpha$, *and*
3. *there is no $\Psi' \subsetneq \Psi$ with $\Psi' \vdash \alpha$.*

For an argument $\mathscr{A} = \langle \Psi, \alpha \rangle$ we say that α is the *claim* of $\mathscr{A}$ and Ψ is the *support* of $\mathscr{A}$.

Thus, in an argument $\mathscr{A} = \langle \Psi, \alpha \rangle$ for α, the set Ψ is a minimal set entailing α. Given a set $\Phi \subseteq \mathcal{L}(\text{At})$ of sentences there may be multiple arguments for α. As in [Besnard and Hunter, 2001] we are interested in arguments that are most cautious.

Definition 5 (Conservativeness). *An argument $\mathscr{A} = \langle \Psi, \alpha \rangle$ is more* conservative *than an argument $\mathscr{B} = \langle \Phi, \beta \rangle$ if and only if $\Psi \subseteq \Phi$ and $\beta \vdash \alpha$.*

In other words, an $\mathscr{A}$ is more conservative than an argument $\mathscr{B}$ if $\mathscr{A}$ has a smaller support (with respect to set inclusion) and a more general conclusion. An argument $\mathscr{A}$ is *strictly more conservative* than an argument $\mathscr{B}$ if and only if $\mathscr{A}$ is more conservative than $\mathscr{B}$ but $\mathscr{B}$ is not more conservative than $\mathscr{A}$. If $\Phi \subseteq \mathscr{L}(\text{At})$ is inconsistent there are arguments with contradictory claims.

Definition 6 (Undercut). An argument $\mathscr{A} = \langle \Psi, \alpha \rangle$ is an *undercut* for an argument $\mathscr{B} = \langle \Phi, \beta \rangle$ if and only if $\alpha = \neg(\phi_1 \wedge \ldots \wedge \phi_n)$ for some $\phi_1, \ldots, \phi_n \subseteq \Phi$.

If $\mathscr{A}$ is an undercut for $\mathscr{B}$ then we also say that $\mathscr{A}$ *attacks* $\mathscr{B}$. In order to consider only those undercuts for an argument that are most general we restrain the notion of undercut as follows.

Definition 7 (Maximally conservative undercut). An argument $\mathscr{A} = \langle \Psi, \alpha \rangle$ is a *maximally conservative undercut* for an argument $\mathscr{B} = \langle \Phi, \beta \rangle$ if and only if $\mathscr{A}$ is an undercut of $\mathscr{B}$ and there is no undercut $\mathscr{A}'$ for $\mathscr{B}$ that is strictly more conservative than $\mathscr{A}$.

Definition 8 (Canonical undercut). An argument $\mathscr{A} = \langle \Psi, \neg(\phi_1 \wedge \ldots \wedge \phi_n) \rangle$ is a *canonical undercut* for an argument $\mathscr{B} = \langle \Phi, \beta \rangle$ if and only if $\mathscr{A}$ is a maximally conservative undercut for $\mathscr{B}$ and $\langle \phi_1, \ldots, \phi_n \rangle$ is the canonical enumeration of Φ.

It can be shown that it suffices to consider only the canonical undercuts for an argument in order to come up with a reasonable argumentative evaluation of some claim α [Besnard and Hunter, 2001]. Having an undercut $\mathscr{B}$ for an argument $\mathscr{A}$ there may also be an undercut $\mathscr{C}$ for $\mathscr{B}$ which *defends* $\mathscr{A}$. In order to give a proper evaluation of some argument $\mathscr{A}$ we have to consider all undercuts for its undercuts as well, and so on. This leads to the notion of an *argument tree*.

Definition 9 (Argument tree). Let $\alpha \in \mathscr{L}(\text{At})$ be some sentence and let $\Phi \subseteq \mathscr{L}(\text{At})$ be a set of sentences. An *argument tree* $\tau_\Phi(\alpha)$ for α in Φ is a tree where the nodes are arguments and that satisfies

1. the root is an argument for α in Φ,
2. for every path $[\langle \Phi_1, \alpha_1 \rangle, \ldots, \langle \Phi_n, \alpha_n \rangle]$ in $\tau_\Phi(\alpha)$ it holds that $\Phi_n \not\subseteq \Phi_1 \cup \ldots \cup \Phi_{n-1}$, and
3. the children $\mathscr{B}_1, \ldots, \mathscr{B}_m$ of a node $\mathscr{A}$ consist of all canonical undercuts for $\mathscr{A}$ such that condition (2) above is not violated when these canonical undercuts are added as children.

Let $\mathscr{T}(\text{At})$ be the set of all argument trees.

An argument tree is a concise representation of the relationships between different arguments that favor or reject some argument $\mathscr{A}$. In order to evaluate whether a claim α can be justified we have to consider all argument trees for α and all argument trees for $\neg \alpha$. For an argument tree τ let $\text{root}(\tau)$ denote the root node of τ. Furthermore, for a node $\mathscr{A} \in \tau$ let $\text{ch}_\tau(\mathscr{A})$ denote the children of $\mathscr{A}$ in τ and $\text{ch}_\tau^{\mathscr{T}}(\mathscr{A})$ denote the set of sub-trees rooted at a child of $\mathscr{A}$.

Definition 10 (Argument structure). Let $\alpha \in \mathscr{L}(\text{At})$ be some sentence and let $\Phi \subseteq \mathscr{L}(\text{At})$ be a set of sentences. The *argument structure* $\Gamma_\Phi(\alpha)$ for α with respect to Φ is the tuple $\Gamma_\Phi(\alpha) = (\mathscr{P}, \mathscr{C})$ such that $\mathscr{P}$ is the set of argument trees for α in Φ and $\mathscr{C}$ is the set of arguments trees for $\neg \alpha$ in Φ.

The argument structure $\Gamma_\Phi(\alpha)$ of a $\alpha \in \mathscr{L}(\text{At})$ gives a complete picture of the reasons for and against α. The argument structure has to be evaluated in order to determine the status of sentences. We introduce the powerful evaluation mechanisms from [Besnard and Hunter, 2001] and give examples of how adequate and simple instantiations can be realized.

Definition 11 (Categorizer). A *categorizer* γ is a function $\gamma: \mathscr{T}(\text{At}) \to \mathbb{R}$.

A categorizer is meant to assign a value to an argument tree τ depending on how strongly this argument tree favors the root argument. In particular, the larger the value of $\gamma(\tau)$ the better justification of believing in the claim of the root argument. For an argument structure $\Gamma_\Phi(\alpha) = (\{\tau_1^p, \ldots, \tau_n^p\}, \{\tau_1^c, \ldots, \tau_m^c\})$ and a categorizer γ we abbreviate

$$\gamma(\Gamma_\Phi(\alpha)) = (\langle \gamma(\tau_1^p), \ldots, \gamma(\tau_n^p) \rangle, \langle \gamma(\tau_1^c), \ldots, \gamma(\tau_m^c) \rangle) \in \mathfrak{PP}(\mathbb{R}) \times \mathfrak{PP}(\mathbb{R}).$$

Definition 12 (Accumulator). An *accumulator* κ is a function $\kappa: \mathfrak{PP}(\mathbb{R}) \times \mathfrak{PP}(\mathbb{R}) \to \mathbb{R}$.

An accumulator is meant to evaluate the categorization of argument trees for or against some sentence α.

Definition 13 (Acceptance). We say that a set of sentences $\Phi \subseteq \mathscr{L}(\text{At})$ *accepts* a sentence α with respect to a categorizer γ and an accumulator κ, denoted by

$$\Phi \mid\sim_{\kappa,\gamma} \alpha \text{ if and and only if } \kappa(\gamma(\Gamma_\Phi(\alpha))) > 0$$

If Φ does not accept α with respect to γ and κ ($\Phi \not\mid\sim_{\kappa,\gamma} \alpha$) we say that Φ *rejects* α with respect to γ and κ.

Some simple instances of categorizers and accumulators are as follows.

Example 3. Let τ be some argument tree. The classical evaluation of an argument tree—as e.g., employed in *Defeasible Logic Programming* [Garcia and Simari, 2004]—is as follows: each leaf of the tree is considered "undefeated"; an inner node is "undefeated" if all its children are "defeated" and "defeated" if there is at least one child that is "undefeated". This evaluation can be formalized by defining the *classical categorizer* γ_0 recursively via

$$\gamma_0(\tau) = \begin{cases} 1 & \text{if } \text{ch}_\tau(\text{root}(\tau)) = \emptyset \\ 1 - \max\{\gamma_0(\tau') \mid \tau' \in \text{ch}_\tau^{\mathscr{T}}(\text{root}(\tau))\} & \text{otherwise} \end{cases}$$

Furthermore, a simple accumulator κ_0 can be defined via

$$\kappa_0(\langle N_1,\ldots,N_n\rangle,\langle M_1,\ldots,M_m\rangle) = N_1 + \ldots + N_n - M_1 - \ldots - M_m.$$

For example, a set of sentences $\Phi \subseteq \mathscr{L}(\mathsf{At})$ accepts a sentence α with respect to γ_0 and κ_0 if and only if there are more argument trees for α where the root argument is undefeated than argument trees for $\neg\alpha$ where the root argument is undefeated. ∎

More examples of categorizers and accumulators can be found in [Besnard and Hunter, 2001]. Using those notions we are able to state for every sentence $\phi \in \Phi$ whether ϕ is accepted in Φ or not, depending on the arguments that favor α and those that reject α.

4 Selective Revision by Deductive Argumentation

Using the deductive argumentation framework presented in the previous section one is able to decide for each sentence $\alpha \in \Phi$ whether α is justifiable with respect to Φ. Note that the framework of deductive argumentation heavily depends on the actual instances of categorizer and accumulator. In the following we only consider categorizers and accumulators that comply with the following minimal requirements.

Definition 14 (Well-behaving categorizer). A categorizer γ is called *well-behaving* if $\gamma(\tau) > \gamma(\tau')$ whenever τ consists only of one single node and τ' consists of at least two nodes.

In other words, a categorizer γ is well-behaving if the argument tree that has no undercuts for its root is considered the best justification for the root.

Definition 15 (Well-behaving accumulator). An accumulator κ is called *well-behaving* if and only if $\kappa((\mathscr{P},\mathscr{C})) > 0$ whenever $\mathscr{P} \neq \emptyset$ and $\mathscr{C} = \emptyset$.

This means that if there are no arguments against a claim α and at least one argument for α in Φ then α should be accepted in Φ. Note that both γ_0 and κ_0 are well-behaving as well as all categorizers and accumulators considered in [Besnard and Hunter, 2001]. Furthermore, if Φ is consistent then every sentence $\alpha \in \Phi$ is accepted by Φ with respect to every well-behaving categorizer and well-behaving accumulator.

Let $\mathscr{K} \subseteq \mathscr{L}(\mathsf{At})$ be a consistent set of sentences, and let γ be some well-behaving categorizer and κ be some well-behaving accumulator. We consider again a selective revision $\circ$ of the form introduced in Definition 3. In order to determine the outcome of the non-prioritized revision $\mathscr{K} \circ \Phi$ for some $\Phi \subseteq \mathscr{L}(\mathsf{At})$ we implement a transformation function f that checks for every sentence $\alpha \in \Phi$ whether α is accepted in $\mathscr{K} \cup \Phi$. Note that although $\mathscr{K}$ is consistent the union $\mathscr{K} \cup \Phi$ is not necessarily consistent which gives rise to an argumentative evaluation. In the following, we consider *two* different transformation functions based on deductive argumentation.

Definition 16 (Skeptical Transformation Function). We define the *skeptical transformation function* $\mathsf{S}_{\mathscr{K}}^{\gamma,\kappa}$ via

$$S_{\mathcal{H}}^{\gamma,\kappa}(\Phi) = \{\alpha \in \Phi \mid \mathcal{H} \cup \Phi \vdash_{\kappa,\gamma} \alpha\}$$

for every $\Phi \subseteq \mathcal{L}(\mathsf{At})$.

Definition 17 (Credulous Transformation Function). We define the *credulous transformation function* $C_{\mathcal{H}}^{\gamma,\kappa}$ via

$$C_{\mathcal{H}}^{\gamma,\kappa}(\Phi) = \{\alpha \in \Phi \mid \mathcal{H} \cup \Phi \not\vdash_{\kappa,\gamma} \neg\alpha\}$$

for every $\Phi \subseteq \mathcal{L}(\mathsf{At})$.

In other words, the value of $S_{\mathcal{H}}^{\gamma,\kappa}(\Phi)$ consists of those sentences of Φ that are accepted in $\mathcal{H} \cup \Phi$ and the value of $C_{\mathcal{H}}^{\gamma,\kappa}(\Phi)$ consists of those sentences of Φ that are not rejected in $\mathcal{H} \cup \Phi$. There is a subtle difference in the behavior of these two transformation functions as the following example shows.

Example 4. Let $\mathcal{H}_1 = \{a\}$ and $\Phi_1 = \{\neg a\}$. Note that there is exactly one argument tree τ_1 for $\neg a$ and one argument tree τ_2 for a in $\mathcal{H}_1 \cup \Phi_1$. In τ_1 the root is the argument $\mathscr{A} = \langle\{\neg a\}, \neg a\rangle$ which has the single canonical undercut $\mathscr{B} = \langle\{a\}, a\rangle$. In τ_2 the situation is reversed and the root of τ_2 is the argument $\mathscr{B}$ which has the single canonical undercut $\mathscr{A}$. Therefore, the argument structure for $\neg a$ is given via $\Gamma_{\mathcal{H} \cup \Phi}(\neg a) = (\{\tau_1\}, \{\tau_2\})$. This implies that $\gamma_0(\tau_1) = \gamma_0(\tau_2) = 0$ and $\kappa_0(\gamma_0(\Gamma_{\mathcal{H} \cup \Phi}(a))) = \kappa_0(\langle 0, 0\rangle) = 0$. Thus, $\mathcal{H} \cup \Phi$ is undecided about both $\neg a$ and a. Consequently, it follows that

$$S_{\mathcal{H}_1}^{\gamma_0,\kappa_0}(\Phi_1) = \emptyset \qquad\qquad C_{\mathcal{H}_1}^{\gamma_0,\kappa_0}(\Phi_1) = \{\neg a\}.$$

∎

Let $*$ be some (prioritized) multiple base revision operator, γ some categorizer, and κ some accumulator. Using the skeptical transformation function we can define the *skeptical argumentative revision* $\circ_S^{\gamma,\kappa}$ following Definition 3 via

$$\mathcal{H} \circ_S^{\gamma,\kappa} \Phi = \mathcal{H} * S_{\mathcal{H}}^{\gamma,\kappa}(\Phi) \qquad (1)$$

for every $\Phi \subseteq \mathcal{L}(\mathsf{At})$ and using the credulous transformation function we can define the *credulous argumentative revision* $\circ_C^{\gamma,\kappa}$ via

$$\mathcal{H} \circ_C^{\gamma,\kappa} \Phi = \mathcal{H} * C_{\mathcal{H}}^{\gamma,\kappa}(\Phi) \qquad (2)$$

for every $\Phi \subseteq \mathcal{L}(\mathsf{At})$.

Example 5. We continue Example 4. Let $*$ be some prioritized multiple base revision. Then it follows that $\mathcal{H}_1 \circ_S^{\gamma_0,\kappa_0} \Phi_1 = \{a\}$ and $\mathcal{H}_1 \circ_C^{\gamma_0,\kappa_0} \Phi_1 = \{\neg a\}$. ∎

We now investigate the formal properties of the transformation functions $S_{\mathcal{H}}^{\gamma,\kappa}$ and $C_{\mathcal{H}}^{\gamma,\kappa}$ and the resulting revision operators $\circ_S^{\gamma,\kappa}$ and $\circ_C^{\gamma,\kappa}$.

Proposition 2. *Let γ be a well-behaving categorizer and κ be a well-behaving accumulator. Then the transformation functions $S_{\mathcal{H}}^{\gamma,\kappa}$ and $C_{\mathcal{H}}^{\gamma,\kappa}$ satisfy inclusion, weak inclusion, weak extensionality, consistency preservation and weak maximality.*

Proof.

Inclusion. This is satisfied by definition as for $\alpha \in S_{\mathcal{H}}^{\gamma,\kappa}(\Phi)$ and each $\alpha \in C_{\mathcal{H}}^{\gamma,\kappa}(\Phi)$ it follows $\alpha \in \Phi$.

Weak Inclusion. This follows directly from the satisfaction of inclusion.

Weak Extensionality. Let $\Phi \cong \Phi'$ and let $\sigma : \Phi \to \Phi'$ be a bijection such that for every $\phi \in \Phi$ it holds that $\phi \equiv \sigma(\phi)$. We extend σ to $\mathcal{H}$ via $\sigma(\psi) = \psi$ for every $\psi \in \mathcal{H}$. If $\Psi \subseteq \mathcal{H} \cup \Phi$ we abbreviate

$$\sigma(\Psi) = \bigcup_{\psi \in \Psi} \{\sigma(\psi)\}.$$

Let $\langle \Psi, \phi \rangle$ be an argument for some $\phi \in \Phi$ with respect to $\mathcal{H} \cup \Phi$. Then $\langle \sigma(\Psi), \sigma(\phi) \rangle$ is an argument for $\sigma(\phi)$ in $\mathcal{H} \cup \Phi'$. It follows that if τ is an argument tree for $\langle \Psi, \phi \rangle$ in $\mathcal{H} \cup \Phi$ then τ' is an argument tree for $\langle \sigma(\Psi), \sigma(\phi) \rangle$ in $\mathcal{H} \cup \Phi'$ where τ' is obtained from τ by replacing each sentence ϕ with $\sigma(\phi)$. This generalizes also to argument structures and it follows that

$$\kappa(\gamma(\Gamma_{\mathcal{H} \cup \Phi}(\phi))) = \kappa(\gamma(\Gamma_{\mathcal{H} \cup \Phi'}(\sigma(\phi)))).$$

Hence, $\phi \in S_{\mathcal{H}}^{\gamma,\kappa}(\Phi)$ if and only if $\sigma(\phi) \in S_{\mathcal{H}}^{\gamma,\kappa}(\Phi')$ for every $\phi \in \Phi$. It follows that $S_{\mathcal{H}}^{\gamma,\kappa}(\Phi) \cong S_{\mathcal{H}}^{\gamma,\kappa}(\Phi')$. The same is true for $C_{\mathcal{H}}^{\gamma,\kappa}$.

Consistency Preservation. Every subset of a consistent set of sentences is consistent and, due to inclusion, it holds that $S_{\mathcal{H}}^{\gamma,\kappa}(\Phi), C_{\mathcal{H}}^{\gamma,\kappa}(\Phi) \subseteq \Phi$ with consistent Φ.

Weak Maximality. If $\mathcal{H} \cup \Phi$ is consistent then for all arguments for a sentence $\alpha \in \Phi$ there do not exist any undercuts as these would have to entail the negation of some sentence of the argument for α which implies inconsistency of $\mathcal{H} \cup \Phi$. The argument structure $\Gamma_\Phi(\alpha) = (\mathcal{P}, \mathcal{C})$ consists of one or more single node trees $\mathcal{P}$ and $\mathcal{C} = \emptyset$. As both γ and κ are well-behaving it follows that $\kappa(\gamma(\Gamma_\Phi(\alpha))) > 0$ for each $\alpha \in \Phi$ and therefore $S_{\mathcal{H}}^{\gamma,\kappa}(\Phi) = \Phi$ and $C_{\mathcal{H}}^{\gamma,\kappa}(\Phi) = \Phi$.

In particular, note that both $S_{\mathcal{H}}^{\gamma,\kappa}$ and $C_{\mathcal{H}}^{\gamma,\kappa}$ do not satisfy either *consistency* or *maximality* in general.

Corollary 1. *Let γ be a well-behaving categorizer and κ be a well-behaving accumulator. Then both $\circ_S^{\gamma,\kappa}$ and $\circ_C^{\gamma,\kappa}$ are non-prioritized multiple base revision operators.*

Proof. This follows directly from Propositions 1 and 2.

Example 6. We continue Example 1 which considered $At = \{s, h, l, m, f, v\}$ with the following informal interpretations.

s : Anna is a surf fanatic
h : Anna travels to Hawaii
f : Anna has financial problems
l : Anna takes a loan
m : Anna has a lot of money
v : There is volcano activity on Hawaii

Now consider Anna's belief base $\mathcal{K}_1$ given via

$$\mathcal{K}_1 = \{s,\ s \Rightarrow h,\ l,\ l \Rightarrow m,\ m \Rightarrow h,\ m \Rightarrow \neg f\}.$$

Note that $\mathcal{K}_1 \vdash h$, i.e. Anna intends to go to Hawaii. As it was mentioned in Example 1, consider the new information $\Phi_1 = \{f,\ f \Rightarrow \neg h,\ v,\ v \Rightarrow \neg h\}$ stemming from communication with Anna's mother. In Φ_1 the mother of Anna tells her not to travel to Hawaii.

As one can see there a several arguments for and against h in $\mathcal{K}_1 \cup \Phi_1$, e.g., $\langle s, s \Rightarrow h, h \rangle$, $\langle f, f \Rightarrow \neg h, \neg h \rangle$. We do not go into details regarding the argumentative evaluation of the sentences in Φ_1 (Example 11 gives a complete description where these kind of details are shown). We only note that $\mathcal{K}_1 \cup \Phi_1$ accepts $f \Rightarrow \neg h$, but rejects f, v, and $v \Rightarrow \neg h$ with respect to γ_0 and κ_0. Furthermore, $\mathcal{K}_1 \cup \Phi_1$ accepts $\neg f$ and rejects $\neg v$ and $\neg(v \Rightarrow \neg h)$ with respect to γ_0 and κ_0 which means that both v and $v \Rightarrow \neg h$ are credulously accepted. Consequently, the values of $S^{\gamma_0, \kappa_0}_{\mathcal{K}_1}(\Phi_1)$ and $C^{\gamma_0, \kappa_0}_{\mathcal{K}_1}(\Phi_1)$ are given via

$$S^{\gamma_0, \kappa_0}_{\mathcal{K}_1}(\Phi_1) = \Phi_1 \setminus \{f, v, v \Rightarrow \neg h\} \quad \text{and} \quad C^{\gamma_0, \kappa_0}_{\mathcal{K}_1}(\Phi_1) = \Phi_1 \setminus \{f\}.$$

Let $*$ be some prioritized multiple base revision operator and define $\circ^{\gamma_0, \kappa_0}_S$ and $\circ^{\gamma_0, \kappa_0}_C$ via (1) and (2), respectively. Then some possible revisions of $\mathcal{K}_1$ with Φ_1 are given via

$$\mathcal{K}_1 \circ^{\gamma_0, \kappa_0}_S \Phi_1 = \{s,\ s \Rightarrow h,\ l,\ l \Rightarrow m,\ m \Rightarrow h,\ m \Rightarrow \neg f,\ f \Rightarrow \neg h\}$$
$$\mathcal{K}_1 \circ^{\gamma_0, \kappa_0}_C \Phi_1 = \{s,\ l \Rightarrow m,\ m \Rightarrow h,\ m \Rightarrow \neg f,\ f \Rightarrow \neg h,\ v \Rightarrow \neg h,\ v\}.$$

Note that it holds that $\mathcal{K}_1 \circ^{\gamma_0, \kappa_0}_S \Phi_1 \vdash h$ and $\mathcal{K}_1 \circ^{\gamma_0, \kappa_0}_C \Phi_1 \vdash \neg h$. ∎

For the evaluation of our approach the sophisticated algorithms presented in [Besnard and Hunter, 2008] can be used. However, the underlying problems of deciding whether a set of propositional formulae is consistent is NP-complete and deciding whether it entails a given formula is co-NP-complete [Garey and Johnson, 1979] such that no generally efficient implementation can be expected.

5 Argumentative Credibility-based Revision

In the previous sections we have shown how argumentation and selective revision can be combined to obtain a non-prioritized revision operator where the incoming information can be evaluated in order to decide if it is accepted or rejected. As shown above, the proposed operator only accepts new information if the information is justifiable with respect to an argumentative evaluation.

Below we will integrate the approach introduced in Section 4 in a multi-agent scenario with information stemming from different agents with different degrees of credibility. We will extend the approach of selective revision by deductive argumentation for this setting by including credibility information in the argumentative process. New information is evaluated based on the credibility of the source in combination with all arguments favoring and opposing the new information. The evaluation process determines which part of the new information is to be accepted for revision and thereupon incorporated into the belief base by an appropriate revision operator.

5.1 Credibility-based Epistemic Models

We continue with developing an epistemic model for an agent in a multi-agent environment that takes the credibilities of other agents into account. Our formalization is based on [Tamargo et al., 2012a]. Let $\mathbb{A} = \{A_1, \ldots, A_n\}$ be a finite set of agents.

Definition 18. If $\phi \in \mathscr{L}(\mathsf{At})$ and $A \in \mathbb{A}$ then $A{:}\phi$ is called an *information object*. Let $\mathfrak{I}(\mathscr{L}(\mathsf{At}), \mathbb{A})$ denote the set of all information objects wrt. $\mathscr{L}(\mathsf{At})$ and $\mathbb{A}$.

An information object $A{:}\phi$ states that ϕ has been uttered by A. For $\mathscr{I} \subseteq \mathfrak{I}(\mathscr{L}(\mathsf{At}), \mathbb{A})$ we abbreviate $\mathsf{Form}(\mathscr{I}) = \{\phi \mid A{:}\phi \in \mathscr{I}\}$. We extend the operator $\mathsf{Cn}()$ to $\mathfrak{I}(\mathscr{L}(\mathsf{At}), \mathbb{A})$ by defining $\mathsf{Cn}(\mathscr{I}) = \mathsf{Cn}(\mathsf{Form}(\mathscr{I}))$. Note that we do not consider nested information objects such as "A said that A' said that ϕ" to keep things simple. We leave this issue for future work.

Remark 1. Although the framework of deductive argumentation from the previous section has been phrased for the language $\mathscr{L}(\mathsf{At})$ we adopt the notions in the same manner for $\mathfrak{I}(\mathscr{L}(\mathsf{At}), \mathbb{A})$ by ignoring the annotated sources. For example, if $\mathscr{I} \subseteq \mathfrak{I}(\mathscr{L}(\mathsf{At}), \mathbb{A})$ and $A{:}\phi \in \mathfrak{I}(\mathscr{L}(\mathsf{At}), \mathbb{A})$ then we say that $\langle \mathscr{I}, \phi \rangle$ is an argument whenever $\langle \mathsf{Form}(\mathscr{I}), \phi \rangle$ is an argument.

For $\mathscr{I} \subseteq \mathfrak{I}(\mathscr{L}(\mathsf{At}), \mathbb{A})$ with $\mathsf{Form}(\mathscr{I}) \nvdash \bot$ and a total preorder $\leq$ on $\mathbb{A}$ (called *credibility order*) the tuple $(\mathscr{I}, \leq)$ is called a *belief base*. If $\mathscr{K}_A = (\mathscr{I}_A, \leq_A)$ is the belief base of an agent A then $A' \leq_A A''$ means that A believes that A'' is at least as credible as A'. The strict relation $<_A$ and the equivalence relation $\equiv_A$ are defined as usual.

Example 7. Let $\mathbb{A} = \{A_1, A_2, A_3\}$ be a set of agents and consider the belief base $\mathscr{K}_{A_1} = (\mathscr{I}_{A_1}, \leq_{A_1})$ of agent A_1 given via

$$\mathscr{I}_{A_1} = \{A_1{:}\neg b, A_2{:}a, A_3{:}a \Rightarrow \neg b, A_3{:}c\}$$
$$\leq_{A_1} = A_1 <_{A_1} A_2 <_{A_1} A_3 \quad .$$

Observe that according to $\mathcal{K}_{A_1}$, A_1 believes that c has been uttered by A_3. Furthermore, A_1 believes that A_2 is less credible than A_3 and that himself is less credible than A_2.

Let $A \in \mathbb{A}$ be an agent and let $\mathcal{K}_A = (\mathcal{I}_A, \leq_A)$ be its belief base. The credibility order $\leq_A$ can be used to specify a preference relation among arguments. Let $\langle \mathcal{I}_1, \phi_1 \rangle, \langle \mathcal{I}_2, \phi_2 \rangle$ be two arguments with $\mathcal{I}_1, \mathcal{I}_2 \subseteq \mathfrak{I}(\mathcal{L}(\mathsf{At}), \mathbb{A})$. Then $\langle \mathcal{I}_1, \phi_1 \rangle$ is as least as preferred as $\langle \mathcal{I}_2, \phi_2 \rangle$ by A, denoted by $\langle \mathcal{I}_2, \phi_2 \rangle \preceq_A \langle \mathcal{I}_1, \phi_1 \rangle$ if and only if for all $B{:}\phi \in \mathcal{I}_1$ there is a $B'{:}\phi' \in \mathcal{I}_2$ such that $B' \leq_A B$. In other words, it holds $\langle \mathcal{I}_2, \phi_2 \rangle \preceq_A \langle \mathcal{I}_1, \phi_1 \rangle$ if and only if the least credible source in $\mathcal{I}_1$ is at least as credible as the least credible source of $\mathcal{I}_2$.

Example 8. Consider $\mathcal{K}_{A_1}$ of Example 7. Let $\langle \mathcal{I}_1, \neg b \rangle$ and $\langle \mathcal{I}_2, c \rangle$ be two arguments with $\mathcal{I}_1 = \{A_2{:}a, A_3{:}a \Rightarrow \neg b\}$ and $\mathcal{I}_2 = \{A_3{:}c\}$. According to $<_{A_1}$, A_2 is less credible than A_3 ($A_2 <_{A_1} A_3$) hence $\langle \mathcal{I}_1, \neg b \rangle \not\preceq_{A_1} \langle \mathcal{I}_2, c \rangle$.

5.2 Credibility-based Revision operation

Consider a multi-agent system with agents $\mathbb{A} = \{A_1, \ldots, A_n\}$ where each agent A_i ($i = 1, \ldots, n$) maintains its own belief base $\mathcal{K}_{A_i} = (\mathcal{I}_{A_i}, \leq_{A_i})$. That is, each agent has some subjective beliefs consisting of individual pieces of information annotated with the source of this information (possibly the agent itself) and some subjective ordering on the credibility of the agents in the system (including itself). When an agent A_j sends some pieces of information $\mathcal{I} \subseteq \mathcal{I}_{A_j}$ to some agent A_i the agent A_i has to deliberate on how to react to receiving $\mathcal{I}$. Clearly, A_i should not blindly— i. e., in a prioritized fashion—revise $\mathcal{I}_{A_i}$ by $\mathcal{I}$ but take into account the credibility of A_j wrt. $\leq_{A_i}$. Furthermore, as $\mathcal{I}$ may contain an information object $A_k{:}\phi$ with $A_k \neq A_j$, i. e., agent A_j forwards some information from A_k to A_i, agent A_i should also consider the credibility of A_k.

Our approach follows the ideas of Section 4 but also incorporates the role of credibilities. On receiving some pieces of information $\mathcal{I} \subseteq \mathcal{I}_{A_j}$ from some agent A_j agent A_i evaluates each $A{:}\phi \in \mathcal{I}$ by an argumentation procedure that results in either accepting or rejecting $A{:}\phi$ for revision. This argumentation procedure is regulated by agent A_i's assessment of the credibilities of the sources of information. In particular, information that comes from a more credible source is preferred to information that comes from a less credible source. For this, we extend our definition of the classical categorizer from Example 3 as follows.

Definition 19. Let $\mathcal{K}_A = (\mathcal{I}_A, \leq_A)$ be the belief base of an agent A, let $\mathcal{I} \subseteq \mathfrak{I}(\mathcal{L}(\mathsf{At}), \mathbb{A})$, and let τ be some argument tree for $A'{:}\phi$ in $\mathcal{I}$. Then define the *credibility categorizer* γ_A^c for A through $\gamma_A^c(\tau) = 1$ if $\mathsf{ch}_\tau(\mathsf{root}(\tau)) = \emptyset$ and through

$$\gamma_A^c(\tau) = 1 - \max\{\gamma_A^c(\tau') \mid \tau' \in \mathsf{ch}_\tau^{\mathcal{I}}(\mathsf{root}(\tau)) \text{ and } \mathsf{root}(\tau) \preceq_A \mathsf{root}(\tau')\}$$

otherwise.

Note that the credibility categorizer extends the classical categorizer as defined in Example 3 as he takes the subjective credibility order of agent A into account by only considering those sub-trees of a node where the root argument is at least as preferred as the node itself.

Example 9. Let $\mathbb{A} = \{A_1, A_2, A_3\}$ be a set of agents and consider the belief base $\mathcal{K}_{A_1} = (\mathcal{I}_{A_1}, \leq_{A_1})$ of agent A_1 where $\mathcal{I}_{A_1} = \{A_2{:}b, A_3{:}c\}$ and $<_{A_1} = A_1 <_{A_1} A_2 <_{A_1} A_3$. Let $\mathcal{I} = \{A_3{:}a \Rightarrow \neg b, A_2{:}a\}$. Note that there is exactly one argument tree τ_1 for $a \Rightarrow \neg b$ and one argument tree τ_2 for $a \wedge b$ in $\mathcal{I}_{A_1} \cup \mathcal{I}$. In τ_1 the root is the argument $\mathscr{A} = \langle \{A_3{:}a \Rightarrow \neg b\}, a \Rightarrow \neg b \rangle$ which has the single canonical undercut $\mathscr{B} = \langle \{A_2{:}a, A_2{:}b\}, a \wedge b \rangle$. In τ_2 the situation is reversed and the root of τ_2 is the argument $\mathscr{B}$ which has the single canonical undercut $\mathscr{A}$. Therefore, the argument structure for $a \Rightarrow \neg b$ is given via $\Gamma_{\mathcal{I}_{A_1} \cup \mathcal{I}}(a \Rightarrow \neg b) = (\{\tau_1\}, \{\tau_2\})$. We can see these argument trees in Figure 1. In τ_1 one can see that the only child of $\mathscr{A}$ is not considered when evaluating with $\gamma^c_{A_1}$ because A_2 is less credible than A_3 according to A_1. For this reason $\gamma^c_{A_1}(\tau_1) = 1$. However, in τ_2 the situation is reversed and $\mathscr{B}$ is considered by $\gamma^c_{A_1}$. For this reason $\gamma^c_{A_1}(\tau_2) = 0$.

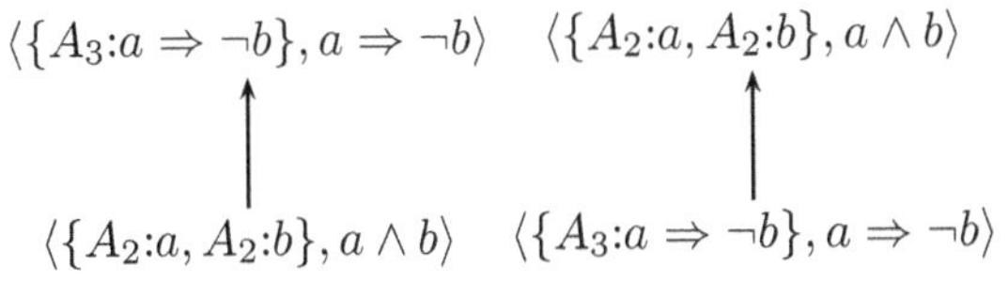

$$\langle\{A_3{:}a \Rightarrow \neg b\}, a \Rightarrow \neg b\rangle \qquad \langle\{A_2{:}a, A_2{:}b\}, a \wedge b\rangle$$
$$\uparrow \qquad\qquad\qquad \uparrow$$
$$\langle\{A_2{:}a, A_2{:}b\}, a \wedge b\rangle \qquad \langle\{A_3{:}a \Rightarrow \neg b\}, a \Rightarrow \neg b\rangle$$

Fig. 1 Argument trees in Example 9

We use the credibility categorizer to evaluate new information $\mathcal{I} \subseteq \mathcal{I}_{A_j}$ received by an agent A_i from agent A_j on an argumentative basis and by taking credibilities into account. As before, we say that an agent A_i with belief base $\mathcal{K}_{A_i} = (\mathcal{I}_{A_i}, \leq_{A_i})$ *accepts* an information object $A{:}\phi \in \mathcal{I}$ wrt. $\mathcal{I}$ if and only if

$$\kappa^c(\mathcal{P}, \mathcal{C}) = \sum_{\tau \in \mathcal{P}} \gamma^c_{A_i}(\tau) - \sum_{\tau \in \mathcal{C}} \gamma^c_{A_i}(\tau) > 0 \qquad (3)$$

where $\Gamma_{\mathcal{I}_{A_j} \cup \mathcal{I}}(A{:}\phi) = (\mathcal{P}, \mathcal{C})$ is the argument structure for $A{:}\phi$ wrt. $\mathcal{I}_{A_j} \cup \mathcal{I}$, cf. the definition of the simple accumulator in Example 3. Equation (3) means that A_i accepts $A{:}\phi$ if there are more reasons to believe in ϕ as there are to believe in $\neg \phi$. Using the notion of acceptance we define transformation functions C_{A_i} and S_{A_i} for agent A_i via

$$C_{A_i}(\mathcal{I}) = \{A{:}\phi \in \mathcal{I} \mid A_i \text{ accepts } A{:}\phi \text{ wrt. } \mathcal{I}\}$$
$$S_{A_i}(\mathcal{I}) = \{A{:}\phi \in \mathcal{I} \mid A_i \text{ does not accept } A'{:}\neg \phi \text{ wrt. } \mathcal{I} \text{ for some } A'\}$$

Note that—in contrast to the transformation functions discussed before—the codomains of C_{A_i} and S_{A_i} are subsets of $\mathfrak{I}(\mathscr{L}(\mathrm{At}), \mathbb{A})$ instead of $\mathscr{L}(\mathrm{At})$.

We now turn to the issue of revising $\mathscr{I}_{A_i}$ in a prioritized fashion by $C_{A_i}(\mathscr{I})$ and $S_{A_i}(\mathscr{I})$, respectively. We do this by exploiting the Levi-identity for belief revision [Alchourron et al., 1985], i. e., by first contracting $\mathscr{I}_{A_i}$ by the complement of $C_{A_i}(\mathscr{I})$ ($S_{A_i}(\mathscr{I})$), which is to be defined, and then expanding by $C_{A_i}(\mathscr{I})$ ($S_{A_i}(\mathscr{I})$). Let $-$ be some belief base contraction—e. g., a kernel contraction [Hansson, 2001]—and define a contraction $-_b$ on $\mathfrak{I}(\mathscr{L}(\mathrm{At}), \mathbb{A})$ for finite $\mathscr{I} \in \mathfrak{I}(\mathscr{L}(\mathrm{At}), \mathbb{A})$ and $\phi \in \mathscr{L}(\mathrm{At})$ through

$$\mathscr{I} -_b \phi = \{A{:}\phi' \in \mathscr{I} \mid \phi' \in \mathrm{Form}(\mathscr{I}) - \phi\} \ .$$

Then, for finite $\mathscr{I}, \mathscr{I}' \in \mathfrak{I}(\mathscr{L}(\mathrm{At}), \mathbb{A})$ with $\mathrm{Form}(\mathscr{I}) \not\vdash \bot$ define a (prioritized) revision $*$ through

$$\mathscr{I} * \mathscr{I}' = (\mathscr{I} -_b \bigvee_{\phi \in \mathrm{Form}(\mathscr{I}')} \neg \phi) \cup \mathscr{I}' \qquad (4)$$

and (non-prioritized) revisions $\circ_A^C$ and $\circ_A^S$ wrt. an agent A through

$$\mathscr{I} \circ_A^C \mathscr{I}' = \mathscr{I} * C_A(\mathscr{I}')$$
$$\mathscr{I} \circ_A^S \mathscr{I}' = \mathscr{I} * S_A(\mathscr{I}')$$

As we stated above, to revise $\mathscr{I}_{A_i}$ for a set of information objects $\mathscr{I}$, we should contract $\mathscr{I}_{A_i}$ by the complement of $\mathscr{I}$. For a given set of information objects $\mathscr{I}$ where $\mathrm{Form}(\mathscr{I}) = \{\phi_1, \ldots, \phi_n\}$, the complement of $\mathscr{I}$ is $\bigvee_{\phi \in \mathrm{Form}(\mathscr{I})} \neg \phi$. However, defining the complement of $\mathscr{I}$ as $\{\neg \phi_1, \ldots, \neg \phi_n\}$ and using a multiple contraction operator as in [Fermé et al., 2003] would not be sufficient as the following example illustrates.

Example 10. Assume $\mathscr{I}_{A_i} = \{\neg a \vee \neg b\}$ and $\mathscr{I} = \{a, b\}$. Any reasonable contraction operator, cf. [Fermé et al., 2003], would change $\mathscr{I}_{A_i}$ in a minimal way such that $\mathscr{I}_{A_i} - \{\neg a, \neg b\} \not\vdash \neg a$ and $\mathscr{I}_{A_i} - \{\neg a, \neg b\} \not\vdash \neg b$. In this case we get $\mathscr{I}_{A_i} - \{\neg a, \neg b\} = \mathscr{I}_{A_i}$, but obviously $\mathscr{I}_{A_i} \cup \mathscr{I} \vdash \bot$.

5.3 Analysis

We first illustrate our approach with an example.

Example 11. Consider again Example 2 given in Section 1 where the agent *Sam* wants to go on vacation. Sam's boss *Bob* doesn't want Sam to go on vacation at this time of the year and tells him that he has to do some work. However, Sam is aware of the fact that *Paul*, a good colleague, can do his work. Now Paul becomes ill—and therefore cannot take Sam's duties—so Sam has to revise his beliefs accordingly.

In this scenario let $\mathbb{A} = \{A_S, A_P, A_B, A_C\}$ where A_S is Sam, A_P is an Sam's colleague Paul, A_B is an Sam's boss, and A_C is his assigned client Carl. Consider the sentences v, w, r and i with the following informal interpretations.

$$v : \text{Sam go on vacation}$$
$$w : \text{There is work to do}$$
$$r : \text{Paul can do Sam's work}$$
$$i : \text{Paul is ill}$$

Now consider Sam's belief base $\mathcal{K}_{A_S}$ given via $\mathcal{I}_{A_S} = \{A_S{:}v, A_C{:}\neg w, A_P{:}r, A_P{:}r \Rightarrow v, A_S{:} \neg w \Rightarrow v\}$ as was shown in Example 2 of Section 1. Furthermore, let the credibility order among agents according to Sam ($<_{A_S}$) be defined via $A_S <_{A_S} A_P <_{A_S} A_B <_{A_S} A_C$.

As was introduced in Example 2 of Section 1, consider the new information $\Phi = \{A_B{:}w, A_P{:}i, A_B{:}i \Rightarrow \neg r\}$ stemming from communication with Sam's boss. As one can see there are some arguments for and against w, i and r in $\mathcal{I}_{A_S} \cup \Phi$, e. g., arguments for and against w are $\langle \{A_B{:}w\}, w \rangle$, $\langle \{A_C{:}\neg w\}, \neg w \rangle$.

We compute the argument structures $\Gamma_{\mathcal{I}_{A_S} \cup \Phi}(\alpha) = (\mathcal{P}, \mathcal{C})$ for each sentence $\alpha \in \text{Form}(\Phi)$ with respect to $\mathcal{I}_{A_S} \cup \Phi$ as follows.

(w). There is exactly one argument tree τ_1 for w and one argument tree τ_2 for $\neg w$ in $\mathcal{I}_{A_S} \cup \Phi$. In τ_1 the root is the argument $\mathcal{A} = \langle \{A_B{:}w\}, w \rangle$ which has the single canonical undercut $\mathcal{B} = \langle \{A_C{:}\neg w\}, \neg w \rangle$. In τ_2 the situation is reversed and the root of τ_2 is the argument $\mathcal{B}$ which has the single canonical undercut $\mathcal{A}$. Therefore, the argument structure for w is given via $\Gamma_{\mathcal{I}_{A_S} \cup \Phi}(w) = (\{\tau_1\}, \{\tau_2\})$. It follows that $\gamma^c_{A_S}(\tau_1) = 0$, $\gamma^c_{A_S}(\tau_2) = 1$ and $\sum_{\tau \in \mathcal{P}} \gamma^c_{A_S}(\tau_1) - \sum_{\tau \in \mathcal{C}} \gamma^c_{A_S}(\tau_2) = -1$ which means that w is *rejected*.

(i). There is exactly one argument tree τ_1 for i and one argument tree τ_2 for $\neg i$ in $\mathcal{I}_{A_S} \cup \Phi$. In τ_1 the root is the argument $\mathcal{A} = \langle \{A_P{:}i\}, i \rangle$ which has the single canonical undercut $\mathcal{B} = \langle \{A_B{:}i \Rightarrow \neg r, A_P{:}r\}, \neg i \rangle$. In τ_2 the situation is reversed and the root of τ_2 is the argument $\mathcal{B}$ which has the single canonical undercut $\mathcal{A}$. Therefore, the argument structure for i is given via $\Gamma_{\mathcal{I}_{A_S} \cup \Phi}(i) = (\{\tau_1\}, \{\tau_2\})$. It follows that $\gamma^c_{A_S}(\tau_1) = \gamma^c_{A_S}(\tau_2) = 0$ and $\sum_{\tau \in \mathcal{P}} \gamma^c_{A_S}(\tau_1) - \sum_{\tau \in \mathcal{C}} \gamma^c_{A_S}(\tau_2) = 0$ which means that the status of i is *undecided*.

($i \Rightarrow \neg r$). There is exactly one argument tree τ_1 for $i \Rightarrow \neg r$ and one argument tree τ_2 for $i \wedge r$ in $\mathcal{I}_{A_S} \cup \Phi$. In τ_1 the root is the argument $\mathcal{A} = \langle \{A_B{:}i \Rightarrow \neg r\}, i \Rightarrow \neg r \rangle$ which has the single canonical undercut $\mathcal{B} = \langle \{A_P{:}i, A_P{:}r\}, i \wedge r \rangle$. In τ_2 the situation is reversed and the root of τ_2 is the argument $\mathcal{B}$ which has the single canonical undercut $\mathcal{A}$. Therefore, the argument structure for $i \Rightarrow \neg r$ is given via $\Gamma_{\mathcal{I}_{A_S} \cup \Phi}(i \Rightarrow \neg r) = (\{\tau_1\}, \{\tau_2\})$. It follows that $\gamma^c_{A_S}(\tau_1) = 1$, $\gamma^c_{A_S}(\tau_2) = 0$ and $\sum_{\tau \in \mathcal{P}} \gamma^c_{A_S}(\tau_1) - \sum_{\tau \in \mathcal{C}} \gamma^c_{A_S}(\tau_2) = 1$ which means that $i \Rightarrow \neg r$ is *accepted*.

Due to the above evaluation the values of $C_{A_i}(\Phi)$ and $S_{A_i}(\Phi)$ can be determined by

$$S_{A_S}(\Phi) = \Phi \setminus \{A_B{:}w, A_P{:}i\} = \{A_B{:}i \Rightarrow \neg r\}$$
$$C_{A_S}(\Phi) = \Phi \setminus \{A_B{:}w\} = \{A_P{:}i, A_B{:}i \Rightarrow \neg r\}$$

If $*$ is defined via (4) we obtain

$$\mathscr{I}_{A_S} * S_{A_S}(\Phi) = \{A_S{:}v, A_C{:}\neg w, A_P{:}r, A_P{:}r \Rightarrow v,$$
$$A_S{:}\neg w \Rightarrow v, A_B{:}i \Rightarrow \neg r\}$$
$$\mathscr{I}_{A_S} * C_{A_S}(\Phi) = \{A_S{:}v, A_C{:}\neg w, A_P{:}r \Rightarrow v,$$
$$A_S{:}\neg w \Rightarrow v, A_P{:}i, A_B{:}i \Rightarrow \neg r\} \quad .$$

The above example illustrates that our approach is quite complex and involves a sophisticated deliberation process for deciding how a non-prioritized revision should be performed. One might ask whether the argumentative decision process is necessary and if the same results could be obtained by a simpler approach that is based on direct comparisons of credibilities. The following definition of a transformation function is suitable to implement this idea.

Definition 20. Let $\mathscr{K}_A = (\mathscr{I}_A, \leq_A)$ and $\mathscr{I} \subseteq \mathfrak{I}(\mathscr{L}(\text{At}), \mathbb{A})$. The function H_A is defined via

$$H_A(\mathscr{I}) = \{A_i{:}\phi \in \mathscr{I} \mid \forall \langle \mathscr{I}', \neg\phi \rangle, \mathscr{I}' \subseteq \mathscr{I}_A \cup \mathscr{I}, \langle \mathscr{I}', \neg\phi \rangle \preceq_A \langle \{A_i{:}\phi\}, \phi \rangle \}.$$

In other words, the function H_A rejects an $A'{:}\phi \in \mathscr{I}$ if there is a proof for $\neg\phi$ in $\mathscr{I}_A \cup \mathscr{I}$ such that the least credible source of this proof is strictly more credible than A'. Therefore, this definition of a transformation function intuitively implements the idea of how a credibility-based revision should be defined. The question arises whether this definition of a transformation is sufficient for realizing a meaningful revision based on credibilities. In Example 12, we show that this is not the case.

Example 12. Let $\mathbb{A} = \{A_1, A_2, A_3\}$ be a set of agents and consider the belief base $(\mathscr{I}_{A_1}, \leq_{A_1})$ of agent A_1 given via

$$\mathscr{I}_{A_1} = \{A_3{:}b, A_3{:}a \Rightarrow \neg b, A_2{:}\neg c\}$$
$$\leq_{A_1} = A_3 <_{A_1} A_2 <_{A_1} A_1 \quad .$$

Assume now that A_1 receives the new information $\mathscr{I}$ given via

$$\mathscr{I} = \{A_3{:}a \Rightarrow c, A_3{:}a\}$$

and consider the revision of $\mathscr{I}_{A_1}$ by $\mathscr{I}$. Observe that

$$C_{A_1}(\mathscr{I}) = \{A_3{:}a \Rightarrow c\}$$
$$H_{A_1}(\mathscr{I}) = \{A_3{:}a \Rightarrow c, A_3{:}a\} = \mathscr{I}$$

If $*$ is defined via (4) we obtain

$$\mathscr{I}_{A_1} * C_{A_1}(\mathscr{I}) = \{A_3{:}b, A_3{:}a \Rightarrow \neg b, A_2{:}\neg c, A_3{:}a \Rightarrow c\}$$
$$\mathscr{I}_{A_1} * H_{A_1}(\mathscr{I}) = \{A_3{:}b, A_3{:}a \Rightarrow c, A_3{:}a\} \quad .$$

As one can see, the revision based on C_{A_1} differs from the revision based on H_{A_1} which stems from $A_3{:}a \in H_{A_1}(\mathscr{I})$ and $A_3{:}a \notin C_{A_1}(\mathscr{I})$. The reason for $A_3{:}a \in H_{A_1}(\mathscr{I})$

is that there are two proofs for $\neg a$ in $\mathscr{I}_{A_1} \cup \mathscr{I} - \{A_3{:}b, A_3{:}a \Rightarrow \neg b\}$ and $\{A_2{:}\neg c, A_3{:} a \Rightarrow c\}$—and the credibility of the least credible agent in both proofs—which is A_3—is not strictly greater than the credibility of $A_3{:}a$—which is A_3 as well. Therefore, H_{A_1} accepts $A_3{:}a$ for revision. For C_{A_1} the situation is different. As $\langle\{A_3{:}a\}, a\rangle$ is the only argument for a and there are two arguments—$\langle\{A_3{:}b, A_3{:}a \Rightarrow \neg b\}, \neg a\rangle$ and $\langle\{A_2{:}\neg c, A_3{:}a \Rightarrow c\}, \neg a\rangle$—for $\neg a$ the argumentative evaluation of a results in the three argument trees depicted in Figure 2 and Figure 3. As all arguments appearing in the argument trees have the same least credible source A_3 no argument is ignored in the evaluation. Therefore the tree for argument $\langle\{A_3{:}a\}, a\rangle$ is categorized to 0 and both trees for $\neg a$ are categorized to 1. By (3) it follows that $A_3{:}a$ is not accepted for revision by C_{A_1}. An implication of this decision is that in $\mathscr{I}_{A_1} * C_{A_1}(\mathscr{I})$ the information $A_2{:}\neg c$—which is the single piece of information that comes from more credible information than any other piece of information—is retained.

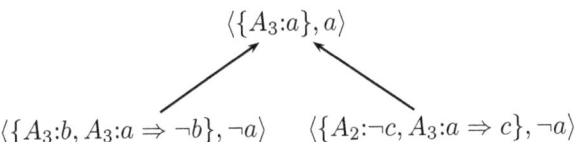

Fig. 2 Argument tree in Example 12

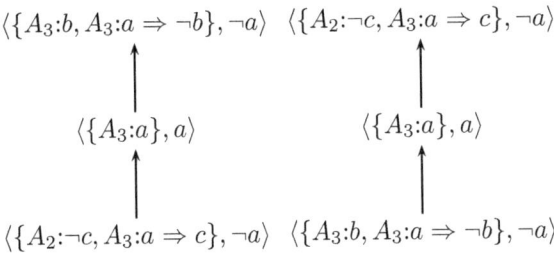

Fig. 3 Argument trees in Example 12

As for formal properties for transformation functions and belief revision our approach behaves well. For the following results source annotations of formulas can be neglected.

Proposition 3. *Let A be some agent. The transformation functions S_A and C_A satisfy* inclusion, weak inclusion, weak extensionality, consistency preservation, *and* weak maximality.

By exploiting Proposition 1, see also [Krümpelmann et al., 2011], we obtain the following result.

Corollary 2. *Let A be some agent. The operators $\circ_A^C$ and $\circ_A^S$ are non-prioritized multiple base revision operators.*

The above corollary shows that argumentative credibility-based revision conforms with expectations to non-prioritized revision.

6 Related Work

This chapter combines works on selected revision with deductive argumentation, and furthermore makes use of credibilities for revision in multiagent systems. It is based on previous work on credibility based multi-source belief revision as well as on the use of argumentation in multi-agent systems. More precisely, our approach to the model for multi-source belief revision has first been presented in [Tamargo *et al.*, 2012a; 2012b] and is combined with the selective revision operator as introduced in [Krümpelmann *et al.*, 2011]. The base approach of Tamargo et al. is similar in its idea to the approaches of [Dragoni *et al.*, 1994] and [Cantwell, 1998].

In [Dragoni *et al.*, 1994], it is considered that agents detect and store in tables the *nogoods*, which are the minimally inconsistent subsets of their knowledge bases. A *good* is a subset of the knowledge base such that: it is not inconsistent (it is not a superset of a *nogood*), and if augmented with whatever else assumption in knowledge base, it becomes inconsistent. In contrast to our approach, they do not remove beliefs to avoid a contradiction, they choose which is the new preferred *good* among them in knowledge base. In [Cantwell, 1998], a *scenario* (set of incoming information) presented by a source is treated as a whole and not sentence by sentence, and therefore, it can be inconsistent. A relation of *trustworthiness* is introduced over sets of sources and not between single sources. Besides, if two sources give the same piece of information α, and a single agent gives $\neg\alpha$, then α will be preferred, that is, the decision is based on majority.

Selective revision is one of the most general non-prioritized revision operator of the type *decision+revision* [Hansson, 1999]. Moreover it allows for partial acceptance of the input, in contrast to most other approaches. Apart from *decision+revision* approaches there are *expansion+consolidation* approaches to non-prioritized belief revision. These perform a simple expansion by the new information, i. e. $\mathcal{K} \cup \Phi$, and then apply a consolidation operator ! that restores consistency, i. e. $\mathcal{K} * \Phi = (\mathcal{K} \cup \Phi)!$. This approach is limited to belief bases since all inconsistent belief sets are equivalent, i. e. $Cn(\bot) = \mathcal{L}(At)$. An instantiation of such an operator that is similar to the setup used in this work has been presented in [Falappa *et al.*, 2002]. The input considered for revision consists of a set of sentences that form an explanation of some claim in the same form as the argument definition used here. However, as with all approaches of the type *expansion+consolidation*, new and old information are completely equivalent for consolidation. In contrast, the approach presented here makes use of two different mechanisms to first decide which part of the input and whether it shall be accepted and then perform prioritized belief revision of the old information.

While there has been some work on the revision of argumentation systems, very little work on the application of argumentation techniques for the revision process has been done so far, cf. [Falappa *et al.*, 2009]. In fact, the work most related to the work presented here makes use of negotiation techniques for belief revision [Booth, 2001; Zhang *et al.*, 2004], without argumentation. In the general setup of [Booth, 2001] a symmetric merging of information from two sources is performed by means of a negotiation procedure that determines which source has to reduce its information in each round. The information to be given up is determined by another function. The negotiation ends when a consistent union of information is reached. While this can be seen as a one step process of merging or consolidation in general, the formalism also allows to differentiate between the information given up from the first source and the second source. In [Booth, 2001], this setting is then successively biased towards prioritizing the second source which leads to representation theorems for operations equivalent to selective revision satisfying *consistent expansion* and for classic AGM operators. However, the negotiation framework used in [Booth, 2001] is very different from the argumentation formalism used here and also very different from the setup of selective revision. Moreover, the functions for the negotiation and concession are left abstract.

In [Zhang *et al.*, 2004] mutual belief revision is considered where two agents revise their respective belief states by information of the other agent. Both agents agree in a negotiation on the information that is accepted by each agent. The revisions of the agents are split into a selection function and two iterated revision functions which leads to operators satisfying *consistent expansion*. The selection function is then a negotiation function on two belief sets that represent the belief sets that each agent is willing to accept from the other agent. This setting has a very different focus as ours and also does not specify the selection function.

There is also work on the use of argumentation to reason about trust, with [Villata *et al.*, 2011] being the most recent work in this area. In [Villata *et al.*, 2011] a meta-argumentation approach is used to not only argue by taking the trustworthiness of information sources into account while evaluating the acceptance of arguments, but also to argue about the trustworthiness itself. In these approaches it is determined for a given set of arguments from different sources which ones are accepted and which ones are not. Dynamics of the system in terms of belief revision are not considered. In contrast, here we treat a belief revision problem of non-argumentative belief bases by employing argumentation in the selection process of belief revision.

While the concepts of trust and reputation are complex, in this approach we have taken the position that they can be seen as a kind of credibility value that the agents assign to each other. In contrast to this work, in [Sabater and Sierra, 2001] a model for reputation is presented that takes into account the social dimension of agents and a hierarchical ontology structure. They show how the model relates to other systems and provide initial experimental results about the benefits of using a social view on the modeling of reputation.

7 Conclusions

In this chapter we showed how argumentation and selective revision can be combined to obtain a non-prioritized revision operator where the incoming information can be evaluated in order to decide if it is accepted or rejected. First, we defined a revision operator that only accepts new information if the information is justifiable with respect to an argumentative evaluation. Then, we went into the details of a multi-agent revision framework based on using deductive argumentation and credibilities for deciding whether new information should be accepted for revision. We used the very general framework of multiple belief base revision and investigated a scenario where an agent has to revise its belief base of propositional formulas with a set propositional formulas. Formulas are annotated with an agent identifier which defines the source of the information. The represent the credibility of each agent by a total preorder over all agent identifiers which carries over to the annotated formulas. We developed an argumentation procedure based on credibility that decides which formulas of the set should be accepted for (prioritized) revision. We investigated the properties of our approach and compared it to a simple approach for multi-agent revision and other related work.

Our approach is concerned with revising the actual content of the belief base of an agent given some *static* credibility assessment. That is, the credibilities of the agents in the system are fixed (subjectively) and must not change. However, this may not be the case in real-world scenarios, see [Tamargo et al., 2011] for a discussion. In particular, information received from an agent may change the subjective assessment of its credibility: if an agent often gives good arguments or his information is confirmed by more credible agents then this agent should be assessed to be more credible as well. The dynamics of credibility assessments can be approached by interpreting credibilities not as annotations but as formulas on the object level and to use traditional revision methods for them as well. Part of future work is on investigating dynamical credibility assessments within our framework of multi-agent revision.

Acknowledgements:

This work is partially supported by PGI-UNS and CONICET Argentina. Patrick Krümpelmann is supported by the DFG, Collaborative Research Center SFB 876, project A5, http://sfb876.tu-dortmund.de.

References

[Alchourron et al., 1985] Carlos E. Alchourron, Peter Gärdenfors, and David Makinson. On the logic of theory change: Partial meet contraction and revision functions. *The Journal of Symbolic*

Logic, 50(2):510–530, 1985.

[Bench-Capon and Dunne, 2007] Trevor J. M. Bench-Capon and Paul E. Dunne. Argumentation in artificial intelligence. *Artificial Intelligence*, 171(10-15):619–641, 2007.

[Besnard and Hunter, 2001] Philippe Besnard and Anthony Hunter. A logic-based theory of deductive arguments. *Artificial Intelligence*, 128(1-2):203–235, 2001.

[Besnard and Hunter, 2008] P. Besnard and A. Hunter. *Elements of Argumentation*. The MIT Press, June 2008.

[Booth, 2001] Richard Booth. A negotiation-style framework for non-prioritised revision. In *Proceedings of TARK'01*, pages 137–150, 2001.

[Cantwell, 1998] John Cantwell. Resolving conflicting information. *Journal of Logic, Language and Information*, 7(2):191–220, 1998.

[Dalal, 1988] Mukesh Dalal. Investigations into a theory of knowledge base revision: Preliminary report. *Seventh National Conference on Artificial Intelligence*, (AAAI–88):475–479, 1988.

[Delgrande and Jin, 2012] James Delgrande and Yi Jin. Parallel belief revision: Revising by sets of formulas. *Artif. Intell.*, 176(1):2223–2245, January 2012.

[Dragoni *et al.*, 1994] Aldo F. Dragoni, Paolo Giorgini, and Paolo Puliti. Distributed belief revision versus distributed truth maintenance. In *Sixth Int. Conf. on Tools with Artificial Intelligence*, 1994.

[Dung, 1995] Phan Minh Dung. On the Acceptability of Arguments and its Fundamental Role in Nonmonotonic Reasoning, Logic Programming and n-Person Games. *Artificial Intelligence*, 77(2):321–358, 1995.

[Falappa *et al.*, 2002] Marcello. A. Falappa, Gabriele Kern-Isberner, and Guillermo R. Simari. Explanations, belief revision and defeasible reasoning. *Artificial Intelligence*, 141(1):1–28, 2002.

[Falappa *et al.*, 2009] Marcello. A. Falappa, Gabriele Kern-Isberner, and Guillermo R. Simari. Belief revision and argumentation theory. In *Argumentation in Artificial Intelligence*, pages 341–360. Springer, 2009.

[Fermé and Hansson, 1999] Eduardo L. Fermé and Sven Ove Hansson. Selective revision. *Studia Logica*, 63:331–342, 1999.

[Fermé *et al.*, 2003] Eduardo Fermé, Karina Saez, and Pablo Sanz. Multiple kernel contraction. *Studia Logica: An International Journal for Symbolic Logic*, 73(2):183–195, March 2003.

[Fuhrmann and Hansson, 1994] André Fuhrmann and Sven Ove Hansson. A survey of multiple contractions. *The Journal of Logic, Language and Information*, 3:39–76, 1994.

[Fuhrmann, 1991] André Fuhrmann. Theory contraction through base contraction. *The Journal of Philosophical Logic*, 20:175–203, 1991.

[Garcia and Simari, 2004] Alejandro J. Garcia and Guillermo R. Simari. Defeasible Logic Programming: An Argumentative Approach. *Theory and Practice of Logic Programming*, 4(1–2):95–138, 2004.

[Gärdenfors, 1978] Peter Gärdenfors. Conditionals and changes of belief. The logic and epistemology of scientific change. *Acta Philosophica Fennica*, 30:381–404, 1978.

[Gärdenfors, 1982] Peter Gärdenfors. Rules for rational changes of belief. Philosophical essays dedicated to Lennart åqvist on his fiftieth birthday. *Philosophical Studies*, 34:88–101, 1982.

[Gärdenfors, 1988] Peter Gärdenfors. *Knowledge in Flux: Modelling the Dynamics of Epistemic States*. The MIT Press, Bradford Books, Cambridge, Massachusetts, 1988.

[Garey and Johnson, 1979] Michael R Garey and David S Johnson. *Computers and intractability*, volume 174. Freeman New York, 1979.

[Hansson, 1993] Sven Ove Hansson. Theory contraction and base contraction unified. *The Journal of Symbolic Logic*, 58(2), 1993.

[Hansson, 1994a] Sven Ove Hansson. Kernel contraction. *The Journal of Symbolic Logic*, 59:845–859, 1994.

[Hansson, 1994b] Sven Ove Hansson. Taking belief bases seriously. *Logic and Philosophy of Science in Uppsala, Kluwer Academic Publishers*, pages 13–28, 1994.

[Hansson, 1999] Sven Ove Hansson. A survey of non-prioritized belief revision. *Erkenntnis*, 50(2-3):413–427, 1999.

[Hansson, 2001] Sven Ove Hansson. *A Textbook of Belief Dynamics*. Kluwer Academic Publishers, Norwell, MA, USA, 2001.

[Katsuno and Mendelzon, 1991] Hirofumi Katsuno and Alberto Mendelzon. Propositional knowledge base revision and minimal change. *Artificial Intelligence*, 52:263–294, 1991.

[Kern-Isberner and Krümpelmann, 2011] G. Kern-Isberner and P. Krümpelmann. A constructive approach to independent and evidence retaining belief revision by general information sets. In T. Walsh, editor, *Proceedings 22nd International Joint Conference on Artificial Intelligence, IJCAI'2011*, pages 937–942, Menlo Park, CA, 2011. AAAI Press.

[Krümpelmann *et al.*, 2011] Patrick Krümpelmann, M. Thimm, Marcello. A. Falappa, Alejandro J. Garcia, Gabriele Kern-Isberner, and Guillermo R. Simari. Selective Revision by Deductive Argumentation. In *Proceedings of the First International Workshop on the Theory and Applications of Formal Argumentation (TAFA'11)*, 2011.

[Levi, 1977] Isaac Levi. Subjunctives, dispositions and chances. *Synthese*, 34:423–455, 1977.

[Sabater and Sierra, 2001] Jordi Sabater and Carles Sierra. Regret: A reputation model for gregarious societies. *Fourth Workshop on Deception, Fraud and Trust in Agent Societies*, pages 61–69, 2001.

[Tamargo *et al.*, 2011] Luciano H. Tamargo, Marcelo A. Falappa, Alejandro J. García, and Guillermo R. Simari. A change model for credibility partial order. In *Proceeding of the 5th International Conference on Scalable Uncertainty Management (SUM)*, pages 317–330, 2011.

[Tamargo *et al.*, 2012a] Luciano H. Tamargo, Alejandro J. García, Marcelo A. Falappa, and Guillermo R. Simari. Modeling knowledge dynamics in multi-agent systems based on informants. *The Knowledge Engineering Review (KER)*, 27(1):87–114, 2012.

[Tamargo *et al.*, 2012b] Luciano H. Tamargo, Alejandro J. García, Matthias Thimm, and Patrick Krümpelmann. Argumentative credibility-based revision in multi-agent systems. In *Proceeding of the 13th Argentine Symposium on Artificial Intelligence (ASAI 2012)*, pages 128–139, 2012.

[Villata *et al.*, 2011] Serena Villata, Guido Boella, Dov M. Gabbay, and Leendert van der Torre. Arguing about the trustworthiness of the information sources. In Weiru Liu, editor, *ECSQARU'11*, volume 6717 of *Lecture Notes in Computer Science*, pages 74–85. Springer, 2011.

[Zhang *et al.*, 2004] Dongmo Zhang, Norman Foo, Thomas Meyer, and Rex Kwok. Negotiation as mutual belief revision. In *Proceedings of the 19th national conference on Artifical intelligence (AAAI'04)*, AAAI'04, pages 317–322. AAAI Press, 2004.

A Glance at Preemption Operators

Philippe Besnard, Éric Grégoire, and Sébastien Ramon

Abstract In contrast to traditional operators for belief revision which insist on dropping no belief when the incoming, new, belief is logically consistent with the current belief base, we discuss a family of operators that regard an incoming belief as a preemptive item so that any former belief subsuming it must be dropped. As such preemption operators cannot conform with the standard postulates for revision, a dedicated series of postulates is provided. Also, a characterization in terms of multiple contraction and expansion is established.

1 Introduction

In belief change, the AGM tradition [Alchourrón et al., 1985] focuses on the case that a belief base must accommodate a new, incoming, belief. As the baseline is that the incoming belief must be part of the resulting belief base, it then happens that the current belief base must be modified inasmuch as it contradicts the incoming belief. The famous AGM postulates express just this principle (for abstract revision operators over belief bases formalized as sets of logical formulas deductively closed under classical logic). However, the AGM postulates go beyond this mere principle by involving a feature dedicated to the case that the current belief base does not contradict the incoming belief: In such a case, the AGM postulates *require* that the belief base remain unaltered and be simply supplemented with the incoming belief.

This additional requirement (which by the way can also be found in the KM postulates [Katsuno and Mendelzon, 1991]) gets much less support from the original

---

Philippe Besnard
IRIT CNRS UMR 5505, 118 route de Narbonne, F-31065 Toulouse, France,
e-mail: besnard@irit.fr

Éric Grégoire · Sébastien Ramon
CRIL CNRS UMR 8188 - Université d'Artois, Rue Jean Souvraz SP18, F-62307 Lens, France,
e-mail: {gregoire, ramon}@cril.fr

idea of a belief base that must *evolve* due to an incoming belief. Here is an illustration.

Consider the situation that "There is a party every Wednesday" is in the current belief base. However, due to disrupted schedule or other reasons, there is a new, incoming, belief stating that "There is a party every week". As an afterthought, the latter belief is expected to take precedence over the original belief. The idea that the weekly party happens just on Wednesdays is dropped and the belief base is to undergo changes so that it no longer entails the subsuming belief "There is a party every Wednesday" but only the belief "There is a party every week". Importantly, changes must take place although the statement "There is a party every week" does *not* contradict by itself the current belief base (indeed, "There is a party every week" can be deduced from "There is a party every Wednesday").

Another illustration is as follows. Assume that the current belief base expresses "If Ruth wins at least one million euros then she'll quit her job". Assume further that a new belief "If Ruth wins at least two million euros then she'll quit her job" arises (from whatever clue). That makes the initial belief superseded, meaning that the new belief is favoured over the former idea about how much gain would prompt Ruth to quit her job. All this happens although there is no logical contradiction between these two beliefs as expressed above. Actually, it is even the case that the statement "If Ruth wins at least one million euros then she'll quit her job" subsumes the new belief "If Ruth wins at least two million euros then she'll quit her job".

This chapter aims at setting out a series of postulates for such operators (that return a new belief base containing the incoming belief but no other belief that would entail it) called *preemption operators*.

Section 2 and Section 3 introduce the technicalities needed throughout. In Section 4, postulates are provided for preemption operators. Section 5 includes a discussion, examples, and elementary properties. In Section 6, relationship with the AGM operators is provided by establishing that a preemption operator can be alternatively defined as multiple contraction [Fuhrmann and Hansson, 1994] (of appropriate formulas) merely followed by expansion [Alchourrón et al., 1985; Gärdenfors, 1988].

2 Formal Preliminaries

We consider classical logic throughout, assuming a propositional language $\mathscr{L}$ defined from a list of propositional variables and the usual connectives $\neg$ (negation), $\wedge$ (conjunction), $\vee$ (disjunction), $\Rightarrow$ (material implication). Also, $\bot$ stands for a contradiction and $\top$ stands for a tautology. A literal is a propositional variable or its negation. A *clausal formula* (called a *clause*) is a finite disjunction of literals. Lowercase letters denote formulas of $\mathscr{L}$. Uppercase letters denote sets of formulas, these being called *belief bases*. $Cn(A)$ denotes the set of all deductive consequences of A. A *theory* A is a deductively closed set of formulas, $A = Cn(A)$. Alternatively, $p \in Cn(A)$ can also be written $A \vdash p$. Thus, $\vdash p$ means that p is tautologous and

$\vdash \neg p$ that p is a contradiction. Two formulas p and q are logically equivalent, written $p \equiv q$, iff $p \in Cn(\{q\})$ and $q \in Cn(\{p\})$. Throughout, it is assumed that belief bases are deductively closed unless stated otherwise. A set of formulas A is consistent iff $\bot \notin Cn(A)$. $K_\bot$ is the trivial belief base, i.e., it consists of all formulas of $\mathscr{L}$ and $K_\top$ is the tautologous belief base, i.e., it consists of the tautologous formulas of $\mathscr{L}$. Lastly, the concept of strict implicant is of paramount importance here: f is a strict implicant of g iff $f \vdash g$ and $g \nvdash f$.

In such a context, a preemption operator $\circledast$ maps a theory K (belief base) and a formula g (incoming belief) to $K \circledast g$ which is meant to represent the resulting belief base. As illustrated above, the problem intuitively consists in turning K such that $K \vdash f$ (where f is some strict implicant of g) into $K' = K \circledast g$ such that $K' \vdash g$ but $K' \nvdash f'$ whenever f' is a strict implicant of g in a relevant class.

Mention of relevant class comes from the fact that not all strict implicants are of interest: Should K and g be taken from clausal propositional logic, only prime implicants f' of g are taken into account [Besnard et al., 2011]. When generalizing the problem to other logics, some more complex notions of strict implicants are needed [Besnard et al., 2013].

3 AGM Postulates

In the AGM paradigm, the operators apply to a base K and a formula g. The most basic of the AGM operators is expansion [Alchourrón et al., 1985; Gärdenfors, 1988].

Expansion [Alchourrón et al., 1985]
The postulates for the expansion of K by g, denoted $K + g$, are:
- (**K+1**) $K + g$ is a theory. (closure)
- (**K+2**) $g \in K + g$. (success)
- (**K+3**) $K \subseteq K + g$. (inclusion)
- (**K+4**) If $g \in K$ then $K + g = K$. (vacuity)
- (**K+5**) If $K \subseteq H$ then $K + g \subseteq H + g$. (monotony)
- (**K+6**) $K+g$ is the smallest set obeying $(K+1) - (K+5)$. **(minimality)**

Based on expansion, the postulates for AGM revision are then as follows.

Revision [Alchourrón et al., 1985]
The postulates for the revision of K by g, denoted $K * g$, are:
- (**K∗1**) $K * g$ is a theory. (closure)
- (**K∗2**) $g \in K * g$. (success)
- (**K∗3**) $K * g \subseteq K + g$. (inclusion)
- (**K∗4**) If $\neg g \notin K$ then $K + g \subseteq K * g$. (vacuity)
- (**K∗5**) $K * g = K_\bot$ iff $\vdash \neg g$. (consistency)
- (**K∗6**) If $g \equiv h$ then $K * g = K * h$. (extensionality)

Intuitively, postulates for preemption operators ⊛ are expected to be rather close to these (with ⊛ replacing ∗) except for (**K**∗**4**). An extra postulate is expected to cope with the fact that strict implicants of g must fail in the belief base after preempting.

4 Postulates for Preemption

Turning to postulates for preemption operators, a restriction is imposed on g being a clause. It is justified by Property 3 in Section 5 below.

(**K**⊛**1**) $K ⊛ g$ is a theory. **(closure)**

I.e., similarly to revision, the outcome of preemption is thus required to be deductived closed.

(**K**⊛**2**) $g \in K ⊛ g$. **(success of insertion)**

I.e., similarly to revision, the new piece of information is meant to be part of the resulting belief base.

(**K**⊛**3**) $K ⊛ g \subseteq K + g$. **(inclusion)**

I.e., similarly to revision, the incoming belief is the only new item (as well as some of its deductive consequences modulo K) in the resulting base.

(**K**⊛**4**) If $G \cap K = \emptyset$ then $K \subseteq K ⊛ g$ else $K \subseteq Cn(K ⊛ g \cup G)$
where $G \stackrel{\text{def}}{=} \{g \Rightarrow f \mid f \text{ is a clausal strict implicant of } g\}$.

(reinstatement)

That is, preemption expels only items that can be reinstated through the formulas $g \Rightarrow f$ (for f clausal strict implicant of g). This means that a preemption operator discards (from the current belief base) nothing more than is needed for g to hold while none of its clausal strict implicants f do.

Why use formulas $g \Rightarrow f$ where the use of (clausal) strict implicants f is expected instead? The reason is that g being in $K ⊛ g$ due to (**K**⊛**2**), and $K ⊛ g$ being deductively closed due to (**K**⊛**1**), the formula $g \Rightarrow f$ is in $K ⊛ g$ if and only if f is in $K ⊛ g$.

Accordingly, (**K**⊛**3**) together with (**K**⊛**4**) mean that $K ⊛ g = K + g$ when no subsuming belief threaten (which is the good case for preemption). This mirrors the fact that (**K**∗**3**) together with (**K**∗**4**) yield $K ⊛ g = K + g$ when g does not contradict K (which is the good case for revision).

The clausal restriction is not specific to (**K**⊛**4**), it merely is a consequence of the general restriction to g being a clause (see next).

(**K**⊛**5**) $f \notin K ⊛ g$ for all clausal strict implicants f of g.

(success of preemption)

(**K ⊛ 5**) is actually the whole point of preemption operators, as they differ from revision operators: No matter what, (clausal) subsuming beliefs are to be expelled from the belief base.

The restriction to g being a clause originates with (**K ⊛ 5**). The formal justification is Property 3, which shows that extending (**K ⊛ 5**) to all strict implicants would make ⊛ to collapse (see details in Section 5).

Lastly, there is no need for a postulate corresponding to (**K ∗ 5**) because it would follow from (**K ⊛ 5**) according to Property 2.

(**K ⊛ 6**) If $g \equiv h$ then $K \circledast g = K \circledast h$. (**extensionality**)

That is, similarly to revision, the outcome of preempting does not depend on any syntactical form of g.

5 Discussion, Examples and Elementary Properties

The first postulate, (**K ⊛ 1**), follows the AGM tradition by identifying a belief base with its deductive closure, even though doing so may seem to cause some confusion. Indeed, let us return to the first motivating example in the introduction. As already mentioned, "There is a party every week" is a consequence of "There is a party every Wednesday". If the belief base is identified with its deductive closure, then the case that the initial belief base contains "There is a party every Wednesday" is a case that this belief base also contains "There is a party every week". Importantly, what happens when the incoming belief is "There is a party every week" is that the belief "There is a party every Wednesday" must definitely be taken into account. Whether the incoming belief is already in the belief base is just secondary, and even more secondary is whether it occurs explicitly or implicitly as a consequence of another belief. The point here is that *whatever* the representation of the belief base (as a deductively closed set of formulas or as a set of representatives for these), the belief "There is a party every Wednesday" must be taken into account. It makes no difference whether its consequences (such as "There is a party every week") are explicitly represented. Indeed, it has been shown [Bienvenu et al., 2008] that it is possible to have recourse to representatives capturing deductively closed sets of formulas and still obtain syntax-insensitive operators for contraction and revision.

In fact, preemption is not for applications where repetition of information (whether previously implicit or explicit) is ineffective. Preemption does mean that repetition of implicit information is significant, being indicative there is something wrong with the belief base as regards the incoming belief and that the belief base must evolve. This of course goes hand to hand with the second postulate, (**K ⊛ 2**), that requires the incoming belief to be part of the resulting belief base.

Postulates (**K ⊛ 4**) and (**K ⊛ 5**) delineate what is to be changed in the belief base in view of the incoming belief. Let us focus on the case that the incoming belief is a consequence of a currently held belief. Certainly, there is a reason for bringing up this *specific* consequence: If a specific consequence of a currently held belief is brought up as relevant information, then it must be "the real thing" and it therefore makes other beliefs it follows from to look bad. That is exactly what $(\mathbf{K} \circledast 5)$ does: discard those beliefs. Again, let us insist that preemption is not universal (see the previous paragraph) and it is not the role of the postulates to determine what cases fall under the umbrella of preemption, the postulates only characterize what is supposed to happen when preemption does apply. Accordingly, $(\mathbf{K} \circledast 5)$ indicates what beliefs must be discarded *provided* preemption is right for the application at hand. Now, $(\mathbf{K} \circledast 5)$ does not indicate to what extent such discarding is to take place (i.e., where discarding must stop). This is what $(\mathbf{K} \circledast 4)$ takes care of. It rules that a belief, if not inferring the incoming belief, must be kept in the resulting belief base. It is actually more precise than that but the technical details are justified in Section 4 and need not be repeated in the current discussion about intuitions underlying our postulates.

$(\mathbf{K} \circledast 6)$ is the counterpart to $(\mathbf{K} \circledast 1)$ for the incoming belief. As we are demanding that our approach be syntax-insensitive with respect to the content of the belief base (this is what $(\mathbf{K} \circledast 1)$ amounts to), we must hold a similar requirement with respect to the incoming belief.

As an illustration, let us now consider all of our postulates through our first example. Let $Mo, Tu, We, \ldots, Su$ be propositional symbols standing for "There is a party every Monday", ..., "There is a party every Sunday". We use another propositional symbol, Any, standing for "There is at least one party every week, not always the same day". Background knowledge includes a formula stating that if the party does not always happen on the same day, then it neither always happens on Monday, nor always on Tuesday and so on:

$$Any \Rightarrow \neg Mo \wedge \cdots \wedge \neg Su \quad (\dagger)$$

Therefore,

$$K = Cn\left(\left\{\begin{array}{c} We, \\ Any \Rightarrow \neg Mo \wedge \cdots \wedge \neg Su \end{array}\right\}\right)$$

which captures the situation that, apart from common sense knowledge $(\dagger)$, the only piece of information available about the dates of our parties is that they are held on Wednesdays. Then,

$$g = Mo \vee Tu \vee We \vee Th \vee Fr \vee Sa \vee Su \vee Any$$

is the incoming belief, expressing "There is a party every week". The fact that it is brought up, although it is already supposed to be held —as a consequence of the former belief "There is a party every Wednesday"— sounds like a correction to the belief that our parties happen on Wednesdays. Accordingly, the effect of $(\mathbf{K} \circledast 5)$ is that for all proper subclauses c of g

$$K \circledast g \not\vdash c.$$

E.g., $K \circledast g \not\vdash Mo \vee Tu \vee We$. More importantly,

$$K \circledast g \not\vdash We.$$

Turning to the effect of (**K**$\circledast$**4**), observe that (†) is in K but not in G (G is defined as $\{g \Rightarrow f \mid f \text{ is a clausal strict implicant of } g\}$, cf Section 4). Applying (**K**$\circledast$**4**), (†) is in $K \circledast g$. In other words, common sense knowledge is preserved in the resulting belief base (this need not be true in general but is true in most applications as common sense knowledge rarely happens to be represented by formulas from G).

Alternative representations are possible, for instance

$$g = We \vee X$$

where X reads "There is a party every week but not always on Wednesdays" while

$$K = Cn\left(\left\{\begin{array}{l} We, \\ X \Rightarrow \neg We \end{array}\right\}\right).$$

According to (**K**$\circledast$**5**),

$$K \circledast g \not\vdash We.$$

Due to (**K**$\circledast$**4**), background knowledge ($X \Rightarrow \neg We$) is preserved:

$$K \circledast g \vdash X \Rightarrow \neg We.$$

Since $K \circledast g \vdash g$ by (**K**$\circledast$**2**), it follows that $K \circledast g \vdash (X \wedge \neg We) \vee (We \wedge \neg X)$ as intuitively expected from preemption (that is, the initial belief that our parties were held on Wednesdays is dropped, and there simply remains the belief that there is a weekly party with no information on what day —possibly the same day every week, possibly not).

There need not be a unique set of beliefs obtained by discarding $g \Rightarrow f$'s from K. For example, let A (for "Availability") be a propositional symbol such that $A \in K$ and $A \Rightarrow We \in K$, if it is believed that availability is the reason for parties to be held on Wednesdays. There are at least two ways to give up We (i.e., parties happen on Wednesdays): Either availability fails ($K \circledast g \not\vdash A$ but $K \circledast g \vdash A \Rightarrow We$), or it fails to be a reason for parties to happen on Wednesdays ($K \circledast g \vdash A$ but $K \circledast g \not\vdash A \Rightarrow We$).

Let us now list a short series of elementary properties that shed some light on the postulates.

The first property states that the outcome of preempting is a tautologous belief base exactly in the case that g is tautologous.

Property 1. Let $\circledast$ satisfy (**K**$\circledast$**1**), together with (**K**$\circledast$**2**) and (**K**$\circledast$**5**). $K \circledast g = K_\top$ iff $\vdash g$.

The next property shows that the only way a trivial belief base results from preempting is by means of preempting by a contradiction.

Property 2. Let $\circledast$ satisfy (**K**$\circledast$**1**), together with (**K**$\circledast$**2**) and (**K**$\circledast$**5**). $K \circledast g = K_\bot$ iff $\vdash \neg g$.

The property below shows that $(\mathbf{K} \circledast 5)$ cannot be extended to all strict implicants of g because, together with $(\mathbf{K} \circledast 1)$ and $(\mathbf{K} \circledast 2)$, it would entail that preemption collapses (expelling all beliefs but g). In brief, Property 3 reveals that not all strict implicants are meaningful when examining what parts of K are to be kept. The restriction to g being a clause comes from the fact that the case is then clear-cut: Strict implicants that are not (equivalent with) subclauses of g do mention atoms unrelated to g, hence unrelated to determining what parts of K are to be kept. Of course, prime implicants happen to be the right notion here [Besnard et al., 2011].

Property 3. Let $\circledast$ satisfy $(\mathbf{K} \circledast 1)$ and $(\mathbf{K} \circledast 2)$. Then, $f \notin K \circledast g$ for all strict implicants f of g iff $K \circledast g$ is logically equivalent with g.

As by Property 4, it is otiose to check in *(reinstatement)* that disjunctions of $g \Rightarrow f_i$ (for distinct clausal strict implicants f_i's of g) are not in K.

Property 4. Let K and g be such that $(g \Rightarrow f) \notin K$ for all clausal strict implicants f of g. Then, there exist no clausal strict implicants i and j of g such that $(i \vee j) \not\equiv g$ and $(g \Rightarrow i) \vee (g \Rightarrow j) \in K$.

6 Characterization

By Levi's identity, revision can be defined [Gärdenfors, 1988] as contraction followed by expansion. Intuitively, preemption amounts to discarding the formulas $g \Rightarrow f$ (for f clausal strict implicant of g) so that g can hold while no such f do. Discarding is what contraction does. A preemption operator could then be captured as contraction followed by expansion. However, preemption would require a set of formulas to be contracted at once whereas contraction as originally introduced in [Alchourrón et al., 1985] applies to a single formula. It is therefore necessary to resort to multiple contraction [Fuhrmann and Hansson, 1994], which permits to contract a belief base K by a set of statements Λ, written $K \ominus \Lambda$, so that no statement in Λ can be inferred from $K \ominus \Lambda$.

Multiple contraction [Fuhrmann and Hansson, 1994]
The postulates for multiple contraction of K by Λ, written $K \ominus \Lambda$, are:
$(\mathbf{K} \ominus 1)$ $K \ominus \Lambda$ is a theory. **(closure)**
$(\mathbf{K} \ominus 2)$ $K \ominus \Lambda \subseteq K$. **(inclusion)**
$(\mathbf{K} \ominus 3)$ If $\Lambda \cap K = \emptyset$ then $K \ominus \Lambda = K$. **(vacuity)**
$(\mathbf{K} \ominus 4)$ If $\Lambda \cap Cn(\emptyset) = \emptyset$ then $\Lambda \cap (K \ominus \Lambda) = \emptyset$. **(success)**
$(\mathbf{K} \ominus 5)$ $K \subseteq Cn((K \ominus \Lambda) \cup \Lambda)$. **(recovery)**
$(\mathbf{K} \ominus 6)$ If $\Lambda \cong \Theta$ then $K \ominus \Lambda = K \ominus \Theta$. **(extensionality)**
$\Lambda \cong \Theta$ means that for each element of Λ there exists a logically equivalent element of Θ, and *vice versa*.
$(\mathbf{K} \ominus 7)$ $(K \ominus \Lambda) \cap (K \ominus \Theta) \subseteq K \ominus (\Lambda \cap \Theta)$. **(intersection)**
$(\mathbf{K} \ominus 8)$ $K \ominus \Theta \subseteq K \ominus (\Theta \cup \{\varphi\})$ if $\varphi \notin K \ominus \Theta$. **(non-deterioration)**
$(\mathbf{K} \ominus 9)$ If $\Lambda \cap (K \ominus \Theta) = \emptyset$ then $K \ominus \Theta \subseteq K \ominus (\Lambda \cup \Theta)$. **(conjunction)**

It happens that none of $(K \ominus 7) - (K \ominus 9)$ is needed below, they are given for exhaustiveness only.

Following the intuition outlined above, next is a general definition of an operator that first contracts K by all clausal stricts implicants of g (and by their equivalent modulo g) and then expands the resulting base by g itself.

Definition 1 ($|||$ operator). Let $\{f_1, f_2, \ldots, f_n, \ldots\}$ be the set of all the clausal strict implicants of g.

$$K \;|||\; g \stackrel{\text{def}}{=} (K \ominus \{g \Rightarrow f_i\}_{i=1,2,\ldots}) + g.$$

Given the definition $G \stackrel{\text{def}}{=} \{g \Rightarrow f \mid f \text{ is a clausal strict implicant of } g\}$ in $(K \circledast 4)$, $K \;|||\; g$ can also be written $(K \ominus G) + g$.

The following two theorems achieve the objective stated above, they show that preemption operators can alternatively be characterized by means of multiple contraction and expansion.

Theorem 1. *If $\ominus$ obeys $(K \ominus 1) - (K \ominus 6)$ and $+$ obeys $(K+1) - (K+6)$, then $|||$ satisfies $(K \circledast 1) - (K \circledast 6)$.*

Theorem 2. *Every $\circledast$ operator satisfying $(K \circledast 1) - (K \circledast 6)$ can be written as an $|||$ operator such that $\ominus$ satisfies $(K \ominus 1) - (K \ominus 6)$, and $+$ satisfies $(K+1) - (K+6)$.*

7 Related Work: Hansson's Replacement Operator

As is given in [Hansson, 2009], replacement permits to replace in a belief base K a proposition p by a proposition q, written $K|_q^p$. Similarly to Levi's identity [Gärdenfors, 1988], it is shown in [Hansson, 2009] that replacement could be captured as a contraction by $q \Rightarrow p$ followed by an expansion by q.

This is just what the characterization in the previous section establishes, when K contains exactly one prime implicant f of g. In such a situation,

$$K \;|||\; g = K|_g^f.$$

Here is an example. Let K be the deductive closure of $(a \vee b) \wedge z$ and let g be $a \vee b \vee c$. Clearly, $a \vee b$ is the only prime implicant of $a \vee b \vee c$ in K. So, it is enough to retract $(a \vee b \vee c) \Rightarrow (a \vee b)$ and to expand by $a \vee b \vee c$. That is, $K|_{a \vee b \vee c}^{a \vee b}$ does the job.

However, in the general case, a multiple replacement operator extending Hansson's would be needed. For instance, let K be the deductive closure of $(a \vee b) \wedge (c \vee d)$ and let g be $a \vee b \vee c \vee d$. Clearly, there are two prime implicants of $a \vee b \vee c \vee d$ in K, i.e., $a \vee b$ and $c \vee d$. Preemption requires both $g \Rightarrow a \vee b$ and $g \Rightarrow c \vee d$ to be retracted *at once* because retracting the disjunction is ineffective and retracting iteratively might cause to retract too much. Lastly, intersecting the replacements

$$\bigcap_{i=1,2,\ldots} K|_g^{f_i}$$

might as well cause to lose too much from K. Hence the need, in general, for a multiple replacement operator

$$K|_g^{\{f_1,f_2,\ldots\}}.$$

8 Conclusion

Preemption is in accordance with revision as they both insist upon the new, incoming, item to hold (in the resulting base) no matter what changes this might require to the current base.

The intuition underlying preemption operators makes them to differ from revision operators when the incoming belief is logically consistent with the current belief base (focusing on the case that K entails at least one $g \Rightarrow f_i$, whereas the opposite case gets preemption operators to behave exactly as revision operators). Also, they may differ when the belief base contradicts the incoming belief. In such a situation, indeed, a revision operator need not throw out all the $(g \Rightarrow f_i)$'s but a preemption operator must. A simple example is $K \vdash \neg a \wedge \neg b \wedge \neg c$ with g being $a \vee b \vee c$. A selection function γ can be picked such that the corresponding revision operator $*$ satisfies $K * g \vdash c \Rightarrow a \vee b$ for instance. In contrast, all preemption operators are such that $K \circledast g \not\vdash c \Rightarrow a \vee b$. This is imposed through $(K \circledast 5)$ and coincides here with the fact that g together with $c \Rightarrow a \vee b$ entail $a \vee b$ which is a clausal strict implicant of g.

As to the interesting case (K entails some clausal strict implicant f of g), a preemption operator is, ultimately, a contraction operator. The reason is that K entails g since K entails f, and the postulates for preemption then make K to be contracted to those current beliefs (including g itself!) that do not entail any $g \Rightarrow f_i$ (or, equivalently, any f_i since g must be in $K \circledast g$). The burden is how to get a systematic formulation capturing those beliefs? This is a topic for future work.

The current set of postulates restricts g to be a clause. A technical reason appears clearly enough from Property 3. However, no such restriction arises from general intuition about preemption: It still makes sense to think that current beliefs entailing the incoming belief should be given up, as common sense makes it quite clear that not all such beliefs are candidates for discard (e.g., $g \wedge x$ would not be such a candidate for x irrelevant with g). Hence, extension to non-clausal g is a due topic for future work.

Lastly, a characterization of preemption operators in terms of multiple contraction and expansion has been provided. Nevertheless, a more direct characterization is still desirable in the form of a representation theorem, again a topic for future work.

9 Acknowledgments

Part of this work has been supported by the *Région Nord/Pas-de-Calais* and the EC under a FEDER project.

References

[Alchourrón *et al.*, 1985] Carlos Alchourrón, Peter Gärdenfors, and David Makinson. On the logic of theory change: partial meet contraction and revision functions. *Journal of Symbolic Logic*, 50(2):510–530, 1985.

[Besnard *et al.*, 2011] Philippe Besnard, Éric Grégoire, and Sébastien Ramon. Enforcing logically weaker knowledge in classical logic. In *5th International Conference on Knowledge Science Engineering and Management (KSEM'11)*, pages 44–55. LNAI 7091, Springer, 2011.

[Besnard *et al.*, 2013] Philippe Besnard, Éric Grégoire, and Sébastien Ramon. Overriding subsuming rules. *International Journal of Approximate Reasoning*, 54(4):452–466, 2013.

[Bienvenu *et al.*, 2008] Meghyn Bienvenu, Andreas Herzig, and Guilin Qi. Prime implicate-based belief revision operators. In 18th *European Conference on Artificial Intelligence (ECAI'08)*, pages 741–742. IOS Press, 2008.

[Fuhrmann and Hansson, 1994] André Fuhrmann and Sven Ove Hansson. A survey of multiple contractions. *Journal of Logic, Language and Information*, 3(1):39–76, 1994.

[Gärdenfors, 1988] Peter Gärdenfors. *Knowledge in Flux: Modeling the Dynamics of Epistemic States*. MIT Press, 1988.

[Hansson, 2009] Sven Ove Hansson. Replacement—a sheffer stroke for belief change. *Journal of Philosophical Logic*, 38(2):127–149, 2009.

[Katsuno and Mendelzon, 1991] Hirofumi Katsuno and Alberto Mendelzon. On the difference between updating a knowledge base and revising it. In 2nd *International Conference on Principles of Knowledge Representation and Reasoning (KR'91)*, pages 387–394. Morgan Kaufmann, 1991.

Representation of basic improvement operators

Mattia Medina Grespan and Ramón Pino Pérez

Abstract Improvement operators, introduced by Konieczny and Pino Pérez, are a generalization of iterated revision operators of Darwiche and Pearl. One important difference is that improvement operators don't satisfy the success postulate (priority of the new piece of information). However, improvement operators have the property of increasing the plausibility of the new piece of information. Moreover, the improvement of beliefs ensures that given an epistemic state, for each new piece of information there is a finite number of iterations of the operator that lead to success. In previous investigations, this is a hypothesis over the operators which has not been characterized semantically. The goal of this paper is to characterize semantically the property of being successful by means of successive iterations, the cornerstone of improvement operators. This is realized by introducing the basic improvement operators and proving a very complete and accurate representation theorem for these operators.

1 Introduction

In belief revision, one of the most important guiding principles is *success*: the new piece of information has to prevail after revision. This can be expressed in the logic model of Alchourrón, Gardenfors and Makinson [Alchourrón *et al.*, 1985] for belief revision. In the most simple logical terms[1] the success postulate can be stated as follows:

Departamento de Matemáticas
Facultad de Ciencias
Universidad de Los Andes
Mérida, Venezuela

e-mail: {mattia,pino}@ula.ve

---

[1] Actually we state the postulate in propositional logic in the style of Katsuno and Mendelzon [Katsuno and Mendelzon, 1991].

(success) $\varphi \circ \alpha \vdash \alpha$

where φ is the belief base representing the "old" information, α is the new information, $\circ$ is a revision operator and $\varphi \circ \alpha$ is the result of revising φ by α. The revision operators obeying this principle are called *prioritized* revision operators.

Of course, there are many situations in which the success principle is desirable. Also, there are situations in which the success principle is not desirable or not totally adequate to model the situation. Many works have been devoted to the non-prioritized revision. We can cite the articles appeared in [Hansson, 1997] and also the Hansson survey [Hansson, 1999]. Among these approaches one can distinguish some families of operators such as the family of operators defined in [Hansson et al., 2001], called *credibility-limited revision operators*, where a successful revision is obtained only if the new information is a formula that belongs to a set of credible formulas[2]. In approaches of this kind there is a crisp situation with respect to the success postulate: after revision, the postulate either holds or does not hold, it depends on the new piece of information. However, in such frameworks there is no adequate way to express that after revision we have a little of success.

In order to be able to express that some small amount of success has occurred after revision, we need to enrich the framework, see for instance [Cantwell, 1997]. Perhaps the most natural way to say that a small amount of success has occurred after revision is through iterations of the operator: after a sufficient number of iterations we have success. In symbols we could put

$$(\cdots(\varphi \circ \alpha) \cdots \circ \alpha) \circ \alpha \vdash \alpha \tag{1}$$

where the number of iterations will depend upon the inputs α, φ of the operator. This is the basic idea which has been developed in a very general framework in [Konieczny and Pino Pérez, 2008; Konieczny et al., 2010]. Actually the language used to express the theory is essentially the framework of epistemic states proposed by Darwiche and Pearl [Darwiche and Pearl, 1997]. This kind of structure is very adequate to express good properties with respect to the iteration process as showed in [Darwiche and Pearl, 1997]. It is interesting to note that the notion of epistemic state we use is very general and can be instantiated to structures such as total preorders (our paradigmatic and preferred notion of concrete epistemic states) but also ordinal conditional functions (OCF's) proposed by Spohn [Spohn, 1988] or even a set of conditionals. The general notion of epistemic state will be also very adequate for our purposes.

The family of operators proposed in [Konieczny and Pino Pérez, 2008; Konieczny et al., 2010], called improvements operators, aims to improve the "plausibility" of the new information. This can be expressed without numbers thanks to the notion of epistemic state. Moreover, this is clearly expressed in the semantic side via the notion of assignments (functions mapping epistemic states into total preorders over the interpretations). In these works some representation theorems are

---

[2] See also the closely-related *screened revision* operators of [Makinson, 1997].

established, some classification of these improvement operators is done and some interesting features about the minimal change are showed.

However, there is a very strong hypothesis in those works. It is assumed that all operators concerned satisfy the postulate of iterated success[3], the postulate roughly described by expression (1). The aim of this work is double. In one hand, we want to put aside this strong hypothesis. On the other hand, we want to introduce new proof techniques for the representation theorems concerning the family of improvement operators.

Actually, the family of operators here considered, called family of basic improvement operators, is a variant of the family of improvement operators considered in [Konieczny and Pino Pérez, 2008; Konieczny et al., 2010]. The postulate expressing that, in certain cases, the "plausibility" of the new information improves (Postulate (I9) in [Konieczny and Pino Pérez, 2008], a postulate very close to property of independence of Jin and Thielscher [Jin and Thielscher, 2007] alias property (P) of Booth and Meyer [Booth and Meyer, 2006]) is modified. Here we adopt a postulate which expresses that at least a part of the new piece of information improves (see Postulate (BI11) in Section 4). However we need to add a couple of postulates saying essentially that there is no part of the new information which worsens. These postulates are like postulates (C3) and (C4) of Darwiche and Pearl.

We have to say a word about our representation theorem (Theorem 7) and our technique of proof. Unlike most of representation theorems in belief revision, going from the semantical side to the syntactical side is not trivial. The difficulty lies in the fact that we cannot suppose in the theorem that the representation equality[4] holds because this equation uses implicitly the postulate of iterated success. Thus we have to prove this postulate and also we have to prove the representation equality.

This work is organized as follows. Section 2 contains the very basic concepts used throughout the work. Section 3 recall some representation results about some classes of improvement operators. Section 4 contains the definition of our basic improvement operators, the semantic concepts and the main representation theorem. Section 5, the longest one, contains the proof of our representation theorem. We have chosen to put it in a section in the corpus of this work and not in an appendix, not only because the technique of proof seems to us very interesting but also because it reveals important aspects about the behavior of our operators. Section 6 contains some links with previous works in the area. In particular we give strong forms of some representation theorems that appeared in [Konieczny et al., 2010], recalled in Section 3. Finally Section 7 contains some concluding remarks and perspectives of work.

---

[3] Postulate (I1) in [Konieczny and Pino Pérez, 2008; Konieczny et al., 2010]. Postulate (BI1) in this work, see Section 4.

[4] The models of the beliefs of the epistemic state Ψ after iterated revision by α are equal to the minimal models of α with respect to the preorder $\leq_\Psi$. For a precise statement of this, see Equation (2).

2 Preliminaries

We consider a propositional language $\mathscr{L}$ defined from a finite set of propositional variables $\mathscr{P}$ and the standard connectives. Let $\mathscr{L}^*$ denote the set of consistent formulae of $\mathscr{L}$.

An interpretation ω is a total function from $\mathscr{P}$ to $\{0,1\}$. The set of all interpretations is denoted $\mathscr{W}$. An interpretation ω is a model of a formula $\phi \in \mathscr{L}$ if and only if it makes it true in the usual truth functional way. $[\![\alpha]\!]$ denotes the set of models of the formula α, i.e., $[\![\alpha]\!] = \{\omega \in \mathscr{W} \mid \omega \models \alpha\}$. When $\{\omega_1,..,\omega_n\}$ is a set of models we denote by $\varphi_{\omega_1,..,\omega_n}$ a formula such that $[\![\varphi_{\omega_1,..,\omega_n}]\!] = \{\omega_1,..,\omega_n\}$.

We will use epistemic states to represent the beliefs of the agent, as usual in iterated belief revision [Darwiche and Pearl, 1997]. An epistemic state Ψ represents the current beliefs of the agent, but also additional conditional information guiding the revision process (usually represented by a pre-order on interpretations, a set of conditionals, a sequence of formulae, etc). Let $\mathscr{E}$ denote the set of all epistemic states. A projection function $B : \mathscr{E} \longrightarrow \mathscr{L}^*$ associates to each epistemic state Ψ a consistent formula $B(\Psi)$, that represents the current beliefs of the agent in the epistemic state Ψ[5].

For simplicity purposes we will only consider in this paper consistent epistemic states and consistent new information. Thus, the change operators we consider are functions $\circ$ mapping an epistemic state and a consistent formula into a new epistemic state, i.e. in symbols, $\circ : \mathscr{E} \times \mathscr{L}^* \longrightarrow \mathscr{E}$. The image of a pair (Ψ, α) under $\circ$ will be denoted by $\Psi \circ \alpha$.

The following definition is very important in our setting:

Definition 1 *Given an operator $\circ$ and a natural number n, we define $\circ^n$ by recursion in the following way:*

$$\Psi \circ^0 \alpha = \Psi$$
$$\Psi \circ^{n+1} \alpha = (\Psi \circ^n \alpha) \circ \alpha$$

Now we define the operator $\star$ in the following way:

$$\Psi \star \alpha = \Psi \circ^n \alpha$$

where n is the first integer such that $B(\Psi \circ^n \alpha) \vdash \alpha$.

Note that $\star$ is undefined if there is no n such that $B(\Psi \circ^n \alpha) \vdash \alpha$. The associated operator $\star$ will be total, when for any pair Ψ, α there will exist n such that $B(\Psi \circ^n \alpha) \vdash \alpha$. Actually, this will be the postulate (BI1) below.

Finally, let $\leq$ be a total pre-order, i.e a reflexive ($x \leq x$), transitive (($x \leq y \wedge y \leq z$) $\rightarrow x \leq z$) and total ($x \leq y \vee y \leq x$) relation over $\mathscr{W}$ (actually the reflexivity follows from totality). Then the corresponding strict relation $<$ is defined as $x < y$ iff $x \leq y$

---

[5] As in most works on iterated revision, we have chosen this very general and abstract framework to define epistemic states (just objects Ψ and their logical belief $B(\Psi)$) because of its simplicity and its flexibility to capture many concrete representations of epistemic states. For a formal definition of epistemic states see [Benferhat et al., 2000].

Representation of basic improvement operators

and $y \not\leq x$, and the corresponding equivalence relation $\simeq$ is defined as $x \simeq y$ iff $x \leq y$ and $y \leq x$. Sometimes we say that x is indifferent to y when $x \simeq y$. We denote $w \ll w'$ when $w < w'$ and there is no w'' such that $w < w'' < w'$. We also use the notation $\min(A, \leq) = \{w \in A \mid \nexists w' \in A \; w' < w\}$, the minimal elements of A with respect to $\leq$. The set of total pre-orders over the interpretations of language will be denoted $\mathscr{TP}$.

When a set $\mathscr{W}$ is equipped with a total pre-order $\leq$, then this set can be split in different levels, giving the ordered sequence of its equivalence classes $\mathscr{W} = \langle S_0, \ldots S_n \rangle$. So $\forall x, y \in S_i \; x \simeq y$. We say in that case that x and y are at the same level of the pre-order. And $\forall x \in S_i \; \forall y \in S_j \; i < j$ implies $x < y$. We say in this case that x is in a lower level than y. We extend straightforwardly these definitions to compare subsets of equivalence classes, i.e if $A \subseteq S_i$ and $B \subseteq S_j$ then we say that A is in a lower level than B if $i < j$.

Definition 1. Let $\circ$ be a change operator such that the operator $\star$ associated is total. Let α, β and Ψ be two formulae and an epistemic state respectively. We say that α is below β with respect to Ψ, given $\circ$, denoted $\alpha \prec_\Psi^\circ \beta$ (or simply $\alpha \prec_\Psi \beta$ if there is no ambiguity about $\circ$) if and only if $\alpha \nvdash \bot$, $\beta \nvdash \bot$, $B(\Psi \star \alpha) \vdash B(\Psi \star (\alpha \vee \beta))$ and $B(\Psi \star \beta) \nvdash B(\Psi \star (\alpha \vee \beta))$.

The pair (α, β) is Ψ-consecutive, denoted $\alpha \lll_\Psi^\circ \beta$ (or simply $\alpha \lll_\Psi \beta$ if there is no ambiguity about $\circ$) if and only if $\alpha \prec_\Psi \beta$ and there is no formula γ such that $\alpha \prec_\Psi \gamma \prec_\Psi \beta$.

Definition 2. Let $\circ$ be a change operator such that the operator $\star$ associated is total. We say that μ is separated in Ψ iff $\forall \beta (B(\Psi \star \beta) \vdash \mu$ or $B(\Psi \star \beta) \vdash \neg \mu)$.

3 Brief recall about improvement operators and their representation

We begin stating the syntactical postulates considered in [Konieczny et al., 2010]:

(I1) There exists n such that $B(\Psi \circ^n \alpha) \vdash \alpha$
(I2) If $B(\Psi) \wedge \alpha \nvdash \bot$, then $B(\Psi \star \alpha) \equiv B(\Psi) \wedge \alpha$
(I3) If $\alpha \nvdash \bot$, then $B(\Psi \circ \alpha) \nvdash \bot$
(I4) For any positive integer n if $\alpha_i \equiv \beta_i$ for all $i \leq n$ then

$$B(\Psi \circ \alpha_1 \circ \ldots \circ \alpha_n) \equiv B(\Psi \circ \beta_1 \circ \ldots \circ \beta_n)$$

(I5) $B(\Psi \star \alpha) \wedge \beta \vdash B(\Psi \star (\alpha \wedge \beta))$
(I6) If $B(\Psi \star \alpha) \wedge \beta \nvdash \bot$, then $B(\Psi \star (\alpha \wedge \beta)) \vdash B(\Psi \star \alpha) \wedge \beta$
(I7) If $\alpha \vdash \mu$ then $B((\Psi \circ \mu) \star \alpha) \equiv B(\Psi \star \alpha)$
(I8) If $\alpha \vdash \neg \mu$ then $B((\Psi \circ \mu) \star \alpha) \equiv B(\Psi \star \alpha)$
(I9) If $B(\Psi \star \alpha) \nvdash \neg \mu$ then $B((\Psi \circ \mu) \star \alpha) \vdash \mu$
(I10) If $B(\Psi \star \alpha) \vdash \neg \mu$ then $B((\Psi \circ \mu) \star \alpha) \nvdash \mu$

(I11) If $B(\Psi \star \alpha) \vdash \neg\mu$ and $\alpha \not\lessdot_\Psi \alpha \wedge \mu$ then $B((\Psi \circ \mu) \star \alpha) \not\vdash \neg\mu$

(H1) If $B(\Psi \star \alpha) \vdash \neg\mu, \alpha \not\lessdot_\Psi \alpha \wedge \mu$ and $\neg\exists\beta(\beta \vdash \neg\mu$ and $\alpha \not\lessdot_\Psi \beta)$, then $B((\Psi \circ \mu) \star \alpha) \not\vdash \neg\mu$

(H2) If $B(\Psi \star \alpha) \vdash \neg\mu$, $\alpha \not\lessdot_\Psi \alpha \wedge \mu$ and $\exists\beta(\beta \vdash \neg\mu$ and $\alpha \not\lessdot_\Psi \beta)$, then $B((\Psi \circ \mu) \star \alpha) \vdash \neg\mu$

(B1) If μ is separated in Ψ, $B(\Psi \star \alpha) \vdash \neg\mu$ and $\alpha \not\lessdot_\Psi \alpha \wedge \mu$, then $B((\Psi \circ \mu) \star \alpha) \not\vdash \neg\mu$

(B2) If μ is not separated in Ψ and $B(\Psi \star \alpha) \vdash \neg\mu$, then $B((\Psi \circ \mu) \star \alpha) \vdash \neg\mu$

Postulates (I2-I6) are very close to postulates (R*2-R*6) of usual belief revision operators [Alchourrón et al., 1985; Katsuno and Mendelzon, 1991; Darwiche and Pearl, 1997]. The important difference lies in postulate (I1) that is weaker that the usual success postulate (R*1). So postulates (I2-I6) hold for sequences of weak improvements (whereas for revision they require only one step). Postulates (I7-I9) correspond to the postulates for iterated revision [Darwiche and Pearl, 1997; Jin and Thielscher, 2007; Booth and Meyer, 2006]. Postulates (I7) and (I8) correspond to postulates (C1) and (C2) of [Darwiche and Pearl, 1997], and postulate (I9) correspond to postulate (P) of [Jin and Thielscher, 2007; Booth and Meyer, 2006]. As for the basic postulates, the difference lies in the fact that they hold for sequences of applications of the operator to same input. Postulate (I10) says literally that a formula μ that is rejected by the agent after several improvements by α, can not be accepted after a improvement by μ and several improvements by α. In fact, the only admissible change of status is that the formula μ that is rejected by the agent after several improvements by α can become undetermined after a improvement by μ and several improvements by α. At least another step of improvement by μ will be required in order to accept this formula by the agent after several improvements by α (in the semantical side this postulate says that the plausibility of the new information can be increase more than one level). Postulate (I11) forces to increase the plausibility of the new information in one level (this is more clear in the semantical side). Postulate (H1) means that when the revision (i.e. sequence of improvements until success) by α implies the negation of μ, if μ is just a little less plausible than its negation given α, then an improvement by μ will be enough to remove its negation from the beliefs of the agent. Note that this postulate is weaker than (I11). Postulate (H2) is very close from (H1), and deals with the case where the revision (i.e. sequence of improvements until success) by α implies the negation of μ, but μ and $\neg\mu$ are both a little less plausible than $\neg\mu$, then an improvement by μ will not be enough to remove its negation from the beliefs of the agent. Postulate (B1) is close to postulates (H1) and (I11), but it holds only when the formula is separated in the epistemic state. Postulate (B2) states that, when the formula is not separated in the epistemic state (which is the general case), the change is the same one than with (H2).

The following definition is taken from [Konieczny et al., 2010]

Definition 3. Let $\circ$ be a change operator.

- If $\circ$ satisfies (I1-I6), it is called a *weak improvement operator*.

Representation of basic improvement operators   201

- If ∘ satisfies (I1-I9), it is called an *improvement operator*.
- If ∘ satisfies (I1-I10), it is called a *soft improvement operator*.
- If ∘ satisfies (I1-I11,) it is called a *one improvement operator*.
- If ∘ satisfies (I1-I10) plus (H1-H2), it is called a *half improvement operator*.
- If ∘ satisfies (I1-I10) plus (B1-B2), it is called a *best improvement operator*.

Now we state the semantical properties of assignments in order to establish the representation theorems appeared in [Konieczny et al., 2010]. Remember that an assignment is a function which maps each epistemic state Ψ in a total preorder over interpretations $\leq_\Psi$.

Definition 4. μ is s-separated in $\leq_\Psi$ iff there is no a pair ω_1, ω_2 such that $\omega_1 \in [\![\mu]\!], \omega_2 \in [\![\neg\mu]\!]$ and $\omega_1 \simeq_\Psi \omega_2$

We consider the following properties:

(1) If $\omega \models B(\Psi)$ and $\omega' \models B(\Psi)$, then $\omega \simeq_\Psi \omega'$.
(2) If $\omega \models B(\Psi)$ and $\omega' \not\models B(\Psi)$, then $\omega <_\Psi \omega'$.
(3) For any positive integer n if $\alpha_i \equiv \beta_i$ for any $i \leq n$ then

$$\leq_{\Psi \circ \alpha_1 \circ \ldots \circ \alpha_n} = \leq_{\Psi \circ \beta_1 \circ \ldots \circ \beta_n}$$

(S1) If $\omega, \omega' \in [\![\alpha]\!]$ then $\omega \leq_\Psi \omega' \Leftrightarrow \omega \leq_{\Psi \circ \alpha} \omega'$.
(S2) If $\omega, \omega' \in [\![\neg\alpha]\!]$ then $\omega \leq_\Psi \omega' \Leftrightarrow \omega \leq_{\Psi \circ \alpha} \omega'$.
(S3) If $\omega \in [\![\alpha]\!], \omega' \in [\![\neg\alpha]\!]$ then $\omega \leq_\Psi \omega' \rightarrow \omega <_{\Psi \circ \alpha} \omega'$.
(S4) If $\omega \in [\![\alpha]\!], \omega' \in [\![\neg\alpha]\!]$ then $\omega' <_\Psi \omega \rightarrow \omega' \leq_{\Psi \circ \alpha} \omega$.
(S5) If $\omega \in [\![\alpha]\!], \omega' \in [\![\neg\alpha]\!]$ then $\omega' \ll_\Psi \omega \rightarrow \omega \leq_{\Psi \circ \alpha} \omega'$.
(SH1) If $\omega \in [\![\mu]\!], \omega' \in [\![\neg\mu]\!], \omega' \ll_\Psi \omega$ and $\not\exists \omega'' \in [\![\neg\mu]\!]$ such that $\omega'' \simeq_\Psi \omega$, then, $\omega \leq_{\Psi \circ \mu} \omega'$.
(SH2) If $\omega \in [\![\mu]\!], \omega' \in [\![\neg\mu]\!], \omega' \ll_\Psi \omega$ and $\exists \omega'' \in [\![\neg\mu]\!]$ such that $\omega'' \simeq_\Psi \omega$ then, $\omega' <_{\Psi \circ \mu} \omega$.
(SB1) If μ is s-separated in $\leq_\Psi$, $\omega \in [\![\mu]\!], \omega' \in [\![\neg\mu]\!]$ and $\omega' \ll_\Psi \omega$ then $\omega \leq_{\Psi \circ \mu} \omega'$.
(SB2) If μ is not s-separated in $\leq_\Psi$, $\omega \in [\![\mu]\!], \omega' \in [\![\neg\mu]\!]$ and $\omega' <_\Psi \omega$ then $\omega' <_{\Psi \circ \mu} \omega$.

Properties (1-2) are fundamental properties saying that the minimal elements of $\leq_\Psi$ are the beliefs of Ψ. Property 3 say that the assignment is independent of the syntax through iterations. S1 and S2 are the properties CR1 and CR2 of Darwiche and Pearl. They correspond to rigidity of evolution of the models of the new information and to rigidity of the models of the negation of the new information. S3 is the semantic condition of Booth-Meyer and Jin-Thielscher (corresponding to postulate P alias I9). S4 says that the plausibility of models of the new information cannot increase more than one degree after revision. S5 says that the plausibility of models of the new information increase exactly one degree after revision. SH1 and SH2 say that the plausibility of the models of the new information increase a half or a one degree. Finally SB1 and SB2 say that in some cases the behavior is to improve one degree the plausibility of the new information (SB1) or in case of not separation only the models at the same level that countermodels are improved.

The following representation theorems were proved in [Konieczny et al., 2010] **for the class of change operators satisfying (I1)**:

Theorem 1. *A change operator* ∘ *is a weak improvement operator if and only if there exists an assignment satisfying conditions (1-3) such that*

$$[\![B(\Psi \star \alpha)]\!] = \min([\![\alpha]\!], \leq_\Psi)$$

Theorem 2. *A change operator* ∘ *is an improvement operator if and only if there exists an assignment satisfying conditions (1-3) plus (S1-S3) such that*

$$[\![B(\Psi \star \alpha)]\!] = \min([\![\alpha]\!], \leq_\Psi)$$

Theorem 3. *A change operator* ∘ *is a soft improvement operator if and only if there exists an assignment satisfying conditions (1-3) plus (S1-S4) such that*

$$[\![B(\Psi \star \alpha)]\!] = \min([\![\alpha]\!], \leq_\Psi)$$

Theorem 4. *A change operator* ∘ *is a one improvement operator if and only if there exists an assignment satisfying conditions (1-3) plus (S1-S5) such that*

$$[\![B(\Psi \star \alpha)]\!] = \min([\![\alpha]\!], \leq_\Psi)$$

Theorem 5. *A change operator* ∘ *is a half improvement operator if and only if there exists an assignment satisfying conditions (1-3) plus (S1-S4) plus (SH1-SH2) such that*

$$[\![B(\Psi \star \alpha)]\!] = \min([\![\alpha]\!], \leq_\Psi)$$

Theorem 6. *A change operator* ∘ *is a best improvement operator if and only if there exists an assignment satisfying conditions (1-3) plus (S1-S4) plus (SB1-SB2) such that*

$$[\![B(\Psi \star \alpha)]\!] = \min([\![\alpha]\!], \leq_\Psi)$$

Actually, when the epistemic states are the total preorders over interpretations, there are a unique one improvement operator, a unique half improvement operator and a unique best improvement operator. They are denoted ⊙, ⊘ and ⊕, one-improvement, half-improvement and best-improvement respectively.

The following figure (taken from [Konieczny et al., 2010]) shows the behavior of these three operators in three different epistemic states:

Representation of basic improvement operators

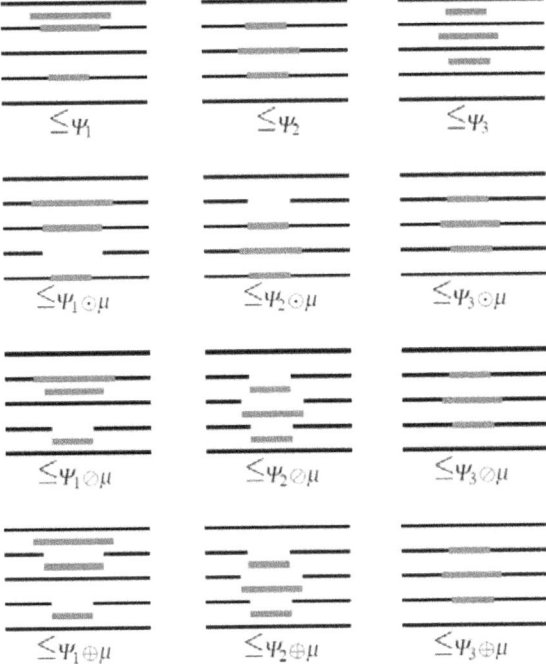

Fig. 1 Example of the behavior of one, half and best improvement

One of the main challenges raised by these theorems is to know if we can suppress the hypothesis that the operators satisfy Postulate I1. In this work we give answer to this question.

The following graphical example shows that Theorems 1, 2 and 3 really need the hypothesis that the operator circ satisfies Postulate I1:

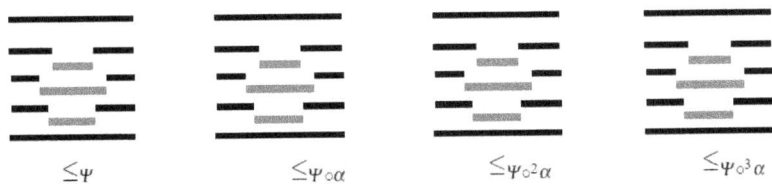

Fig. 2 The models of α are in gray.

As one easily check, the properties (S1-S4) of assignments are satisfied. And we can extend this assignment in such a way that the properties (1-3) of assignments are satisfied.

However Theorems 4, 5 and 6 don't need the hypothesis that the considered operators satisfy I1. This will be seen in Section 6. Actually we will state these theorems in a stronger form.

4 Basic improvement operators

As we explain in the introduction, in the operators we are going to consider we change the postulate (I9) (corresponding to Property (P) in [Booth and Meyer, 2006]) which says essentially that after one step of revision the whole new piece information has improved. We are going to replace that by postulates which say that at least a part of the new piece of information improves and the whole new piece of information does not worsen. In some cases the new postulate will be weaker than (I9) and in some cases the new postulate will be stronger than (I9). More precisely the postulates defining our operators appear in the following definition:

Definition 2 *Let $\circ$ be an change operator and $\star$ its associated change operator. The operator circ is called a basic improvement operator if the following postulates hold* [6].

(BI1)   There exists n such that $B(\Psi \circ^n \alpha) \vdash \alpha$.
(BI2)   If $B(\Psi) \wedge \alpha \not\vdash \bot$, then $B(\Psi \star \alpha) \equiv B(\Psi) \wedge \alpha$.
(BI3)   If $\alpha \not\vdash \bot$, then $B(\Psi \star \alpha) \not\vdash \bot$.
(BI4)   For all positive integer n if $\alpha_i \equiv \beta_i$ for any $i \leq n$ and $\gamma \equiv \mu$ then

$$B((\Psi \circ \alpha_1 \circ \cdots \circ \alpha_n) \star \gamma) \equiv B((\Psi \circ \beta_1 \circ \cdots \circ \beta_n) \star \mu)$$

(BI5) $B(\Psi \star \alpha) \wedge \beta \vdash B(\Psi \star (\alpha \wedge \beta))$
(BI6)   If $B(\Psi \star \alpha) \wedge \beta \not\vdash \bot$, then $B(\Psi \star (\alpha \wedge \beta)) \vdash B(\Psi \star \alpha) \wedge \beta$
(BI7)   If $\alpha \vdash \mu$ then $B((\Psi \circ \mu) \star \alpha) \equiv B(\Psi \star \alpha)$
(BI8)   If $\alpha \vdash \neg \mu$ then $B((\Psi \circ \mu) \star \alpha) \equiv B(\Psi \star \alpha)$
(BI9)   If $B(\Psi \circ \alpha) \vdash \mu$, then $B((\Psi \circ \mu) \star \alpha) \vdash \mu$
(BI10)   If $B(\Psi \circ \alpha) \not\vdash \neg \mu$, then $B((\Psi \circ \mu) \star \alpha) \not\vdash \neg \mu$
(BI11)   If there exists β such that β is consistent with α, β is consistent with $\neg \alpha$ and $B(\Psi \star \beta) \not\vdash \alpha$, then at least one of the following conditions holds:

     (a) $\exists \gamma$ such that $B(\Psi \star \gamma)$ is consistent with α, $B(\Psi \star \gamma)$ is consistent with $\neg \alpha$ and $B((\Psi \circ \alpha) \star \gamma)$ is inconsistent with $\neg \alpha$ but consistent with α
     (b) $\exists \gamma$ such that $B(\Psi \star \gamma) \vdash \neg \alpha$ and $B((\Psi \circ \alpha) \star \gamma)$ is consistent with α

---

[6] As usual, in all the postulates is implicit the universal quantification over the logical parameters. For instance postulate BI1 have to be read as follows: for all Ψ and for all α there exists n such that $B(\Psi \circ^n \alpha) \vdash \alpha$. However, in order to make the notation less cumbersome, we omit the universal quantifiers.

Most of these postulates have been considered in [Konieczny and Pino Pérez, 2008; Konieczny et al., 2010]. Therein they were called (In), where n is a number. We have recalled these postulates in Section 3. To avoid giving the same name to different postulates and for the sake of uniformity we have called the postulates (BIn), as basic improvement, where n is a number. We explain below their correspondences to postulates in the literature and their meaning.

Postulate(BI1) is the postulate of success through iteration. It allows to define the $\star$ operator in terms of which the other postulates are stated. It is, actually, Postulate (I1) in [Konieczny and Pino Pérez, 2008; Konieczny et al., 2010]. Postulate (BI2) is the postulate corresponding to vacuity postulate in the AGM framework. This postulate is exactly Postulate (I2) in [Konieczny and Pino Pérez, 2008; Konieczny et al., 2010]. Postulate (BI3) is the postulate of coherence. It is the postulate (I3) in [Konieczny and Pino Pérez, 2008; Konieczny et al., 2010]. Postulate (BI4) is the postulate of irrelevance of the syntax. It corresponds to Postulate (I4) in [Konieczny and Pino Pérez, 2008; Konieczny et al., 2010]. We refer to [Konieczny and Pino Pérez, 2008] for an explanation of the necessity of this strong iterated form of the postulate. Postulates (BI5) and (BI6) are the postulates corresponding to postulates (R5) and (R6) of the AGM revision as presented by Katsuno and Mendelzon [Katsuno and Mendelzon, 1991]. They express the good behavior of the iteration with respect to conjunction. They are called (I5) and (I6) respectively in [Konieczny and Pino Pérez, 2008; Konieczny et al., 2010]. Postulates (BI7) and (BI8) are the postulates expressing that there are no changes in the structure of the new information and in the structure of negation of the new information. These postulates correspond to postulates (C1) and (C2) of Darwich and Pearl [Darwiche and Pearl, 1997]. As they have explained, they are necessary for good iterative behavior. Postulates (BI7) and (BI8) are Postulates (I7) and (I8) respectively in [Konieczny and Pino Pérez, 2008; Konieczny et al., 2010]. Postulates (BI9) and (BI10) are the postulates corresponding to the fact that the new piece of information does not worsen after revision. They are new in the framework of improvement. Actually, they correspond to postulates (C3) and (C4) of Darwiche and Pearl [Darwiche and Pearl, 1997]. They were not considered in previous works of improvement operators because they are a consequence of Postulate (I9) considered in [Konieczny and Pino Pérez, 2008; Konieczny et al., 2010]. Postulate (BI11) is also new. It expresses the fact that at least a part of the new information improves after revision if there is something to improve. The disjunctive form of the conclusion of the postulate corresponds to two kinds of possible improvements. Note that this postulate is in some sense more precise than postulate (I9) (corresponding to Property (P)) considered in [Konieczny and Pino Pérez, 2008; Konieczny et al., 2010]: it captures the idea that whenever there is something to improve in the new piece of information, there will be something improved after revision.

The following notion of basic assignment will be the semantic key notion for the representation theorem.

Definition 3 *Let $\circ$ be a change operator. A mapping $\Psi \mapsto \leq_\Psi$ which sends each epistemic state Ψ into a total preorder $\leq_\Psi$ over the interpretations is called a* basic assignment *for $\circ$ if the following conditions[7] hold:*

(BS1) If $\omega \models B(\Psi)$ and $\omega' \models B(\Psi)$, then $\omega \simeq_\Psi \omega'$
(BS2) If $\omega \models B(\Psi)$ and $\omega' \not\models B(\Psi)$, then $\omega <_\Psi \omega'$
(BS3) For all positive integers n if $\alpha_i \equiv \beta_i$ for any $i \leq n$ then

$$\leq_{\Psi \circ \alpha_1 \circ \cdots \circ \alpha_n} = \leq_{\Psi \circ \beta_1 \circ \cdots \circ \beta_n}$$

(BS4) If $\omega, \omega' \in [\![\mu]\!]$ then $\omega \leq_\Psi \omega' \Leftrightarrow \omega \leq_{\Psi \circ \mu} \omega'$
(BS5) If $\omega, \omega' \in [\![\neg\mu]\!]$ then $\omega \leq_\Psi \omega' \Leftrightarrow \omega \leq_{\Psi \circ \mu} \omega'$
(BS6) If $\omega \models \mu$ and $\omega' \models \neg\mu$ then $\omega <_\Psi \omega' \Rightarrow \omega <_{\Psi \circ \mu} \omega'$
(BS7) If $\omega \models \mu$ and $\omega' \models \neg\mu$ then $\omega \leq_\Psi \omega' \Rightarrow \omega \leq_{\Psi \circ \mu} \omega'$
(BS8) If $\omega \in [\![\mu]\!]$, $\omega' \in [\![\neg\mu]\!]$ and $\omega' \leq_\Psi \omega$ then at least one of the following conditions holds:

(i) $\exists \omega_1, \omega_2$ such that $\omega_1 \models \mu$, $\omega_2 \models \neg\mu$, $\omega_1 \simeq_\Psi \omega_2$ and $\omega_1 <_{\Psi \circ \mu} \omega_2$
(ii) $\exists \omega_1, \omega_2$ such that $\omega_1 \models \mu$, $\omega_2 \models \neg\mu$, $\omega_2 <_\Psi \omega_1$ and $\omega_1 \leq_{\Psi \circ \mu} \omega_2$

Most of these conditions have been considered in [Konieczny and Pino Pérez, 2008; Konieczny et al., 2010] (see Section 3). Therein they were numbered or called (Sn), where n is a number. To avoid giving the same name to different conditions and for the sake of uniformity we have called the conditions (BSn), as basic semantics, where n is a number. We explain below the correspondences of these conditions to previous conditions considered in the literature and their meaning.

Conditions (BS1) and (BS2) correspond to conditions (1) and (2) for strong faithful assignments considered in [Konieczny and Pino Pérez, 2008; Konieczny et al., 2010]. They express the fact that the models of the beliefs of an epistemic state are the minimal models of the total preorder corresponding to it. Condition (BS3) is a very natural condition that links preorders associated to iteration of improvements: two sequences of improvements of the same pre-order by equivalent formulas lead to the same pre-order. This condition corresponds to condition (3) considered in [Konieczny and Pino Pérez, 2008; Konieczny et al., 2010] for strong faithful assignments. Condition (BS4) expresses that the models of μ maintain the same relation before and after revision by μ. They correspond to Condition (S1) in [Konieczny and Pino Pérez, 2008; Konieczny et al., 2010] which is in fact a form of the condition (CR1) of Darwiche and Pearl [Darwiche and Pearl, 1997]. Condition (BS5) expresses that the models of $\neg\mu$ maintain the same relation before and after revision by μ. They correspond to Condition (S2) in [Konieczny and Pino Pérez, 2008; Konieczny et al., 2010] which is in fact a form of the condition (CR2) of Darwiche and Pearl [Darwiche and Pearl, 1997]. Conditions (BS6) and (BS7) express that the models of μ do not worsen with respect the models of $\neg\mu$ after revision by μ. They correspond to Conditions (CR3) and (CR4) of Darwiche and Pearl

---

[7] Like in Definition 2, we omit the universal quantifiers in the conditions.

Representation of basic improvement operators

[Darwiche and Pearl, 1997]. Conditions (BS6) and (BS7) are new in the framework of improvement. Finally Condition (BS8) expresses the fact that if there is the possibility of improving a model of μ there will be a model which will improve after revision by μ.

Now we are ready to state our main result:

Theorem 7. *If $\circ$ is a basic improvement operator, that is, it satisfies the postulates (BI1)-(BI11), then there exists a basic assignment for $\circ$ mapping each epistemic state Ψ into a total preorder $\leq_\Psi$ such that*

$$[\![B(\Psi \star \alpha)]\!] = min([\![\alpha]\!], \leq_\Psi) \qquad (2)$$

Conversely, assume that $\circ$ is a change operator for which there is a basic assignment mapping each epistemic state Ψ into a total preorder $\leq_\Psi$, then the operator $\circ$ is a basic improvement operator and the equation (2) holds for this basic assignment.

The proof of this theorem will be done in the next section but before we give some examples of such operators.

Example 1 *The operators (defined in [Konieczny et al., 2010]) $\odot$, $\oslash$ and $\oplus$ called one improvement, half improvement and best improvement respectively are basic improvement operators.*

5 The proof

We begin by stating a very useful lemma for the proof of Theorem 7.

Lemma 1 *Let $\circ$ be an operator satisfying (BI1). Suppose there exists a function mapping each epistemic state Ψ into a total preorder $\leq_\Psi$ over $\mathcal{W}$ satisfying the properties (BS1), (BS2) and (BS3) of a basic assignment. Moreover, suppose that Equation (2) holds. Then, for all formulae μ and all consistent formulae α, we have $B(\Psi \star \alpha) \vdash \mu$ if, and only if, there exists a interpretation ω such that $\omega \in [\![\mu \wedge \alpha]\!]$ and $\omega <_\Psi \omega'$ for all $\omega' \in [\![\alpha \wedge \neg\mu]\!]$.*

Proof: (If) Suppose we have $\omega \in [\![\mu \wedge \alpha]\!]$ such that $\omega <_\Psi \omega'$ for all $\omega' \in [\![\alpha \wedge \neg\mu]\!]$. We want to see $B(\Psi \star \alpha) \vdash \mu$ which by the equation (2) is equivalent to show $min([\![\alpha]\!], \leq_\Psi) \subset [\![\mu]\!]$. Towards a contradiction, suppose there exists $\omega' \in min([\![\alpha]\!], \leq_\Psi)$ and $\omega' \notin [\![\mu]\!]$. Since $\omega \in [\![\alpha]\!]$, we have $\omega' \leq_\Psi \omega$. But $\omega' \in [\![\alpha \wedge \neg\mu]\!]$, a contradiction.

(Only if) Suppose α consistent such that $B(\Psi \star \alpha) \vdash \mu$. By the equation (2) we have $min([\![\alpha]\!], \leq_\Psi) \subseteq [\![\mu]\!]$. Take $\omega \in min([\![\alpha]\!], \leq_\Psi)$, then $\omega \in [\![\alpha \wedge \mu]\!]$. We claim that $\omega <_\Psi \omega'$ for all $\omega' \in [\![\alpha \wedge \neg\mu]\!]$. Towards a contradiction, suppose there exists

$\omega' \in [\![\alpha \wedge \neg \mu]\!]$ such that $\omega' \leq_\Psi \omega$. Necessarily $\omega' \in min([\![\alpha]\!], \leq_\Psi)$. But this contradicts the fact that $min([\![\alpha]\!], \leq_\Psi) \subseteq [\![\mu]\!]$. ∎

Now, we are ready to begin with the proof of our main theorem.

Proof of Theorem 7: **(Only if)** Suppose that ∘ is an operator satisfying postulates (BI1)-(BI11). For each epistemic state Ψ define a relation $\leq_\Psi$ in the following way:

$$\omega \leq_\Psi \omega' \Leftrightarrow \omega \models B(\Psi \star \varphi_{\omega,\omega'})$$

where, we recall, $\varphi_{\omega,\omega'}$ denotes a formula such that $[\![\varphi_{\omega,\omega'}]\!] = \{\omega, \omega'\}$.
We have to show that $\leq_\Psi$ is a total preorder.

Totality: Let ω, ω' be worlds. Since $[\![\varphi_{\omega,\omega'}]\!] \neq \emptyset$, by (BI3) we have $[\![B(\Psi \star \varphi_{\omega,\omega'})]\!] \neq \emptyset$. By (BI1) we have $[\![B(\Psi \star \varphi_{\omega,\omega'})]\!] \subseteq [\![\varphi_{\omega,\omega'}]\!]$, therefore either $\omega \models B(\Psi \star \varphi_{\omega,\omega'})$ or $\omega' \models B(\Psi \star \varphi_{\omega,\omega'})$. That is to say, either $\omega \leq_\Psi \omega'$ or $\omega' \leq_\Psi \omega$. The totality of $\leq_\Psi$ is proved in this way.

Transitivity: Let $\omega_1, \omega_2, \omega_3$ be worlds such that $\omega_1 \leq_\Psi \omega_2$ and $\omega_2 \leq_\Psi \omega_3$. We want to show that $\omega_1 \leq_\Psi \omega_3$. We consider the following three cases:

Case 1: $\omega_1 \models B(\Psi)$.
In this case $B(\Psi)$ and $\varphi_{\omega_1,\omega_3}$ are mutually consistent; then, by (BI2) we have $B(\Psi \star \varphi_{\omega_1,\omega_3}) \equiv B(\Psi) \wedge \varphi_{\omega_1,\omega_3}$. Therefore $\omega_1 \models B(\Psi \star \varphi_{\omega_1,\omega_3})$, thus $\omega_1 \leq_\Psi \omega_3$.

Case 2: $\omega_1 \not\models B(\Psi)$ and $\omega_2 \models B(\Psi)$.
By the hypothesis we have $[\![B(\Psi) \wedge \varphi_{\omega_1,\omega_2}]\!] = \{\omega_2\}$. Thus, by (BI2), we have $[\![B(\Psi) \star \varphi_{\omega_1,\omega_2}]\!] = \{\omega_2\}$, but $\omega_1 \leq_\Psi \omega_2$ i.e. $\omega_1 \models B(\Psi \star \varphi_{\omega_1,\omega_2})$, contradiction. Therefore this case is impossible.

Caso 3: $\omega_1, \omega_2 \not\models B(\Psi)$.
Since $[\![\varphi_{\omega_1,\omega_2,\omega_3}]\!] \neq \emptyset$, by (BI3) we have $[\![B(\Psi \star \varphi_{\omega_1,\omega_2,\omega_3})]\!] \neq \emptyset$. By (BI1) we have $[\![B(\Psi \star \varphi_{\omega_1,\omega_2,\omega_3})]\!] \subseteq \{\omega_1, \omega_2, \omega_3\}$. We consider now two subcases.

Case 3.1: $[\![B(\Psi \star \varphi_{\omega_1,\omega_2,\omega_3})]\!] \cap \{\omega_1, \omega_2\} = \emptyset$.
In this case we have necessarily $[\![B(\Psi \star \varphi_{\omega_1,\omega_2,\omega_3})]\!] = \{\omega_3\}$. Then

$$[\![B(\Psi \star \varphi_{\omega_1,\omega_2,\omega_3})]\!] \cap [\![\varphi_{\omega_2,\omega_3}]\!] = \{\omega_3\} \neq \emptyset.$$

It follows, from (BI5) and (BI6),

$$\begin{aligned}[\![B(\Psi \star (\varphi_{\omega_1,\omega_2,\omega_3} \wedge \varphi_{\omega_2,\omega_3}))]\!] &= [\![B(\Psi \star \varphi_{\omega_1,\omega_2,\omega_3}) \wedge \varphi_{\omega_2,\omega_3}]\!] \\ &= [\![B(\Psi \star \varphi_{\omega_1,\omega_2,\omega_3})]\!] \cap [\![\varphi_{\omega_2,\omega_3}]\!] \\ &= \{\omega_3\}.\end{aligned}$$

Representation of basic improvement operators

Therefore $[\![B(\Psi \star (\varphi_{\omega_2,\omega_3}))]\!] = \{\omega_3\}$, this entails in particular $\omega_2 \not\leq_\Psi \omega_3$, contradicting our initial hypothesis.

Case 3.2: $[\![B(\Psi \star \varphi_{\omega_1,\omega_2,\omega_3})]\!] \cap \{\omega_1, \omega_2\} \neq \emptyset$.
In this case $B(\Psi \star \varphi_{\omega_1,\omega_2,\omega_3})$ and $\varphi_{\omega_1,\omega_2}$ are mutually consistent. Then, by (BI5) and (BI6), we have $B(\Psi \star (\varphi_{\omega_1,\omega_2,\omega_3} \wedge \varphi_{\omega_1,\omega_2})) \equiv B(\Psi \star \varphi_{\omega_1,\omega_2,\omega_3}) \wedge \varphi_{\omega_1,\omega_2}$ i.e. $[\![B(\Psi \star \varphi_{\omega_1,\omega_2})]\!] = [\![B(\Psi \star \varphi_{\omega_1,\omega_2,\omega_3})]\!] \cap \{\omega_1, \omega_2\}$. Note that, by hypothesis, $\omega_1 \models B(\Psi \star \varphi_{\omega_1,\omega_2})$, so $\omega_1 \models B(\Psi \star \varphi_{\omega_1,\omega_2,\omega_3}) \wedge \varphi_{\omega_1,\omega_2}$. In particular $\omega_1 \models B(\Psi \star \varphi_{\omega_1,\omega_2,\omega_3})$ thus $\omega_1 \in [\![B(\Psi \star \varphi_{\omega_1,\omega_2,\omega_3})]\!] \cap [\![\varphi_{\omega_1,\omega_3}]\!]$ i.e. $B(\Psi \star \varphi_{\omega_1,\omega_2,\omega_3})$ and $\varphi_{\omega_1,\omega_3}$ are mutually consistent. Again by (BI5) an (BI6) we have $[\![B(\Psi \star \varphi_{\omega_1,\omega_3})]\!] = [\![B(\Psi \star \varphi_{\omega_1,\omega_2,\omega_3})]\!] \cap \{\omega_1, \omega_3\}$. Since $\omega_1 \in \{\omega_1, \omega_3\}$ and $\omega_1 \models B(\Psi \star \varphi_{\omega_1,\omega_2,\omega_3})$ we have $\omega_1 \models B(\Psi \star \varphi_{\omega_1,\omega_3})$ i.e. $\omega_1 \leq_\Psi \omega_3$.

Now we prove that the assignment $\Psi \mapsto \leq_\Psi$ is indeed a basic assignment:

Condition (BS1): It is enough to see that if $\omega \models B(\Psi)$ then $\omega \leq_\Psi \omega'$ for all $\omega' \in \mathcal{W}$. If ω is a model of $B(\Psi)$ then $\omega \models B(\Psi) \wedge \varphi_{\omega,\omega'}$. By (BI2) we have $B(\Psi \star \varphi_{\omega,\omega'}) \equiv B(\Psi) \wedge \varphi_{\omega,\omega'}$ thus $\omega \models B(\Psi \star \varphi_{\omega,\omega'})$ that is $\omega \leq_\Psi \omega'$.

Condition (BS2): Suppose $\omega \models B(\Psi)$ and $\omega' \not\models B(\Psi)$. We want to show $\omega <_\Psi \omega'$. By (BI2), $[\![B(\Psi \star \varphi_{\omega,\omega'})]\!] = \{\omega\}$ i.e. $\omega' \not\leq_\Psi \omega$. By the previous condition $\omega \leq_\Psi \omega'$. Thus $\omega <_\Psi \omega'$.

Condition (BS3):
Let n be a positive integer such that $\alpha_i \equiv \beta_i$ for all $i \leq n$. We want to see that $\leq_{\Psi \circ \alpha_1 \circ \cdots \circ \alpha_n} = \leq_{\Psi \circ \beta_1 \circ \cdots \circ \beta_n}$.
Note that for any ω and ω' we have

$$\omega \leq_{\Psi \circ \alpha_1 \circ \cdots \circ \alpha_n} \omega' \Leftrightarrow \omega \models B((\Psi \circ \alpha_1 \circ \cdots \circ \alpha_n) \star \varphi_{\omega,\omega'})$$

and

$$\omega \leq_{\Psi \circ \beta_1 \circ \cdots \circ \beta_n} \omega' \Leftrightarrow \omega \models B((\Psi \circ \beta_1 \circ \cdots \circ \beta_n) \star \varphi_{\omega,\omega'})$$

Moreover by (BI4) $B((\Psi \circ \alpha_1 \circ \cdots \circ \alpha_n) \star \varphi_{\omega,\omega'}) \equiv B((\Psi \circ \beta_1 \circ \cdots \circ \beta_n) \star \varphi_{\omega,\omega'})$. From this follows that $\leq_{\Psi \circ \alpha_1 \circ \cdots \circ \alpha_n} = \leq_{\Psi \circ \beta_1 \circ \cdots \circ \beta_n}$

With the conditions proved until now we can show Equation (2) holds, i.e.

$$[\![B(\Psi \star \alpha)]\!] = min([\![\alpha]\!], \leq_\Psi)$$

Remember that we work under the hypothesis that the epistemic states and the new piece of information are consistent. Thus, suppose α is consistent. In order to prove Equation (2), we show first $[\![B(\Psi \star \alpha)]\!] \subseteq min([\![\alpha]\!], \leq_\Psi)$. Towards a contradiction, assume there is ω such that $\omega \models B(\Psi \star \alpha)$ and $\omega \notin min(\llbracket\alpha\rrbracket, \leq_\Psi)$. By (BI1) we have $\omega \models \alpha$. Since $\omega \notin min(\llbracket\alpha\rrbracket, \leq_\Psi)$ there is $\omega' \models \alpha$ such that $\omega' <_\Psi \omega$. We consider the following two cases:

Case 1: $\omega' \models B(\Psi)$.
Since $\omega' \models \alpha$, we have $\omega' \models B(\Psi) \wedge \alpha$. Then, by (BI2), $\omega' \models B(\Psi \star \alpha) \equiv B(\Psi) \wedge \alpha$. But $\omega \models B(\Psi \star \alpha)$. Thus $\omega \models B(\Psi)$. By condition (BS1), $\omega \leq_\Psi \omega'$, contradicting the fact $\omega' <_\Psi \omega$.

Case 2: $\omega' \not\models B(\Psi)$.
Since $\omega' <_\Psi \omega$ we have $\llbracket\Psi \star \varphi_{\omega,\omega'}\rrbracket = \{\omega'\}$. Note that ω and ω' are in $\llbracket\alpha\rrbracket$, thus $\alpha \wedge \varphi_{\omega,\omega'} \equiv \varphi_{\omega,\omega'}$. Then, by (BI5) and (BI4) we have $B(\Psi \star \alpha) \wedge \varphi_{\omega,\omega'} \vdash B(\Psi \star \varphi_{\omega,\omega'})$ i.e. $\llbracket B(\Psi \star \alpha)\rrbracket \cap \{\omega, \omega'\} \subseteq \llbracket B(\Psi \star \varphi_{\omega,\omega'})\rrbracket$. But $\llbracket B(\Psi \star \varphi_{\omega,\omega'})\rrbracket = \{\omega'\}$, thus $\omega \not\models B(\Psi \star \alpha)$, contradicting the hypothesis.

We turn now to prove $min(\llbracket\alpha\rrbracket, \leq_\Psi) \subseteq \llbracket B(\Psi \star \alpha)\rrbracket$. Let ω be in $min(\llbracket\alpha\rrbracket, \leq_\Psi)$. Towards a contradiction, suppose $\omega \not\models B(\Psi \star \alpha)$. Since α is consistent, by (BI3), there exists ω' in $\llbracket B(\Psi \star \alpha)\rrbracket$. By (BI1) $\omega' \models \alpha$. Since $\omega, \omega' \in \llbracket\alpha\rrbracket$, we have $\alpha \wedge \varphi_{\omega,\omega'} \equiv \varphi_{\omega,\omega'}$. Note that $\omega' \in \llbracket B(\Psi \star \alpha)\rrbracket \cap \{\omega, \omega'\}$ thus $B(\Psi \star \alpha) \wedge \varphi_{\omega,\omega'}$ is consistent. Then, by (BI5), (BI6) and (BI4), we have $B(\Psi \star \alpha) \wedge \varphi_{\omega,\omega'} \equiv B(\Psi \star \varphi_{\omega,\omega'})$ i.e. $\llbracket B(\Psi \star \alpha)\rrbracket \cap \{\omega, \omega'\} = \llbracket B(\Psi \star \varphi_{\omega,\omega'})\rrbracket$. Since $\omega \not\models B(\Psi \star \alpha)$ we have $\{\omega'\} = \llbracket B(\Psi \star \varphi_{\omega,\omega'})\rrbracket$, i.e. $\omega' <_\Psi \omega$, contradicting the minimality of ω in $\llbracket\alpha\rrbracket$ with respect to $\leq_\Psi$.

We have proved that Equation (2) holds. This equation will be useful in proving the other conditions of basic assignment.

Now we can carry on the proof of the other conditions of basic assignment:

Condition(BS4): Assume that $\omega, \omega' \in \llbracket\mu\rrbracket$. We want to show $\omega \leq_\Psi \omega' \Leftrightarrow \omega \leq_{\Psi \circ \mu} \omega'$. Let α be a formula such that $\llbracket\alpha\rrbracket = \{\omega, \omega'\}$. Then $\alpha \vdash \mu$. By (BI7) we have $B((\Psi \circ \mu) \star \alpha) \equiv B(\Psi \star \alpha)$ i.e. $min(\llbracket\alpha\rrbracket, \leq_{\Psi \circ \mu}) = \llbracket B(\Psi \circ \mu) \star \alpha\rrbracket = \llbracket B(\Psi \star \alpha)\rrbracket = min(\llbracket\alpha\rrbracket, \leq_\Psi)$. In particular $min(\llbracket\alpha\rrbracket, \leq_{\Psi \circ \mu}) = min(\llbracket\alpha\rrbracket, \leq_\Psi)$. From this equality we get easily $\omega \leq_\Psi \omega' \Leftrightarrow \omega \leq_{\Psi \circ \mu} \omega'$.

Condition (BS5): This condition is proven in an analogous way to the previous condition.

Condition (BS6): Let ω and ω' be worlds such that $\omega \models \mu$ and $\omega' \models \neg\mu$. Suppose $\omega <_\Psi \omega'$. We want to see that $\omega <_{\Psi \circ \mu} \omega'$. Let α be a formula such that $\llbracket\alpha\rrbracket = \{\omega, \omega'\}$. Note that $\omega \models \alpha \wedge \mu$, $\omega <_\Psi \omega'$ and $\llbracket\alpha\rrbracket \cap \llbracket\neg\mu\rrbracket = \{\omega'\}$. Then, by Lemma 1, $B(\Psi \star \alpha) \vdash \mu$. Thus, by (BI9), $B((\Psi \circ \mu) \star \alpha) \vdash \mu$. Again, by Lemma 1, we have $\omega <_{\Psi \circ \mu} \omega'$, because $[\![\alpha \wedge \mu]\!] = \{\omega\}$ and $[\![\alpha \wedge \neg \mu]\!] = \{\omega'\}$.

Condition (BS7): Note that this condition is equivalent (under the assumption that the relations $\leq_\Phi$ are total preorders for any epistemic state Φ) to the following condition:

(S7') If $\omega \models \mu$ and $\omega' \models \neg \mu$ then $\omega <_{\Psi \circ \mu} \omega' \Rightarrow \omega' <_\Psi \omega$.

In order to prove (S7') suppose that $\omega \models \mu$, $\omega' \models \neg \mu$ and $\omega <_{\Psi \circ \mu} \omega'$. We want to show $\omega' <_\Psi \omega$. Let α be a formula such that $[\![\alpha]\!] = \{\omega, \omega'\}$. Note that $[\![\alpha \wedge \mu]\!] = \{\omega\}$ and $[\![\alpha \wedge \neg \mu]\!] = \{\omega'\}$. Since $\omega' <_{\Psi \circ \mu} \omega$, by Lemma 1, $B((\Psi \circ \mu) \star \alpha) \vdash \neg \mu$. Using the contrapositive form of (BI10), i.e.

(I10') if $B((\Psi \circ \mu) \star \alpha) \vdash \neg \mu$, then $B(\Psi \star \alpha) \vdash \neg \mu$,

we get $B(\Psi \star \alpha) \vdash \neg \mu$. Again, by Lemma 1, we have $\omega' <_\Psi \omega$, because $[\![\alpha]\!] \cap [\![\neg \mu]\!] = \{\omega'\}$ and $[\![\alpha]\!] \cap [\![\mu]\!] = \{\omega\}$.

Condition (BS8): Assume that there are worlds ω and ω' such that $\omega \models \mu$, $\omega' \models \neg \mu$ and $\omega' \leq_\Psi \omega$. We want to see that at least one of the following conditions holds:

(i) $\exists \omega_1, \omega_2$ such that $\omega_1 \models \mu$, $\omega_2 \models \neg \mu$, $\omega_1 \simeq_\Psi \omega_2$ and $\omega_1 <_{\Psi \circ \mu} \omega_2$

(ii) $\exists \omega_1, \omega_2$ such that $\omega_1 \models \mu$, $\omega_2 \models \neg \mu$, $\omega_2 <_\Psi \omega_1$ and $\omega_1 \leq_{\Psi \circ \mu} \omega_2$

Consider the formula $\varphi_{\omega, \omega'}$ having ω and ω' as the sole models. Since $\omega' \leq_\Psi \omega$ we have $\omega' \in min([\![\varphi_{\omega, \omega'}]\!], \leq_\Psi)$. By Equation (2), $\omega' \models B(\Psi \star \varphi_{\omega, \omega'})$. Since $\omega' \models \neg \mu$ we have $[\![B(\Psi \star \varphi_{\omega, \omega'})]\!] \not\subseteq [\![\mu]\!]$, i.e. $B(\Psi \star \varphi_{\omega, \omega'}) \not\vdash \mu$. Thus, by (BI11), one of the following conditions holds:

(a) $\exists \gamma$ such that $B(\Psi \star \gamma)$ is consistent with μ and also consistent with $\neg \mu$ and $B((\Psi \circ \mu) \star \gamma)$ is inconsistent with $\neg \mu$ but consistent with μ.

(b) $\exists \gamma$ such that $B(\Psi \star \gamma) \vdash \neg \mu$ and $B((\Psi \circ \mu) \star \gamma)$ is consistent with μ.

In the case in which (a) holds, we have $\exists \omega_1, \omega_2$ such that $\omega_1 \models \mu$, $\omega_2 \models \neg \mu$ and $\omega_1, \omega_2 \models B(\Psi \star \gamma)$. By Equation (2), $\omega_1, \omega_2 \in min([\![\gamma]\!], \leq_\Psi)$. Thus $\omega_1 \simeq_\Psi \omega_2$. By condition (BS7) we have $\omega_1 \leq_{\Psi \circ \mu} \omega_2$. We claim that $\omega_2 \not\leq_{\Psi \circ \mu} \omega_1$. Towards a contradiction suppose $\omega_2 \leq_{\Psi \circ \mu} \omega_1$. Then $\omega_1 \simeq_{\Psi \circ \mu} \omega_2$. Since $[\![B(\Psi \circ \mu) \star \gamma]\!] = min([\![\gamma]\!], \leq_{\Psi \circ \mu})$ does not contain models of $\neg \mu$, necessarily $\omega_1 \notin min([\![\gamma]\!], \leq_{\Psi \circ \mu})$ (otherwise ω_2 would be minimal too). But $B(\Psi \circ \mu) \star \gamma$ is consistent with μ, so there exists $\omega_3 \models \mu$ such that $\omega_3 \models min([\![\gamma]\!], \leq_{\Psi \circ \mu})$. Therefore $\omega_3 <_{\Psi \circ \mu} \omega_1$. Since $\omega_1 \in min([\![\gamma]\!], \leq_\Psi)$ and $\omega_3 \models \gamma$ we have $\omega_1 \leq_\Psi \omega_3$. Since $\omega_1, \omega_3 \models \mu$, by (BS4), $\omega_1 \leq_{\Psi \circ \mu} \omega_3$, a contradiction. Thus, $\omega_1 <_{\Psi \circ \mu} \omega_2$. In this way, we have seen that (i) holds.

In case (b) holds, $\exists \omega_1$ such that $\omega_1 \models \mu$ and $\omega_1 \models B((\Psi \circ \mu) \star \gamma)$. By Equation (2), $\omega_1 \in min([\![\gamma]\!], \leq_{\Psi \circ \mu})$. Since $B(\Psi \star \gamma) \vdash \neg \mu$, we have

$min([\![\gamma]\!], \leq_\Psi) \subseteq [\![\neg\mu]\!]$. Therefore there exists $\omega_2 \models \gamma \wedge \neg\mu$ such that $\omega_2 <_\Psi \omega_1$. Since $\omega_1 \in min([\![\gamma]\!], \leq_{\Psi \circ \mu})$ and $\omega_2 \models \gamma$ we have $\omega_1 \leq_{\Psi \circ \mu} \omega_2$. That is, condition (ii) holds.

We have finished the *only if* part of the proof of Theorem 7. ∎

Before continuing with the *if* part of the proof of Theorem 7 we would like to make some comments about how it is developed.

First of all, note that this part is quite different from similar parts in other representations theorems. The difference lies in the fact that we do not assume that Equation (2) holds. Actually we cannot assume such an equation before establishing that the operator ⋆ has full meaning. In order to do that we need to prove that Postulate (BI1) holds. Thus our first task will be to prove this postulate. Then, as a side product, we will get that Equation (2) holds. Finally, with the help of this equation, we will prove the remaining postulates, *i.e.* (BI2)-(BI11).

In order to prove that Postulate (BI1) holds, we need to develop some techniques and establish some results which will be very useful for our purpose. From now on we suppose that ∘ is an operator for which $\Psi \mapsto \leq_\Psi$ is a basic assignment. We begin with the following observation

Observation 1 *By successive iterations of condition (BS4) it is easy to see that for all $i \in \mathbb{N}$ if $\omega, \omega' \models \alpha$ then, $\omega \leq_\Psi \omega' \Leftrightarrow \omega \leq_{\Psi \circ^i \alpha} \omega'$.*

Analogously, by successive iterations of condition (BS5), for all $i \in \mathbb{N}$ if $\omega, \omega' \models \neg\alpha$ then, $\omega \leq_\Psi \omega' \Leftrightarrow \omega \leq_{\Psi \circ^i \alpha} \omega'$.

Given a formula α and the total preorder $\leq_\Psi$ we define a relation $\sim_\alpha$ over $\mathcal{W}$ in the following way.

Definition 4 $\omega \sim_\alpha \omega'$ *if and only if either* $\omega, \omega' \models \alpha$ *and* $\omega \simeq_\Psi \omega'$ *or* $\omega, \omega' \models \neg\alpha$ *and* $\omega \simeq_\Psi \omega'$.

This relation $\sim_\alpha$ is obviously an equivalence relation because it inherits symmetry, reflexivity and transitivity from relation $\simeq_\Psi$. Also, note that each equivalence class is either totally contained in $[\![\alpha]\!]$ or totally contained in $[\![\neg\alpha]\!]$ because, by definition, a model of α is never in relation (under $\sim_\alpha$) with a model of $\neg\alpha$.

This observation allows us to partition the set of equivalence classes $\mathcal{W}/\sim_\alpha$ into two disjoint sets: the set of classes whose elements are models of α, denoted $C(\alpha)$ and the set of classes whose elements are models of $\neg\alpha$, denoted $C(\neg\alpha)$.

We consider the relation $\leq_{\widetilde{\Psi}}$ over $\mathcal{W}/\sim_\alpha$ defined in the following way:

$$[\omega] \leq_{\widetilde{\Psi}} [\omega'] \Leftrightarrow \omega \leq_\Psi \omega'$$

It is clear that $\leq_{\widetilde{\Psi}}$ defines a total preorder over $\mathcal{W}/\sim_\alpha$ because it inherits totality and transitivity from $\leq_\Psi$.

The indifference relation associated to $\leq_{\widetilde{\Psi}}$ will be denoted $\simeq_{\widetilde{\Psi}}$.

Observation 2 *The relation $\leq_{\tilde{\psi}}$ restrained to $C(\alpha)$ is a linear order.*

Proof: We know that $\leq_{\tilde{\psi}}$ is transitive and total. It remains to prove that $\leq_{\tilde{\psi}}$ satisfies antisymmetry. Let $[\omega], [\omega']$ be elements of $C(\alpha)$. Suppose $[\omega] \leq_{\tilde{\psi}} [\omega']$ and $[\omega'] \leq_{\tilde{\psi}} [\omega]$. Then, by definition, $\omega \leq_{\psi} \omega'$ and $\omega' \leq_{\psi} \omega$, i.e. $\omega \simeq_{\psi} \omega'$. But $\omega, \omega' \models \alpha$, thus $\omega \sim_{\alpha} \omega'$, i.e. $[\omega] = [\omega']$. ∎

With an argument analogous to the previous one we can prove the following:

Observation 3 *The relation $\leq_{\tilde{\psi}}$ restrained to $C(\neg\alpha)$ is a linear order.*

As in Definition 4, we define a relation $\sim_{\alpha^i}$ for each positive integer i. More precisely, given a formula α and the total preorder $\leq_{\psi \circ^i \alpha}$ we define a relation $\sim_{\alpha^i}$ over $\mathcal{W}$ in the following way.

Definition 5 $\omega \sim_{\alpha^i} \omega'$ if and only if either $\omega, \omega' \models \alpha$ and $\omega \simeq_{\psi \circ^i \alpha} \omega'$ or $\omega, \omega' \models \neg\alpha$ and $\omega \simeq_{\psi \circ^i \alpha} \omega'$.

Like the relation $\sim_{\alpha}$, the relations $\sim_{\alpha^i}$ are equivalence relations because they inherit their properties from $\simeq_{\psi \circ^i \alpha}$.

As before, we can partition the set $\mathcal{W}/\sim_{\alpha^i}$ in two disjoint sets defined as follows:

$$C^i(\alpha) = \{[\omega] : \omega \in [\![\alpha]\!]\}$$
$$C^i(\neg\alpha) = \{[\omega] : \omega \in [\![\neg\alpha]\!]\}$$

We define, as before, the relation $\leq_{\tilde{\psi} \circ^i \alpha}$ over $\mathcal{W}/\sim_{\alpha^i}$ by letting

$$[\omega] \leq_{\tilde{\psi} \circ^i \alpha} [\omega'] \Leftrightarrow \omega \leq_{\psi \circ^i \alpha} \omega'$$

And again, with similar arguments to the previous ones in the case of relation $\leq_{\tilde{\psi}}$, we can see that $\leq_{\tilde{\psi} \circ^i \alpha}$ is a total preorder over $\mathcal{W}/\sim_{\alpha^i}$ and, moreover, for each positive integer i the relation restrained to $C^i(\alpha)$ is a linear order and the relation restrained to $C^i(\neg\alpha)$ is also a linear order.

The relation of indifference associated to $\leq_{\tilde{\psi} \circ^i \alpha}$ is denoted $\simeq_{\tilde{\psi} \circ^i \alpha}$.

Observation 4 *For all $i \in \mathbb{N}$ we have $\sim_{\alpha} = \sim_{\alpha^i}$. Moreover, $C(\alpha) = C^i(\alpha)$, $C(\neg\alpha) = C^i(\neg\alpha)$, the linear orders $(C(\alpha), \leq_{\tilde{\psi}})$ and $(C^i(\alpha), \leq_{\tilde{\psi} \circ^i \alpha})$ coincide and also the linear orders $(C(\neg\alpha), \leq_{\tilde{\psi}})$ and $(C^i(\neg\alpha), \leq_{\tilde{\psi} \circ^i \alpha})$ coincide.*

Proof: From definitions of $\sim_{\alpha}$ and $\sim_{\alpha^i}$ (Definitions 4 and 5) and from Observation 1, we get easily $\sim_{\alpha} = \sim_{\alpha^i}$, $C^i(\alpha) = C(\alpha)$ and $C(\neg\alpha) = C^i(\neg\alpha)$ for all $i \in \mathbb{N}$. Again by Observation 1 and the definitions of $\leq_{\tilde{\psi} \circ^i \alpha}$ and $\leq_{\tilde{\psi}}$ we have for all $[\omega], [\omega'] \in C(\alpha)$ that $[\omega] \leq_{\tilde{\psi} \circ^i \alpha} [\omega'] \Leftrightarrow \omega \leq_{\psi \circ^i \alpha} \omega' \Leftrightarrow \omega \leq_{\psi} \omega' \Leftrightarrow [\omega] \leq_{\tilde{\psi}} [\omega']$. Therefore $\leq_{\tilde{\psi} \circ^i \alpha} = \leq_{\tilde{\psi}}$ over $C(\alpha)$.

Using an analogous reasoning to the previous one we can see that $\leq_{\tilde{\psi} \circ^i \alpha} = \leq_{\tilde{\psi}}$ over $C(\neg\alpha)$. ∎

The next definition is very important. It allows to associate a weight to an element of $C(\alpha)$ (alias $C^i(\alpha)$) at the iteration i.

Definition 6 *Let $[\omega]$ be in $C(\alpha)$. For each positive integer $i \in \mathbb{N}$, we define $D^i([\omega])$ as the set of classes $[\omega'] \in C(\neg\alpha)$ such that $[\omega'] \leq_{\widetilde{\psi}\circ i_\alpha} [\omega]$, that is to say,*

$$D^i([\omega]) = \{[\omega'] \in C(\neg\alpha) : [\omega'] \leq_{\widetilde{\psi}\circ i_\alpha} [\omega]\}$$

For each positive integer $i \in \mathbb{N}$ we define a function $\alpha^i : C(\alpha) \to \mathbb{N}$ by letting

$$\alpha^i([\omega]) = |D^i([\omega])|$$

Actually, $\alpha^i([\omega])$ is counting the number of levels containing models of $\neg\alpha$ starting to count from the level of ω until the minimal level in the total preorder $\leq_{\psi\circ i_\alpha}$.

Essentially, the following proposition tells us that this weight does not increase as we iterate.

Observation 5 *For each positive integer $i \in \mathbb{N}$, if $[\omega] \in C(\alpha)$, then $D^{i+1}([\omega]) \subseteq D^i([\omega])$.*

Proof: Let $[\omega']$ be in $D^{i+1}([\omega])$. We want to see that $[\omega'] \in D^i([\omega])$, that is to say $[\omega'] \leq_{\widetilde{\psi}\circ i_\alpha} [\omega]$. By hypothesis $[\omega'] \in C(\neg\alpha)$, $[\omega] \in C(\alpha)$ and $[\omega'] \leq_{\widetilde{\psi}\circ i+1_\alpha} [\omega]$. Thus, by definition of $\leq_{\widetilde{\psi}\circ i+1_\alpha}$ we get $\omega' \leq_{\psi\circ i+1_\alpha} \omega$. From this, using the contrapositive form of condition (BS6) we obtain $\omega' \leq_{\psi\circ i_\alpha} \omega$. This implies, by definition, that $[\omega'] \leq_{\widetilde{\psi}\circ i_\alpha} [\omega]$. ∎

As a straightforward consequence of the previous observation we have the following:

Observation 6 *For each positive integer $i \in \mathbb{N}$, $\alpha^{i+1}([\omega]) \leq \alpha^i([\omega])$.*

The following observation is very intuitive: it says essentially that the lower the model of α is in the total preorder $\leq_{\psi\circ i_\alpha}$, the lower its weight is. More precisely:

Observation 7 *For each positive integer $i \in \mathbb{N}$, if $[\omega], [\omega'] \in C(\alpha)$ and $[\omega] \leq_{\widetilde{\psi}\circ i_\alpha} [\omega']$ then $\alpha^i([\omega]) \leq \alpha^i([\omega'])$.*

Proof: If $\alpha^i([\omega]) = 0$, then $\alpha^i([\omega]) \leq \alpha^i([\omega'])$. Now suppose that $\alpha^i([\omega]) \neq 0$. Thus, $D^i([\omega]) \neq \emptyset$. Let $[\omega'']$ in $D^i([\omega])$. By definition, $[\omega''] \leq_{\widetilde{\psi}\circ i_\alpha} [\omega]$. Then, by transitivity, $[\omega''] \leq_{\widetilde{\psi}\circ i_\alpha} [\omega']$ i.e. $[\omega''] \in D^i([\omega'])$.
Thus we have proved $D^i([\omega]) \subseteq D^i([\omega'])$. Therefore $\alpha^i([\omega]) \leq \alpha^i([\omega'])$. ∎

We know, by Observation 2, that $\leq_{\widetilde{\psi}}$ is a linear order over $C(\alpha)$. Note that $|C(\alpha)|$ is finite because $\mathscr{W}$ is finite. Put $m = |C(\alpha)|$. Thus, we can enumerate the elements of $C(\alpha)$ in such a way that $C(\alpha) = \{[\omega'_1], \ldots, [\omega'_m]\}$ and $[\omega'_j] <_{\widetilde{\psi}} [\omega'_{j+1}]$ for all $j \in \{1, \ldots, m-1\}$.
In an analogous way, by Observation 3, we can enumerate the elements of $C(\neg\alpha) =$

$\{[\omega_1''], \ldots, [\omega_n'']\}$, where $n = |C(\neg\alpha)|$, in such a way that $[\omega_j''] <_{\tilde{\Psi}} [\omega_{j+1}'']$ for all $j \in \{1, \ldots, n-1\}$.

By the fact that the linear orders between the elements $C(\alpha)$ coincide for all $\leq_{\tilde{\Psi} \circ^i \alpha}$ (Observation 4), we have $[\omega_j'] <_{\tilde{\Psi} \circ^i \alpha} [\omega_{j+1}']$ for all $j \in \{1, \ldots, m-1\}$. Thus, by Observation 7 we have $\alpha^i([\omega_j']) \leq \alpha^i([\omega_{j+1}'])$ for all $j \in \{1, \ldots, m-1\}$.

The following two observations will be useful later.

Observation 8 *For all $i \in \mathbb{N}$, $min(\leq_{\tilde{\Psi} \circ^i \alpha})$ is either $\{[\omega_1']\}$ or $\{[\omega_1'']\}$ or $\{[\omega_1'], [\omega_1'']\}$.*

Proof: Because of Observation 4 there are at most two elements in $min(\leq_{\tilde{\Psi} \circ^i \alpha})$. By the same observation the only possibilities are $\{[\omega_1']\}$ or $\{[\omega_1'']\}$ or $\{[\omega_1'], [\omega_1'']\}$. ∎

Observation 9 *For all $i \in \mathbb{N}$, $min(\leq_{\Psi \circ^i \alpha}) = \bigcup min(\leq_{\tilde{\Psi} \circ^i \alpha})$.*

By the previous remark, for each $i \in \mathbb{N}$, the vector $(\alpha^i([\omega_1']), \alpha^i([\omega_2']), \ldots, \alpha^i([\omega_m']))$ is an array of non decreasing natural numbers. These vectors will be decisive in the rest of the proof.

Before continuing to establish the results which lead us to the proof of Postulate (BI1), we introduce an example whose behavior obeys the postulates. It will illustrate the concepts and will show the iterative behavior of the basic assignment.

Example 2 *Assume that $|\mathcal{W}| = 16$. Let α be a formula such that $|[\![\neg\alpha]\!]| = 8$ and $|[\![\alpha]\!]| = 8$. Let Ψ be an epistemic state which is mapped by the basic assignment into the total preorder $\leq_\Psi$ over $\mathcal{W}$ represented in the figure below. The models of α are in black circles and the models of $\neg\alpha$ are in circles marked with $\times$. The lower a model is in this representation, the more preferred the model is. Thus the minimal models of $\leq_\Psi$ are the four models of $\neg\alpha$ at the bottom line.*

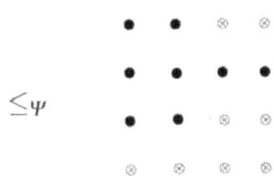

Now we consider $\mathcal{W}/\sim_\alpha$. The equivalence classes are formed by the sets of models of α which are $\leq_\Psi$-indifferent between them (at the same line in the figure) and by the sets of models of $\neg\alpha$ which are $\leq_\Psi$-indifferent between them. The first set of equivalence classes is $C(\alpha)$ and the second set of equivalent classes is $C(\neg\alpha)$.

The following figure represents the total preorder $\leq_{\tilde{\psi}}$ over $\mathcal{W}/\sim_\alpha$.

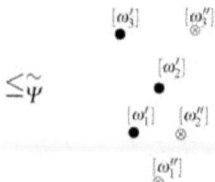

From this figure, using Definition 6, we have $\alpha^0([\omega'_1]) = 2$, $\alpha^0([\omega'_2]) = 2$ and $\alpha^0([\omega'_3]) = 3$. Thus, at the beginning of the process the array is the following

$$(\alpha^0([\omega_1]), \alpha^0([\omega_2]), \alpha^0([\omega_3])) = (2, 2, 3)$$

Suppose that after one iteration the total preorder $\leq_{\tilde{\psi} \circ \alpha}$ corresponding to the total preorder $\leq_{\Psi \circ \alpha}$ is represented by the figure below:

Note that in this case ω'_1 and ω'_3 have improved, so the property BS8 is satisfied. The postulates BS4 and BS5 of "rigidity" are satisfied because we are working with the equivalent classes. The postulates BS6 and BS7 of "not worsening" are also clearly satisfied. Postulates BS1 and BS2 are satisfied by definition. In this structure we have $\alpha^1([\omega'_1]) = 1$, $\alpha^1([\omega'_2]) = 2$ and $\alpha^1([\omega'_3]) = 2$. Thus, after one iteration the array is the following

$$(\alpha^1([\omega'_1]), \alpha^1([\omega'_2]), \alpha^1([\omega'_3])) = (1, 2, 2)$$

Representation of basic improvement operators

Suppose that after two iterations the total preorder $\tilde{\leq}_{\Psi \circ^2 \alpha}$ corresponding to the total preorder $\leq_{\Psi \circ^2 \alpha}$ is represented by the figure below:

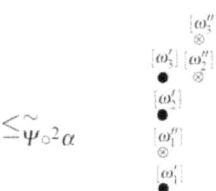

Here is also easy to check that properties BS1-BS8 are satisfied. In this structure we have $\alpha^2([\omega_1']) = 0$, $\alpha^2([\omega_2']) = 1$ and $\alpha^2([\omega_3']) = 2$. Thus, after two iterations the array is the following

$$(\alpha^2([\omega_1']), \alpha^2([\omega_2']), \alpha^2([\omega_3'])) = (0, 1, 2)$$

What this example suggests is that after a certain number k of iterations we will have, for the operators satisfying BS1-BS8, that $\alpha^k([\omega_1']) = 0$ (in the example $k = 2$). In such a case, we will prove that $[\![B(\Psi \circ^k \alpha)]\!] = [\omega_1']$ and therefore (BI1) holds because $[\omega_1'] \subset [\![\alpha]\!]$.

Note that from Conditions (BS1) and (BS2), the next observation follows straightforwardly:

Observation 10 *For every epistemic state Ψ, $[\![B(\Psi)]\!] = min(\leq_\Psi)$. In particular, for all $i \in \mathbb{N}$ $[\![B(\Psi \circ^i \alpha)]\!] = min(\leq_{\Psi \circ^i \alpha})$.*

Observation 11 *If there exists $k \in \mathbb{N}$ such that $\alpha^k([\omega_1']) = 0$ then*

$$[\![B(\Psi \circ^k \alpha)]\!] = [\omega_1']$$

Proof: Since $\alpha^k([\omega_1']) = 0$ there are no classes in $C(\neg \alpha)$ below or at the level of $[\omega_1']$ with respect to the preorder $\tilde{\leq}_{\Psi \circ^k \alpha}$. This says that $min(\tilde{\leq}_{\Psi \circ^k \alpha}) = [\omega_1']$. From this, by definition of $\tilde{\leq}_{\Psi \circ^k \alpha}$, we obtain $min(\leq_{\Psi \circ^k \alpha}) = [\omega_1']$. By Observation 10, we have $min(\leq_{\Psi \circ^k \alpha}) = [\![B(\Psi \circ^k \alpha)]\!]$, that is to say $[\![B(\Psi \circ^k \alpha)]\!] = [\omega_1]$. ∎

Observation 12 *If for all Ψ and for all α there exists $k \in \mathbb{N}$ such that $\alpha^k([\omega_1']) = 0$ then Postulate (BI1) holds.*

Proof: Assume there is a k such that $\alpha^k([\omega_1']) = 0$. By Observation 11, $[\![B(\Psi \circ^k \alpha)]\!] = [\omega_1']$. Note that $[\omega_1'] \subset [\![\alpha]\!]$, thus $[\![B(\Psi \circ^k \alpha)]\!] \subset [\![\alpha]\!]$. That is, there is a k such that $B(\Psi \circ^k \alpha) \vdash \alpha$, which is exactly what (BI1) expresses. ∎

Given the previous observation, our goal now is to prove that, given Ψ and α, there exists an integer k such that $\alpha^k([\omega'_1]) = 0$. In order to do that we will prove first that our arrays eventually decrease with respect to the lexicographical order. More precisely:

Proposition 1 *If $\alpha^i([\omega'_1]) \neq 0$ then there exists $j \in \mathbb{N}$ such that*

$$(\alpha^i([\omega'_1]), \ldots, \alpha^i([\omega'_m])) >_{lex} (\alpha^{i+j}([\omega'_1]), \ldots, \alpha^{i+j}([\omega'_m]))$$

This crucial result will be proved by induction on the number of equivalence classes in $C(\alpha)$ which are not indifferent to the elements of $C(\neg\alpha)$ with respect to the relation $\lesssim_{\Psi \circ^i \alpha}$. More precisely we define the set S^i_α of such classes as follows:

$$S^i_\alpha = \{[\omega] \in C(\alpha) : \not\exists [\omega'] \in C(\neg\alpha) \text{ and } [\omega'] \simeq_{\Psi \circ^i \alpha} [\omega]\}$$

When a class of $C(\alpha)$ is in S^i_α we say that it is *isolated* in the total preorder $\lesssim_{\Psi \circ^i \alpha}$.

The induction will be done on the cardinality s^i_α of this set ($s^i_\alpha = |S^i_\alpha|$), but first, we are going to illustrate the previous concepts through the total preorders of Example 2.

Example 3 *Let α, $\lesssim_{\Psi}$, $\lesssim_{\Psi \circ \alpha}$ and $\lesssim_{\Psi \circ^2 \alpha}$ be as in Example 2. We have $s^0_\alpha = 1$ because $[\omega'_2]$ is the only isolated class in $\lesssim_{\Psi}$; $s^1_\alpha = 3$ because $[\omega'_1]$, $[\omega'_2]$ and $[\omega'_3]$ are the isolated classes in $\lesssim_{\Psi \circ \alpha}$. Finally $s^2_\alpha = 2$ because $[\omega'_1]$ and $[\omega'_2]$ are the isolated classes in $\lesssim_{\Psi \circ^2 \alpha}$.*

We need some more observations and results before the proof of Proposition 1.

Observation 13 *If $\alpha^i([\omega'_1]) \neq 0$ then at least one of the following conditions holds:*

(I) $\exists [\omega], [\omega']$ *such that* $\omega \in C(\alpha)$, $[\omega'] \in C(\neg\alpha)$, $[\omega] \simeq_{\Psi \circ^i \alpha} [\omega']$ *and* $[\omega] <_{\Psi \circ^{i+1} \alpha} [\omega']$

(II) $\exists [\omega], [\omega']$ *such that* $[\omega] \in C(\alpha)$, $[\omega'] \in C(\neg\alpha)$, $[\omega'] <_{\Psi \circ^i \alpha} [\omega]$ *and* $[\omega] \lesssim_{\Psi \circ^{i+1} \alpha} [\omega']$

Proof: By hypothesis $\alpha^i([\omega_1]) \neq 0$, that is to say there exists $[\omega'] \in C(\neg\alpha)$ such that $[\omega'] \lesssim_{\Psi \circ^i \alpha} [\omega_1]$. From this it follows that there exists $\omega' \models \neg\alpha$ such that $\omega' \lesssim_{\Psi \circ^i \alpha} \omega_1$. Thus the hypothesis of Condition (BS8) holds. Therefore, at least one of the following two conditions holds:

(i) $\exists \omega, \omega'$ such that $\omega \models \alpha$, $\omega' \models \neg\alpha$, $\omega \simeq_{\Psi \circ^i \alpha} \omega'$ and $\omega <_{\Psi \circ^{i+1} \alpha} \omega'$

(ii) $\exists \omega, \omega'$ such that $\omega \models \alpha$, $\omega' \models \neg\alpha$, $\omega' <_{\Psi \circ^i \alpha} \omega$ and $\omega \lesssim_{\Psi \circ^{i+1} \alpha} \omega'$

But by definition of the equivalence classes and the relations $\lesssim_{\Psi \circ^i \alpha}$, $\simeq_{\Psi \circ^i \alpha}$ and $<_{\Psi \circ^i \alpha}$, the previous conditions are easily translated in

Representation of basic improvement operators 219

(I) $\exists [\omega], [\omega']$ such that $\omega \in C(\alpha), [\omega'] \in C(\neg\alpha), [\omega] \simeq_{\widetilde{\Psi} \circ i_\alpha} [\omega']$ and $[\omega] <_{\widetilde{\Psi} \circ i+1_\alpha} [\omega']$

(II) $\exists [\omega], [\omega']$ such that $[\omega] \in C(\alpha), [\omega'] \in C(\neg\alpha), [\omega'] <_{\widetilde{\Psi} \circ i_\alpha} [\omega]$ and $[\omega] \leq_{\widetilde{\Psi} \circ i+1_\alpha} [\omega']$

∎

Definition 7 *We say that a change of type (I) takes place (in $\leq_{\widetilde{\Psi} \circ i_\alpha}$) if after one more iteration Condition (I) holds. We say that a change of type (II) takes place (in $\leq_{\widetilde{\Psi} \circ i_\alpha}$) if after one more iteration Condition (II) holds.*

Lemma 2 *If there is no change of type (I) on $\leq_{\widetilde{\Psi} \circ i_\alpha}$, then $S_\alpha^{i+1} \subseteq S_\alpha^i$ and therefore $s_\alpha^{i+1} \leq s_\alpha^i$.*

Proof: Note that the classes in S_α^{i+1} can be partitioned into two sorts: the classes of $C(\alpha)$ which were in S_α^i, that is to say those classes in $C(\alpha)$ which are not indifferent to any class in $C(\neg\alpha)$ with respect to $\leq_{\widetilde{\Psi} \circ i_\alpha}$ and which continue to be isolated with respect to $\leq_{\widetilde{\Psi} \circ i+1_\alpha}$ and the new ones. Note that the new classes were classes which were not isolated in $\leq_{\widetilde{\Psi} \circ i_\alpha}$ and now they are isolated. Suppose $[\omega]$ is such a new class. Thus, necessarily, there is $[\omega']$ in $C(\neg\alpha)$ such that $[\omega] \simeq_{\widetilde{\Psi} \circ i_\alpha} [\omega']$ and since $[\omega]$ is isolated in $\leq_{\widetilde{\Psi} \circ i+1_\alpha}$ we have $[\omega] <_{\widetilde{\Psi} \circ i+1_\alpha} [\omega']$ or $[\omega'] <_{\widetilde{\Psi} \circ i+1_\alpha} [\omega]$. But the last option can not occur because of Condition (BS7). Thus, the only way for us to have new elements in S_α^{i+1} is that a change of type (I) occurs and this is impossible by hypothesis. Therefore there are no new classes in S_α^{i+1}. Therefore, $s_\alpha^{i+1} \leq s_\alpha^i$. ∎

Now we are ready to start the proof of Proposition 1.

Proof of Proposition 1: We proceed by induction on s_α^i. We start with the base case of our induction that is $s_\alpha^i = 0$. This means there are no isolated elements in $\leq_{\widetilde{\Psi} \circ i_\alpha}$, i.e.

$$\forall [\omega] \in C(\alpha), \exists [\omega'] \in C(\neg\alpha) \text{ such that } [\omega] \simeq_{\widetilde{\Psi} \circ i_\alpha} [\omega']$$

We claim that

$$(\alpha^i([\omega'_1]), \ldots, \alpha^i([\omega'_m])) >_{lex} (\alpha^{i+1}([\omega'_1]), \ldots, \alpha^{i+1}([\omega'_m]))$$

Note that by virtue of Observation 13 there is a change of type (I) or there is a change of type (II). Thus we consider two cases.

Case 1: There is a change of type (I).
Let $[\omega'_c]$ be the lowest class in $C(\alpha)$ producing a change of type (I). Thus, there exists $[\omega''] \in C(\neg\alpha)$ such that $[\omega_c] \simeq_{\widetilde{\Psi} \circ i_\alpha} [\omega'']$ and $[\omega_c] <_{\widetilde{\Psi} \circ i+1_\alpha} [\omega'']$. From this, it

follows clearly that $[\omega''] \in D^i([\omega'_c])$ and $[\omega''] \notin D^{i+1}([\omega'_c])$. Putting together this and Observation 5 we have $D^{i+1}([\omega'_c]) \subsetneq D^i([\omega'_c])$. Therefore $|D^{i+1}(\omega'_c)| < |D^i(\omega'_c)|$, i.e. $\alpha^{i+1}([\omega'_c]) < \alpha^i([\omega'_c])$. By Observation 6, we have $\alpha^{i+1}([\omega'_p]) \leq \alpha^i([\omega'_p])$ for all $p < c$. Thus, necessarily

$$(\alpha^i([\omega'_1]),...,\alpha^i([\omega'_c]),...,\alpha^i([\omega'_m])) >_{lex} (\alpha^{i+1}([\omega'_1]),...,\alpha^{i+1}([\omega'_c]),...,\alpha^{i+1}([\omega'_m])).$$

Case 2: There is a change of type (II).
Let $[\omega'_c]$ be the lowest class in $C(\alpha)$ producing a change of type (II). Thus, there exists $[\omega''] \in C(\neg\alpha)$ such that, $[\omega''] <_{\widetilde{\psi_{\circ i}\alpha}} [\omega'_c]$ and $[\omega'_c] \leq_{\widetilde{\psi_{\circ i+1}\alpha}} [\omega'']$. Since $s^i_\alpha = 0$ necessarily there exists $[\omega'''] \in C(\neg\alpha)$ such that $[\omega_c] \simeq_{\widetilde{\psi_{\circ i}\alpha}} [\omega''']$. Then, by definition, $\omega'_c \simeq_{\psi_{\circ i}\alpha} \omega'''$. By hypothesis $[\omega''] <_{\widetilde{\psi_{\circ i}\alpha}} [\omega'_c]$ then $[\omega''] <_{\widetilde{\psi_{\circ i}\alpha}} [\omega''']$. Thus, $\omega'' <_{\psi_{\circ i}\alpha} \omega'''$. Then, by Condition (BS5) $\omega'' <_{\psi_{\circ i+1}\alpha} \omega'''$. Note that $\omega'_c \leq_{\psi_{\circ i+1}\alpha} \omega''$ because $[\omega'_c] \leq_{\widetilde{\psi_{\circ i+1}\alpha}} [\omega'']$. Then, by transitivity, $\omega'_c <_{\psi_{\circ i+1}\alpha} \omega'''$, so $[\omega'_c] <_{\widetilde{\psi_{\circ i+1}\alpha}} [\omega''']$. Thus a change of type (I) occurs. We know that in this case the claim is true.
This completes the proof in the case $s^i_\alpha = 0$.

Now suppose that the proposition is true when $s^i_\alpha \leq k$, that is if $s^i_\alpha \leq k$ there exists $j \in \mathbb{N}$ such that

$$(\alpha^i([\omega'_1]),\ldots,\alpha^i([\omega'_m])) >_{lex} (\alpha^{i+j}([\omega'_1]),\ldots,\alpha^{i+j}([\omega'_m]))$$

We want to see that the result holds when $s^i_\alpha = k+1$. By Observation 13 we know that there is a change of type (I) or there is a change of type (II) or both. We are going to consider two cases: there is a change of type (I) or there is no change of type (I).

Case 1: There is a change of type (I).
Let $[\omega'_c]$ be the lowest class in $C(\alpha)$ producing a change of type (I) in $\leq_{\widetilde{\psi_{\circ i}\alpha}}$. With an analogous reasoning to the base case (Case 1) we get $\alpha^{i+1}([\omega'_c]) < \alpha^i([\omega'_c])$. By Observation 6, we have $\alpha^{i+1}([\omega'_p]) \leq \alpha^i([\omega'_p])$ for all $p < c$. Thus, necessarily

$$(\alpha^i([\omega'_1]),...,\alpha^i([\omega'_c]),...,\alpha^i([\omega'_m])) >_{lex} (\alpha^{i+1}([\omega'_1]),...,\alpha^{i+1}([\omega'_c]),...,\alpha^{i+1}([\omega'_m]))$$

Case 2: There is no change of type (I).
By Observation 13 necessarily there is a change of type (II). Let $[\omega'_c]$ be the lowest class in $C(\alpha)$ producing a change of type (II) in $\leq_{\widetilde{\psi_{\circ i}\alpha}}$. Thus, there exists $[\omega''] \in C(\neg\alpha)$ such that $[\omega''] <_{\widetilde{\psi_{\circ i}\alpha}} [\omega'_c]$ and $[\omega'_c] \leq_{\widetilde{\psi_{\circ i+1}\alpha}} [\omega'']$. We claim that $[\omega'_c]$ is isolated in $\leq_{\widetilde{\psi_{\circ i}\alpha}}$, i.e. there is no $[\omega'''] \in C(\neg\alpha)$ such that $[\omega'''] \simeq_{\widetilde{\psi_{\circ i}\alpha}} [\omega'_c]$. Towards a contradiction, suppose there is $[\omega'''] \in C(\neg\alpha)$ such that $[\omega'''] \simeq_{\widetilde{\psi_{\circ i}\alpha}} [\omega'_c]$. Thus, $[\omega''] <_{\widetilde{\psi_{\circ i}\alpha}} [\omega''']$. Then $\omega'' <_{\psi_{\circ i}\alpha} \omega'''$. By Condition (BS5) $\omega'' <_{\psi_{\circ i+1}\alpha} \omega'''$. Then $[\omega''] <_{\widetilde{\psi_{\circ i+1}\alpha}} [\omega''']$. By hypothesis $[\omega'_c] \leq_{\widetilde{\psi_{\circ i+1}\alpha}} [\omega'']$. Thus, by transitivity, $[\omega'_c] <_{\widetilde{\psi_{\circ i+1}\alpha}} [\omega''']$. Therefore $[\omega'_c]$ produces a change of type (I), a contradiction.

Representation of basic improvement operators

As $[\omega'_c] \leq_{\tilde{\psi} \circ^{i+1} \alpha} [\omega'']$ we have two possibilities: either $[\omega'_c] <_{\tilde{\psi} \circ^{i+1} \alpha} [\omega'']$ or $[\omega'_c] \simeq_{\tilde{\psi} \circ^{i+1} \alpha} [\omega'']$. They are analyzed separately.

Case 2.1: $[\omega'_c] <_{\tilde{\psi} \circ^{i+1} \alpha} [\omega'']$.
Since $[\omega''] <_{\tilde{\psi} \circ^i \alpha} [\omega'_c]$, we have $[\omega''] \in D^i([\omega'_c])$. By assumption $[\omega'_c] <_{\tilde{\psi} \circ^{i+1} \alpha} [\omega'']$, so $[\omega''] \notin D^{i+1}([\omega'_c])$. Putting together this and Observation 5 we have $D^{i+1}([\omega'_c]) \subsetneq D^i([\omega'_c])$, that is $\alpha^{i+1}([\omega'_c]) < \alpha^i([\omega'_c])$. By Observation 6, we have $\alpha^{i+1}([\omega'_p]) \leq \alpha^i([\omega'_p])$ for all $p < c$. Thus, necessarily

$$(\alpha^i([\omega'_1]), \ldots, \alpha^i([\omega'_c]), \ldots, \alpha^i([\omega'_m])) >_{lex} (\alpha^{i+1}([\omega'_1]), \ldots, \alpha^{i+1}([\omega'_c]), \ldots, \alpha^{i+1}([\omega'_m]))$$

Case 2.2: $[\omega'_c] \simeq_{\tilde{\psi} \circ^{i+1} \alpha} [\omega'']$.
Since there is no change of type (I), we can apply Lemma 2 and obtain $s_\alpha^{i+1} \leq s_\alpha^i$. We claim that $s_\alpha^{i+1} \leq k$. In order to see this, note that $[\omega'_c]$ is isolated in $\leq_{\tilde{\psi} \circ^i \alpha}$. Therefore

$$[\omega'_c] \in S_\alpha^i = \{[\omega] \in C(\alpha) : \nexists [\omega'] \in C(\neg \alpha) \text{ and } [\omega'] \simeq_{\tilde{\psi} \circ^i \alpha} [\omega]\}$$

However, by hypothesis, $[\omega'_c] \simeq_{\tilde{\psi} \circ^{i+1} \alpha} [\omega'']$. Therefore

$$[\omega'_c] \notin S_\alpha^{i+1} = \{[\omega] \in C(\alpha) : \nexists [\omega'] \in C(\neg \alpha) \text{ and } [\omega'] \simeq_{\tilde{\psi} \circ^{i+1} \alpha} [\omega]\}.$$

From this and together with Lemma 2, $S_\alpha^{i+1} \subsetneq S_\alpha^i$. Then $s_\alpha^{i+1} < s_\alpha^i = k+1$. Therefore $s_\alpha^{i+1} \leq k$. Thus, by induction hypothesis, there exists $j \in \mathbb{N}$ such that

$$(\alpha^{i+1}([\omega'_1]), \ldots, \alpha^{i+1}([\omega'_m])) >_{lex} (\alpha^{i+1+j}([\omega'_1]), \ldots, \alpha^{i+1+j}([\omega'_m]))$$

By Observation 5, we have

$$(\alpha^i([\omega'_1]), \ldots, \alpha^i([\omega'_m])) \geq_{lex} (\alpha^{i+1}([\omega'_1]), \ldots, \alpha^{i+1}([\omega'_m]))$$

Then, by transitivity of $\geq_{lex}$ we have

$$(\alpha^i([\omega'_1]), \ldots, \alpha^i([\omega'_m])) >_{lex} (\alpha^{i+1+j}([\omega'_1]), \ldots, \alpha^{i+1+j}([\omega'_m]))$$

which completes the proof of Proposition 1. ∎

Now we are ready to prove the *if* part of Theorem 7.

Proof of Theorem 7 (*if* part):
First we prove that Postulate (BI1) holds.
Note that Observation 12 says that Postulate (BI1) holds whenever there exists k such that $\alpha^k([\omega'_1]) = 0$. We claim that such a k exists. If not, by Proposition 1, we can construct an infinite sequence of arrays $(\alpha^{r_n}([\omega'_1]), \ldots, \alpha^{r_n}([\omega'_m]))$ (where $(r_n)_{n \in \mathbb{N}}$ is a increasing sequence of integers) such that

$$(\alpha^{r_n}([\omega'_1]),\ldots,\alpha^{r_n}([\omega'_m])) >_{lex} (\alpha^{r_{n+1}}([\omega'_1]),\ldots,\alpha^{r_{n+1}}([\omega'_m])).$$

But this contradicts the fact that the lexicographical order is well founded over the m-tuples of natural numbers.

Now we are going to prove that Equation (2) holds.
Let k be the minimum natural number such that $B(\Psi \circ^k \alpha) \vdash \alpha$, in particular $\Psi \star \alpha = \Psi \circ^k \alpha$. Note that
$$k = min\{n : B(\Psi \circ^n \alpha) \vdash \alpha\}$$
$$= min\{n : [\![B(\Psi \circ^n \alpha)]\!] \subseteq [\![\alpha]\!]\}$$
$$= min\{n : min(\leq_{\Psi \circ^n \alpha}) \subseteq [\![\alpha]\!]\}$$

the last equality because of Observation 10. By Observation 9,

$$min(\leq_{\Psi \circ^n \alpha}) = \bigcup min(\leq_{\widetilde{\Psi} \circ^n \alpha}).$$

From this we have that the expression

$$min(\leq_{\Psi \circ^n \alpha}) \subseteq [\![\alpha]\!]$$

is equivalent to the expression

$$\bigcup min(\leq_{\widetilde{\Psi} \circ^n \alpha}) \subseteq [\![\alpha]\!].$$

By Observation 8, the last expression is equivalent to

$$min(\leq_{\widetilde{\Psi} \circ^n \alpha}) = [\omega'_1].$$

But, by definition, $[\omega'_1]$ is exactly $min([\![\alpha]\!], \leq_\Psi)$.
Therefore $[\![B(\Psi \star \alpha)]\!] = min([\![\alpha]\!], \leq_\Psi)$, i.e. Equation (2) holds.

Now we are going to prove Postulates (BI2)-(BI11).

Postulate (BI2):
Suppose $B(\Psi) \wedge \alpha$ is consistent. We want to see that $B(\Psi \star \alpha) \equiv B(\Psi) \wedge \alpha$. By Equation (2) it is enough to see that $[\![B(\Psi) \wedge \alpha]\!] = min([\![\alpha]\!], \leq_\Psi)$. In order to establish this equation, we show first that $[\![B(\Psi) \wedge \alpha]\!] \subseteq min([\![\alpha]\!], \leq_\Psi)$. Let ω be an element of $[\![B(\Psi)]\!] \cap [\![\alpha]\!]$. By Conditions (BS1) and (BS2), we have $\omega \leq_\Psi \omega'$ for all interpretations ω'. Thus, in particular ω is minimal in $[\![\alpha]\!]$ with respect to $\leq_\Psi$, i.e. $\omega \in min([\![\alpha]\!], \leq_\Psi)$.
Now we show that $min([\![\alpha]\!], \leq_\Psi) \subseteq [\![B(\Psi) \wedge \alpha]\!]$. Assume that $\omega \in min([\![\alpha]\!], \leq_\Psi)$. Towards a contradiction, suppose that $\omega \notin [\![B(\Psi) \wedge \alpha]\!]$. Since $\omega \models \alpha$ necessarily $\omega \not\models B(\Psi)$. Since $[\![B(\Psi) \wedge \alpha]\!] \neq \emptyset$, there exists ω' such that $\omega' \in [\![B(\Psi) \wedge \alpha]\!]$. Then, by the inclusion previously proved, $\omega' \in min([\![\alpha]\!], \leq_\Psi)$. Since $\omega' \models B(\Psi)$ and $\omega \not\models B(\Psi)$, by Condition (BS2) we have $\omega' <_\Psi \omega$, contradicting the minimality

of ω.

Postulate (BI3):
Suppose $[\![\alpha]\!] \neq \emptyset$. We want to see that $B(\Psi \star \alpha)$ is consistent. Since we are in the finite case, $[\![\alpha]\!]$ is finite. Thus $min([\![\alpha]\!], \leq_\Psi) \neq \emptyset$. Then, by Equation (2) $B(\Psi \star \alpha)$ is consistent.

Postulate (BI4):
Let n be a positive integer and let α_i, β_i, γ and μ be formulas such that $\gamma \equiv \mu$ and for all $i \leq n$ $\alpha_i \equiv \beta_i$. We want to show that

$$B((\Psi \circ \alpha_1 \circ \cdots \circ \alpha_n) \star \gamma) \equiv B((\Psi \circ \beta_1 \circ \cdots \circ \beta_n) \star \mu)$$

By condition (BS3) we have $\leq_{\Psi \circ \alpha_1 \circ \cdots \circ \alpha_n} = \leq_{\Psi \circ \beta_1 \circ \cdots \circ \beta_n}$. By hypothesis $[\![\gamma]\!] = [\![\mu]\!]$ then $min([\![\gamma]\!], \leq_{\Psi \circ \alpha_1 \circ \cdots \circ \alpha_n}) = min([\![\mu]\!], \leq_{\Psi \circ \beta_1 \circ \cdots \circ \beta_n})$. From this, using Equation (2), we get $B((\Psi \circ \alpha_1 \circ \cdots \circ \alpha_n) \star \gamma) \equiv B((\Psi \circ \beta_1 \circ \cdots \circ \beta_n) \star \mu)$.

Postulate (BI5):
We want to show $B(\Psi \star \alpha) \wedge \beta \vdash B(\Psi \star (\alpha \wedge \beta))$ or equivalently $[\![B(\Psi \star \alpha) \wedge \beta]\!] \subseteq [\![B(\Psi \star (\alpha \wedge \beta))]\!]$. If $B(\Psi \star (\alpha \wedge \beta))$ is inconsistent the result is obvious. Suppose now $B(\Psi \star \alpha) \wedge \beta$ is consistent. Assume $\omega \in [\![B(\Psi \star \alpha) \wedge \beta]\!]$; we want to show that $\omega \in [\![B(\Psi \star (\alpha \wedge \beta))]\!]$. Towards a contradiction, suppose $\omega \notin [\![B(\Psi \star (\alpha \wedge \beta))]\!]$. By Equation (2), $\omega \in min([\![\alpha]\!], \leq_\Psi)$, $\omega \in [\![\beta]\!]$ and $\omega \notin min([\![\alpha \wedge \beta]\!], \leq_\Psi)$. Thus there exists $\omega' \in [\![\alpha]\!] \cap [\![\beta]\!]$, such that $\omega' <_\Psi \omega$. But in particular $\omega' \models \alpha$, contradicting the fact $\omega \in min([\![\alpha]\!], \leq_\Psi)$.

Postulate (BI6):
Suppose that $B(\Psi \star \alpha) \wedge \beta$ is consistent. We want to see that $B(\Psi \star (\alpha \wedge \beta)) \vdash B(\Psi \star \alpha) \wedge \beta$. It is enough to prove the equivalent semantical counterpart of this postulate: $[\![B(\Psi \star (\alpha \wedge \beta))]\!] \subseteq [\![B(\Psi \star \alpha) \wedge \beta]\!]$. Towards a contradiction, suppose that the inclusion does not hold, i.e. there exists $\omega \models B(\Psi \star (\alpha \wedge \beta))$ such that $\omega \not\models B(\Psi \star \alpha) \wedge \beta$. By Equation (2), we have $\omega \in min([\![\alpha \wedge \beta]\!], \leq_\Psi)$ and $\omega \notin min([\![\alpha]\!], \leq_\Psi) \cap [\![\beta]\!]$. Since $\omega \models \beta$, necessarily $\omega \notin min([\![\alpha]\!], \leq_\Psi)$. Since $[\![B(\Psi \star \alpha) \wedge \beta]\!] \neq \emptyset$, there exists $\omega' \in min([\![\alpha]\!], \leq_\Psi) \cap [\![\beta]\!]$. Since $\omega \in min([\![\alpha \wedge \beta]\!], \leq_\Psi)$ and $\omega' \models \alpha \wedge \beta$, we have $\omega \leq_\Psi \omega'$. But $\omega' \in min([\![\alpha]\!], \leq \Psi)$ thus $\omega \in min([\![\alpha]\!], \leq_\Psi)$, contradiction.

Postulate (BI7):
Assume $\alpha \vdash \mu$. We want to see that $B(\Psi \star \alpha) \equiv B((\Psi \circ \mu) \star \alpha)$. Condition (BS4) says that $\leq_\Psi$ and $\leq_{\Psi \circ \mu}$ coincide over $[\![\mu]\!]$. Since $[\![\alpha]\!] \subseteq mod(\mu)$, we have

$$min([\![\alpha]\!], \leq_\Psi) = min([\![\alpha]\!], \leq_{\Psi \circ \mu})$$

Then, by Equation (2), we get $B(\Psi \star \alpha) \equiv B((\Psi \circ \mu) \star \alpha)$.

Postulate (BI8):
The proof of this postulate is completely analogous to the proof of the previous postulate.

Postulate (BI9):
Suppose $B(\Psi \star \alpha) \vdash \mu$. We want to see $B((\Psi \circ \mu) \star \alpha) \vdash \mu$. From the assumption $B(\Psi \star \alpha) \vdash \mu$, by Lemma 1, we get that there exists $\omega \in [\![\alpha]\!] \cap mod(\mu)$ such that $\omega <_\Psi \omega'$ for all $\omega' \in [\![\alpha]\!] \cap [\![\neg\mu]\!]$. Then, by condition (BS6), $\omega \leq_{\Psi \circ \mu} \omega'$ for all $\omega' \in [\![\alpha]\!] \cap [\![\neg\mu]\!]$. From this, using again Lemma 1, we get $B((\Psi \circ \mu) \star \alpha) \vdash \alpha$.

Postulate (BI10):
Note that Postulate (BI10) is equivalent to the following postulate called (BI10'):
(I10') If $B(\Psi \circ \mu) \star \alpha \vdash \neg\mu$ then $B(\Psi \star \alpha) \vdash \neg\mu$.
Thus it is enough to prove (BI10'). Note that Condition (BS7) is equivalent to the following condition:
(S7') If $\omega \models \mu$ and $\omega' \models \neg\mu$, then $\omega' <_{\Psi \circ \mu} \omega \Rightarrow \omega' <_\Psi \omega$.
Assume $B((\Psi \circ \mu) \star \alpha) \vdash \neg\mu$. We want to see $B(\Psi \star \alpha) \vdash \neg\mu$. By Lemma 1, there exists $\omega \in mod(\alpha \wedge \neg\mu)$ such that $\omega <_{\Psi \circ \mu} \omega'$ for all $\omega' \in [\![\alpha \wedge \mu]\!]$. From this and Condition (S7') we get $\omega <_\Psi \omega'$ for all $\omega' \in [\![\alpha \wedge \mu]\!]$. From this, using again Lemma 1, we have $B(\Psi \star \alpha) \vdash \neg\mu$.

Postulate (BI11):
Suppose that there exists β consistent with α and consistent with $\neg\alpha$ such that $B(\Psi \star \beta) \not\vdash \alpha$. We want to see that at least one of the following conditions holds:

(a) $\exists \gamma$ such that $B(\Psi \star \gamma)$ is consistent with α, $B(\Psi \star \gamma)$ is consistent with $\neg\alpha$ and $B((\Psi \circ \alpha) \star \gamma)$ is inconsistent with $\neg\alpha$ but consistent with α

(b) $\exists \gamma$ such that $B(\Psi \star \gamma) \vdash \neg\alpha$ and $B((\Psi \circ \alpha) \star \gamma)$ is consistent with α

By the assumptions it is clear that β is consistent. Then, by (BI3) (already proved), $B(\Psi \star \beta) \not\vdash \bot$, i.e. $[\![B(\Psi \star \beta)]\!] \neq \emptyset$. By hypothesis $[\![B(\Psi \star \beta)]\!] \not\subseteq [\![\alpha]\!]$. Thus $\exists \omega' \models B(\Psi \star \beta) \wedge \neg\alpha$. Then, by Equation (2), $\omega' \in min([\![\beta]\!], \leq_\Psi)$. Therefore $\omega' \leq_\Psi \omega$. Thus the hypotheses of Condition (BS8) hold (taking α instead of μ). Therefore at least one of the following conditions holds:

(i) $\exists \omega_1, \omega_2$ such that $\omega_1 \models \alpha$, $\omega_2 \models \neg\alpha$, $\omega_1 \simeq_\Psi \omega_2$ and $\omega_1 <_{\Psi \circ \alpha} \omega_2$

(ii) $\exists \omega_1, \omega_2$ such that $\omega_1 \models \alpha$, $\omega_2 \models \neg\alpha$, $\omega_2 <_\Psi \omega_1$ and $\omega_1 \leq_{\Psi \circ \alpha} \omega_2$

Suppose condition (i) holds. Consider the formula $\varphi_{\omega_1,\omega_2}$.
Then $min([\![\varphi_{\omega_1,\omega_2}]\!], \leq_\Psi) = \{\omega_1, \omega_2\}$. From this and Equation (2), it follows $B(\Psi \star \varphi_{\omega,\omega'})$ is consistent with α and also consistent with $\neg\alpha$.
Moreover $min([\![\varphi_{\omega_1,\omega_2}]\!], \leq_{\Psi \circ \alpha}) = \{\omega_1\}$. Thus, by Equation (2), $B((\Psi \circ \alpha) \star \varphi_{\omega_1,\omega_2})$

is consistent with α and inconsistent with $\neg \alpha$. Thus letting $\gamma = \varphi_{\omega_1,\omega_2}$ condition (a) holds.

Suppose condition (ii) holds. Consider the formula $\varphi_{\omega_1,\omega_2}$. Then $min(\llbracket \varphi_{\omega_1,\omega_2} \rrbracket, \leq_\Psi) = \{\omega_2\} \subseteq \llbracket \neg \alpha \rrbracket$. Thus, by Equation (2), $B(\Psi \star \varphi_{\omega,\omega'}) \vdash \neg \alpha$. Also by Equation (2), $\omega \in min(\llbracket \varphi_{\omega,\omega'} \rrbracket, \leq_{\Psi \circ \alpha})$. Thus $B((\Psi \circ \alpha) \star \varphi_{\omega,\omega'})$ is consistent with α. Therefore by letting $\gamma = \varphi_{\omega_1,\omega_2}$ condition (b) holds. ∎

6 Implications on related works

Remember that the operators defined in [Konieczny et al., 2010] and considered in Example 1, ⊙, ⊘ and ⊕, are actually basic improvement operators. Thus we can get more powerful representation theorem for these operators by observing two facts:

The first fact is that Condition (S3) entails Conditions (BS6) and (BS7).

The second fact is that the other conditions of the assignments for these operators entail Condition (BS8)

Thus the assignments associated to these operators (systematic enhancement, half gradual assignment and best gradual assignment) are indeed basic assignments satisfying other specific conditions. Then we have the following three results:

Theorem 1 *If ○ is a one improvement operator (in the sense of [Konieczny et al., 2010]), that is, it satisfies the postulates (I1)-(I11), then there exists a systematic enhancement (an assignment satisfying conditions 1, 2, 3, S1, S2, S3, S4 and S5) mapping each epistemic state Ψ into a total preorder $\leq_\Psi$ such that*

$$\llbracket B(\Psi \star \alpha) \rrbracket = min(\llbracket \alpha \rrbracket, \leq_\Psi) \quad (3)$$

Conversely, assume that ○ is a change operator for which there is a systematic enhancement mapping each epistemic state Ψ into a total preorder $\leq_\Psi$, then the operator ○ is an improvement operator and the equation (3) holds for this gradual assignment.

Theorem 2 *If ○ is a half improvement operator (in the sense of [Konieczny et al., 2010]), that is, it satisfies the postulates (I1)-(I10),(H1),(H2) then there exists a half gradual assignment (an assignment satisfying conditions 1, 2, 3, S1, S2, S3, S4, SH1 and SH2) mapping each epistemic state Ψ into a total preorder $\leq_\Psi$ such that*

$$\llbracket B(\Psi \star \alpha) \rrbracket = min(\llbracket \alpha \rrbracket, \leq_\Psi) \quad (4)$$

Conversely, assume that ○ is a change operator for which there is a half gradual assignment mapping each epistemic state Ψ into a total preorder $\leq_\Psi$, then the operator ∘ is an improvement operator and the equation (4) holds for this gradual assignment.

Theorem 3 *If ∘ is a best improvement operator (in the sense of [Konieczny et al., 2010]), that is, it satisfies the postulates (I1)-(I1),(B1),(B2) then there exists a best gradual assignment (an assignment satisfying conditions 1, 2, 3, S1, S2, S3, S4, SB1 and SB2) mapping each epistemic state Ψ into a total preorder $\leq_\Psi$ such that*

$$[\![B(\Psi \star \alpha)]\!] = min([\![\alpha]\!], \leq_\Psi) \qquad (5)$$

Conversely, assume that ∘ is a change operator for which there is a best gradual assignment mapping each epistemic state Ψ into a total preorder $\leq_\Psi$, then the operator ∘ is an improvement operator and the equation (5) holds for this gradual assignment.

7 Concluding remarks

We have found conditions on the assignment, in particular the *non worsening* conditions (Conditions (BS6) and (BS7)) and the condition saying that *something improves* if there is any possibility of improvement (Condition (BS8)) which allow to show that the property of iterated success, *i.e.* Postulate (BI1), is satisfied. This suggests that we have to enrich the conditions of gradual assignment to allow improvement in all the situations (even when the models of μ are isolated) if we want to get a full representation theorem for the improvement operators (those satisfying postulates (I1)-(I9)), for the weak improvement operators (those satisfying postulates (I1)-(I6)) and for the soft improvement operators (those satisfying postulates (I1)-(I10)) considered in [Konieczny et al., 2010]. Full in the sense that we do not make *a priori* the hypothesis that the operators considered satisfy Postulate (BI1).

A future work is exploring the minimality of the set of postulates defining the basic improvement operators. We conjecture that this set is minimal.

Akcnowledgements

We will thank the referees by the helpful remarks contributing to improve the first version of this work. The second author thanks the CDCHTA-ULA for the financial support through the project C-1855-13-05-AA.

References

[Alchourrón et al., 1985] C. E. Alchourrón, P. Gärdenfors, and D. Makinson. On the logic of theory change: Partial meet contraction and revision functions. *Journal of Symbolic Logic*, 50:510–530, 1985.

[Benferhat et al., 2000] S. Benferhat, S. Konieczny, O. Papini, and R. Pino Pérez. Iterated revision by epistemic states: axioms, semantics and syntax. In *Fourteenth European Conference on Artificial Intelligence (ECAI'2000)*, pages 13–17, 2000.

[Booth and Meyer, 2006] R. Booth and T. Meyer. Admissible and restrained revision. *Journal of Artificial Intelligence Research*, 26:127–151, 2006.

[Cantwell, 1997] J. Cantwell. On the logic of small changes in hypertheories. *Theoria*, 63(1-2):54–89, 1997.

[Darwiche and Pearl, 1997] A. Darwiche and J. Pearl. On the logic of iterated belief revision. *Artificial Intelligence*, 89:1–29, 1997.

[Hansson et al., 2001] S. O. Hansson, E. Fermé, J. Cantwell, and M. Falappa. Credibility limited revision. *Journal of Symbolic Logic*, 66:1581–1596, 2001.

[Hansson, 1997] S. O. Hansson, editor. *Theory*, Special issue on non-prioritized belief revision, vol 63, number 1-2. 1997.

[Hansson, 1999] S. O. Hansson. A survey of non-prioritized belief revision. *Erkenntnis*, 50:413–427, 1999.

[Jin and Thielscher, 2007] Y. Jin and M. Thielscher. Iterated belief revision, revised. *Artificial Intelligence*, 171:1–18, 2007.

[Katsuno and Mendelzon, 1991] H. Katsuno and A. O. Mendelzon. Propositional knowledge base revision and minimal change. *Artificial Intelligence*, 52:263–294, 1991.

[Konieczny and Pino Pérez, 2008] S. Konieczny and R. Pino Pérez. Improvement operators. In *Proceedings of the Eleventh International Conference on Principles of Knowledge Representation And Reasoning (KR 2008).*, pages 177–187, 2008.

[Konieczny et al., 2010] S. Konieczny, M. Medina Grespan, and R. Pino Pérez. Taxonomy of improvement operators and the problem of minimal change. In *Proceedings of the Twelfth International Conference on Principles of Knowledge Representation And Reasoning (KR 2010).*, pages 161–170, 2010.

[Makinson, 1997] D. Makinson. Screened revision. *Theoria*, 63(1-2):14–23, 1997.

[Spohn, 1988] W. Spohn. Ordinal conditional functions: A dynamic theory of epistemic states. In W. L. Harper and B. skyrms, editors, *Causation in Decision: Belief Change and Statistics*, pages 105–134. Kluwer, 1988.

Argument Revision as a means of supporting dishonesty

Mark Snaith and Chris Reed

Abstract In this paper, we demonstrate how our previous work on Argument Revision can be used to assist a participant in a dialogue to be dishonest. We first provide answers to the questions of why a participant would choose to lie, and what constitutes a lie in terms of structured argumentation. We then go on to show how Argument Revision can be used not only in selecting a "minimal" lie, but also in maintaining that lie in order to avoid detection.

1 Introduction

Dishonesty is a common human behaviour. Types of dishonesty range from innocent and trivial "white lies" (such as telling a child about the existence of Father Christmas), to those with more serious consequences (such as a politician lying while in office). In either case, it is necessary for the speaker of the lie to maintain it, or face negative consequences (an upset child, or charges of corruption).

There are clear connections between dishonesty and belief revision, the process of updating beliefs to accommodate new or changed information. At the most basic level, dishonesty involves a speaker communicating something they do not believe, with the intention that the hearer believe that they (the speaker) do believe it; in other words, the speaker wishes to present an external belief state that is different to their actual beliefs. As a simple example, assume that Alice believes $\{\neg P, Q, P \wedge Q \vdash \bot\}$. If Alice lies through uttering P, she cannot outwardly appear to believe Q, because to do so would present an inconsistency. Further, if it is later established that $P \vdash S$, she needs to be prepared to appear, dialogically, to believe S, even if she does not actually believe it.

Further connections exist between dishonesty and belief revision. For instance, a main idea in the logical account of lying specified by Sakama et al. [2010], is

---

Mark Snaith · Chris Reed
School of Computing, University of Dundee, Dundee, UK,
e-mail: marksnaith@computing.dundee.ac.uk

keeping dishonest acts as small as possible, because doing so makes them easier to maintain. This is strikingly similar to the concept of "minimal change" in the AGM model of belief revision, where a revision process is carried out with the smallest effect (with respect to epistemic entrenchment) on remaining beliefs. [Alchourrón et al., 1985]

Inspired by the recent momentum in the exploration of connections between argumentation and belief revision [Falappa et al., 2009, 2011], Snaith and Reed [2012] present a model for Argument Revision using structured argumentation. The model, subsequently expanded by Snaith [2012], removes the need for an entrenchment ordering and instead determines minimal change on the basis of measurable effects on the system, providing an answer the question of uniquely specifying revision functions [Gärdenfors, 1988, 1992]. These measurable effects have as their basis sets of arguments (and, by extension, formulae — the conclusions of arguments) that are impacted when a specified change is made to the system. It is through these sets that this model lends itself well to supporting dishonesty — not only can a minimal lie be identified, the effects on other beliefs are also recorded and can be used as a means of maintaining the lie.

The purpose of the paper is to show how the model of Argument Revision presented in [Snaith and Reed, 2012] and [Snaith, 2012] can support lying in dialogue, from initially choosing to lie, to maintaining it.

2 Background

In this section, we briefly introduce previous work upon which the present paper builds.

2.1 ASPIC$^+$ framework

The ASPIC$^+$ framework [Prakken, 2010] extends the original ASPIC framework [Amgoud et al., 2006] and instantiates Dung's [1995] abstract approach to argumentation by adding structure to arguments. The basic notion of the framework is an *argumentation system*:

Definition 1. (Argumentation system)
An argumentation system $\mathscr{AS} = \langle \mathscr{L}, \bar{\ }, \mathscr{R}, \leq \rangle$ where $\mathscr{L}$ is a logical language, $\bar{\ }$ is a contrariness function from $\mathscr{L}$ to $2^{\mathscr{L}}$, $\mathscr{R} = \mathscr{R}_s \cup \mathscr{R}_d$ is a set of strict $\mathscr{R}_s$ and defeasible[1] $\mathscr{R}_d$ inference rules such that $\mathscr{R}_s \cap \mathscr{R}_d = \emptyset$ and $\leq$ is a partial preorder on $\mathscr{R}_d$.

---

[1] From [Pollock, 1987], a rule is strict iff it holds without exception; otherwise, it is strict.

Remark 1. The contrariness function is represented as $p \in \bar{q}$, which means "p is a contrary of q", or $\bar{q} = \{p, r\}$, which means "p and r are contraries of q". Where p and q are contraries of each other (i.e. $p \in \bar{q}$ and $q \in \bar{p}$) they are said to be contradictory, which is represented as $p = -q$.

An argumentation system contains a knowledge base, $\langle \mathcal{K}, \leq' \rangle$ where $\mathcal{K} \subseteq \mathcal{L}$ and $\leq'$ is a partial preorder on $\mathcal{K} \setminus \mathcal{K}_n$. $\mathcal{K} = \mathcal{K}_n \cup \mathcal{K}_p \cup \mathcal{K}_a \cup \mathcal{K}_i$, where $\mathcal{K}_n$ is a set of (necessary) axioms, $\mathcal{K}_p$ is a set of ordinary premises, $\mathcal{K}_a$ is a set of assumptions and $\mathcal{K}_i$ is a set of issues.

¿From the knowledge base ($\mathcal{K}$) and rules ($\mathcal{R}$), arguments are constructed. For an argument $\mathcal{A}$, $Prem(\mathcal{A})$ is a function that returns all the premises in $\mathcal{A}$; $Conc(\mathcal{A})$ is a function that returns the conclusion of $\mathcal{A}$; $Sub(\mathcal{A})$ is a function that returns all the sub-arguments of $\mathcal{A}$; $DefRules(\mathcal{A})$ is a function that returns all defeasible rules in $\mathcal{A}$; and $TopRule(\mathcal{A})$ is a function that returns the last inference rule use in $\mathcal{A}$.

On the basis of these functions, $\mathcal{A}$ is:

1. φ if $\varphi \in \mathcal{K}$ with:
   - $Prem(\mathcal{A}) = \{\varphi\}$
   - $Conc(\mathcal{A}) = \varphi$
   - $Sub(\mathcal{A}) = \varphi$
   - $DefRules(\mathcal{A}) = \emptyset$
   - $TopRule(\mathcal{A}) =$ undefined.

2. $\mathcal{A}_1, \ldots, \mathcal{A}_n \to \psi$ if $\mathcal{A}_1, \ldots, \mathcal{A}_n$ are arguments such that there exists a strict rule $Conc(\mathcal{A}_1), \ldots Conc(\mathcal{A}_n) \to \psi$ in $\mathcal{R}_s$ with:
   - $Prem(\mathcal{A}) = Prem(\mathcal{A}_1) \cup \ldots \cup Prem(\mathcal{A}_n)$
   - $Conc(\mathcal{A}) = \psi$
   - $Sub(\mathcal{A}) = Sub(\mathcal{A}_1) \cup \ldots \cup Sub(\mathcal{A}_n) \cup \{\mathcal{A}\}$
   - $DefRules(\mathcal{A}) = DefRules(\mathcal{A}_1) \cup \ldots \cup DefRules(\mathcal{A}_n)$
   - $TopRule(\mathcal{A}) = Conc(\mathcal{A}_1), \ldots, Conc(\mathcal{A}_n) \to \psi$.

3. $\mathcal{A}_1, \ldots, \mathcal{A}_n \Rightarrow \psi$ if $\mathcal{A}_1, \ldots, \mathcal{A}_n$ are arguments such that there exists a defeasible rule $Conc(\mathcal{A}_1), \ldots Conc(\mathcal{A}_n) \Rightarrow \psi$ in $\mathcal{R}_d$ with
   - $Prem(\mathcal{A}) = Prem(\mathcal{A}_1) \cup \ldots \cup Prem(\mathcal{A}_n)$
   - $Conc(\mathcal{A}) = \psi$
   - $Sub(\mathcal{A}) = Sub(\mathcal{A}_1) \cup \ldots \cup Sub(\mathcal{A}_n) \cup \{\mathcal{A}\}$
   - $DefRules(\mathcal{A}) = DefRules(\mathcal{A}_1) \cup \ldots \cup DefRules(\mathcal{A}_n)$
   - $TopRule(\mathcal{A}) = Conc(\mathcal{A}_1), \ldots, Conc(\mathcal{A}_n) \Rightarrow \psi$.

¿From [Prakken, 2010], we write $S \vdash \varphi$ if there exists a strict argument for φ with all premises taken from S and $S \mid\!\sim \varphi$ if there exists a defeasible argument for φ with all premises taken from S.

An argument can be attacked in three ways: on a (non-axiom) premise (undermine), on a defeasible inference rule (undercut) or on a conclusion (rebuttal).

Given an argumentation system $\mathcal{AS}$ and a knowledge base $\mathcal{KB}$, an argumentation theory is $\mathcal{AT} = \langle \mathcal{AS}, \mathcal{KB}, \preceq \rangle$ where $\preceq$ is an argument ordering on the set of all arguments that can be constructed from $\mathcal{KB}$ in $\mathcal{AT}$. An argumentation theory is *well-formed* iff: if ϕ is a contrary of ψ then ψ is not in $\mathcal{K}_n$ and ψ is not the consequent of a strict rule.

In this paper, we will use the following notations: $Args(\mathcal{AT})$ to represent the set of all arguments in an argumentation theory; $e(\mathcal{AT})$ to represent an extension of the derived abstract framework under some unspecified, single-extension[2] semantics; $\mathcal{K}(\mathcal{AT})$ is the knowledge base of the argumentation system in $\mathcal{AT}$.

2.2 Argument Revision

The theory of Argument Revision in [Snaith and Reed, 2012] has two main principles — the process of removing, or rendering unacceptable, an argument, and the subsequent determination as to exactly how to perform that process. This theory was subsequently updated by Snaith [2012] to extend the process to sets of arguments: given a set of arguments $\mathcal{S}$, when every argument in $\mathcal{S}$ is either no longer present or acceptable in an argumentation theory $\mathcal{AT}$, the process is complete. This is denoted as $\mathcal{AT} \dot{-} \mathcal{S}$ and is called the *contraction* of $\mathcal{AT}$ by $\mathcal{S}$. The contraction of an argumentation theory by a set of arguments is a series of modifications to the knowledge base, such that the arguments no longer exist, or are rendered unacceptable, and the resultant argumentation theory is well-formed. Knowledge base modification involves adding and/or removing formulae until the goal of the Argument Revision is achieved.

When a formula is removed from a knowledge base, the result is a new argumentation system that is identical to the original argumentation system, except for the knowledge base which lacks the removed formula.

Formally, this is captured by the *formula removal function*, which takes as input an argumentation system and some formula, and provides as output the new argumentation system whose knowledge base does not contain that formula:

Definition 2. Formula removal function
$\mathcal{AS} - \phi : \mathcal{K}(\mathcal{AS} - \phi) = \mathcal{K}(\mathcal{AS}) \setminus \{\phi\}$

Remark 2. The notation $\mathcal{AT} - (\phi, \mathcal{AS})$ is used to represent the new argumentation theory yielded by removing a formula ϕ from the knowledge base in $\mathcal{AS}$.

The *formula addition function* governs the addition of a formula to the knowledge base of an argumentation system.

One issue of adding information to the knowledge base is knowing which subset it should be placed in. The nature of the formula addition function is that it adds formulae simply to introduce new arguments, with no justification (where justification could, for instance, come from another agent). We therefore classify added formulae as assumptions.

---

[2] We limit the present work to single-extension semantics because to examine the nature of belief derivation from multiple extensions is beyond the scope of this paper.

Definition 3. Formula addition function
$\mathcal{AS} + \phi: \mathcal{K}_a(\mathcal{AS} + \phi) = \mathcal{K}_a(\mathcal{AS}) \cup \{\phi\}$

Remark 3. Note also that because $\mathcal{K}_a(\mathcal{AS}) \subseteq \mathcal{K}(\mathcal{AS})$, $\mathcal{K}(\mathcal{AS} + \phi) = \mathcal{K}(\mathcal{AS}) \cup \{\phi\}$

Remark 4. The notation $\mathcal{AT} + (\phi, \mathcal{AS})$ is used to represent the new argumentation theory yielded by adding a formula ϕ to the knowledge base in $\mathcal{AS}$.

There may be multiple possible ways to contract an argumentation system; these are modelled using a *change graph*.

Definition 4. (Change Graph)
A change graph $CG(\mathcal{AT}, \Pi')$ for the revisions of $\mathcal{AT}$ to a set of argumentation theories $\Pi' = \{\mathcal{AT}_1^{\pm}, \ldots, \mathcal{AT}_n^{\pm}\}$ is a directed acyclic graph $\langle \Upsilon, \Omega \rangle$ where:

- $\Upsilon \subseteq \Pi$ (where Π is the set of all possible argumentation theories)
- $\Pi' \subseteq \Upsilon$ is minimal (w.r.t. set inclusion) in that $\forall \mathcal{AT}' \in \Pi'$, $\neg \exists \mathcal{AT}'' \in \Pi'$, $\mathcal{K}(\mathcal{AT}) \Delta \mathcal{K}(\mathcal{AT}'') \subset \mathcal{K}(\mathcal{AT}) \Delta \mathcal{K}(\mathcal{AT}')$
- **(Atomic change)** $\Omega \subseteq \Upsilon \times \Upsilon$ where $\forall \omega \in \Omega$ such that $\omega = (\mathcal{AT}', \mathcal{AT}'')$, we have that $| \mathcal{K}(\mathcal{AT}') \Delta \mathcal{K}(\mathcal{AT}'') | = \pm 1$

In order to choose which contraction method to perform, a cost function is used, based on four criteria of minimality — **argument loss** (those previously acceptable arguments that are lost in the resultant argumentation theory), **acceptability loss** (those previously acceptable arguments that remain in the resultant theory, but are now unacceptable), **argument gain** (new arguments in the theory) and **acceptability gain** (previously unacceptable arguments that are acceptable in the new theory). These effects are captured by a set of functions, where $\mathcal{AT}^{\pm}$ is an argumentation theory that has had an unspecified revision process carried out on it.

Definition 5. The *argument drop function* Δ_A for an extension E (under some single-extension semantics subsumed by complete semantics) in an argumentation theory $\mathcal{AT}$:

$\Delta_A: \Pi \times \Pi \to 2^{\mathcal{A}rgs(\mathcal{AT})}$,
$\Delta_A(\mathcal{AT}, \mathcal{AT}^{\pm}) = \{\mathcal{A} \mid \mathcal{A} \in e(\mathcal{AT}), \mathcal{A} \notin \mathcal{A}rgs(\mathcal{AT}^{\pm})\}$

Definition 6. The *acceptability drop function* Δ_S for an extension E (under some single-extension semantics subsumed by complete semantics) in an argumentation theory $\mathcal{AT}$:

$\Delta_S: \Pi \times \Pi \to 2^{\mathcal{A}rgs(\mathcal{AT})}$,
$\Delta_S(\mathcal{AT}, \mathcal{AT}^{\pm}) = \{\mathcal{A} \in e(\mathcal{AT}), \mathcal{A} \in \mathcal{A}rgs(\mathcal{AT}^{\pm}), \mathcal{A} \notin e(\mathcal{AT}^{\pm})\}$

Definition 7. The *argument gain function* Γ_A:

$\Gamma_A: \Pi \times \Pi \to 2^{\mathcal{A}rgs(\mathcal{AT}^{\pm})}$,
$\Gamma_A(\mathcal{AT}, \mathcal{AT}^{\pm}) = \{\mathcal{A} \mid \mathcal{A} \notin \mathcal{A}rgs(\mathcal{AT}), \mathcal{A} \in \mathcal{A}rgs(\mathcal{AT}^{\pm})\}$

Definition 8. The *acceptability gain function* Γ_S for an extension E (under some single-extension semantics subsumed by complete semantics):

$$\Gamma_S: \Pi \times \Pi \to 2^{Args(\mathcal{AT})},$$
$$\Gamma_S(\mathcal{AT}, \mathcal{AT}^\pm) = \{\mathcal{A} \mid \mathcal{A} \notin e(\mathcal{AT}), \mathcal{A} \in e(\mathcal{AT}^\pm)\}$$

These functions are used to determine a *path cost*:

Definition 9. (Path cost)
Given a change graph $CG(\mathcal{AT}, \Pi') = \langle \Upsilon, \Omega \rangle$, the *path cost* from $\mathcal{AT}$ to $\mathcal{AT}' \in \Pi'$ is
$$V(\mathcal{AT}, \mathcal{AT}') = |\Delta_A(\mathcal{AT}, \mathcal{AT}') \cup \Delta_S(\mathcal{AT}, \mathcal{AT}') \cup \Gamma_A(\mathcal{AT}, \mathcal{AT}') \cup \Gamma_S(\mathcal{AT}, \mathcal{AT}')|$$

While at present the four measures of change are given equal weighting in determining a path cost, it is our intention in future work to investigate the role of preferences between them.

2.3 Dishonesty

While dishonesty is common in everyday communication, few attempts have been made to formalise it in terms of inter-agent communication.

Castelfranchi [2000] made one of the first attempts, categorising deception into two forms: *strong deception*, where an agent utters p despite explicitly believing $\neg p$; and *weak deception*, where an agent utters p but does not believe p. These characterisations are similar to two of the three types of dishonesty specified by Sakama et al. [2010]: *lying* is analogous to strong deception, in that it requires explicit belief in the contrary of what is stated and *bullshit* (henceforth referred to as "BS") is analogous to weak deception, in that the statement is grounded neither in truth nor falsity. A third form of dishonesty, *deception* is also specified by Sakama et al., where an agent utters a statement they believe to be true, but with the intention the hearer derives a false conclusion.

For the purposes of this paper, we use an adapted version of the Sakama et al. [2010] definition of lying to show how Argument Revision can support the dishonesty process.

3 Lying with structured argumentation

In this section, we explore the nature of lying when using the ASPIC$^+$ framework as a basis for beliefs. We focus on a single act, lying, because the purpose of this paper is not to provide a formal specification of lying, but instead to show how Argument Revision can support dishonesty and can be used by a participant in a dialogue to update the epistemic state they wish to externally present.

Each participant in a dialogue possesses a personal argumentation theory ($\mathcal{PAT}$) from which they derive beliefs from its associated abstract framework.

Definition 10. (Beliefs)
The beliefs of a participant α with a personal argumentation theory $\mathcal{PAT}_\alpha$ are:

$$\mathcal{B}_\alpha = \{Conc(\mathcal{A}) \mid \mathcal{A} \in e(\mathcal{PAT}_\alpha)\}$$

In addition, we do not use an existing or specify a new dialogue protocol, but will instead focus on a single locution, $utter_\alpha(\phi)$, to denote the utterance of a formula $\phi \in \mathcal{L}$ by a participant α.

3.1 Why lie?

Before showing how Argument Revision can support lying, we first answer the question of why a participant in a dialogue would choose to lie.

When choosing how to respond to a statement F made by an opponent in a persuasion dialogue, a participant will carry out a three-stage process: i) determining whether or not they believe the statement; if not, ii) determining whether or not they have a contrary or contradictory argument against the statement; if not, iii) conceding the statement.

A significant drawback of this process is that unless the participant has a counter-argument against F, they must concede, regardless of the impact of doing so. In certain protocols, conceding a statement causes a participant to incur commitment to it [Walton and Krabbe, 1995] which in turn will affect current and potential future commitments; current commitments could become inconsistent, while future commitments will themselves have to be consistent with the conceded statement.

This presents one scenario in which a participant might benefit from being dishonest — where doing so is less costly (in terms of minimal change, with respect to impact on commitments) than conceding. Accommodating dishonesty may bring about fewer changes, both in terms of current and future commitments, than conceding the statement to their opponent.

A second scenario relies on another intrinsic human behaviour — keeping certain beliefs private, no matter what. However, this can lead to a situation where a participant is able to honestly defeat their opponent's argument, but cannot do so without sharing information they wish to keep private. They might therefore be dishonest in order to being about the same result (i.e. defeating their opponent's argument), but without revealing private information.

It is beyond the scope of this paper to explore the exact nature of private beliefs, so we will assume that having applied some unspecified criteria, a participant has arrived at a set $P_\mathcal{L} \subseteq \mathcal{L}$ of formulae that they wish to keep private. Using this set, they can determine a set of private arguments in an argumentation theory. An argument is private if it contains a sub-argument whose conclusion is in the set of private formulae.

Definition 11. (Private arguments)
Given an argumentation system $\mathscr{AS}$ and a set of private formulae $P_{\mathscr{L}}$, the set of private arguments is:

$$P_{\mathscr{A}rgs} = \{\mathscr{A} \in \mathscr{A}rgs(\mathscr{AS}) \mid \exists \mathscr{A}' \in Sub(\mathscr{A}) \text{ s.t. } Conc(\mathscr{A}') \in P_{\mathscr{L}}\}$$

By extension, we can define a set of private beliefs.

Definition 12. (Private beliefs)
Given a set of private arguments, $P_{\mathscr{A}rgs}$, a participant's private beliefs are:

$$P_B = \{\phi \mid \phi \in B, \phi \in \bigcup_{\mathscr{A} \in P_{\mathscr{A}rgs}} Conc(\mathscr{A})\}$$

The concept of private arguments and beliefs could leave a participant in a situation of believing their opponent's statement is false, but because the argument that leads them to that belief is private, they are powerless to respond without revealing this private information.

The process of (honest) dialogue move selection, incorporating private arguments, is shown in Figure 1.

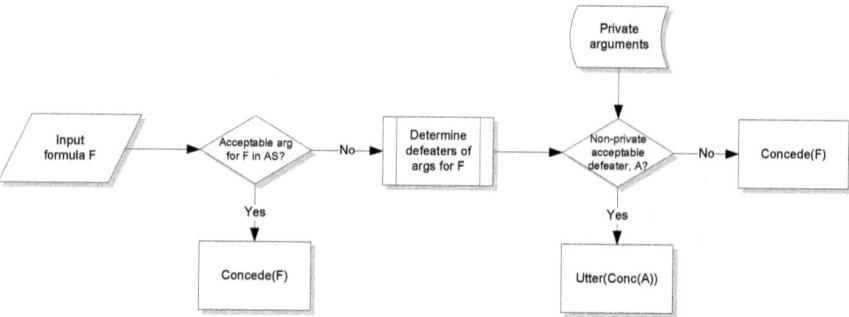

Fig. 1 Dialogue move selection

As an example of a situation where a dialogue participant will want to keep certain arguments private, consider the following knowledge base and rules in an argumentation system, where a government minister is attempting to convince a political opponent that they should not spend money on welfare. The real reason is that the government thinks a war is about to happen that will need to be paid for, but they do not want to share this. Instead, the position to not spend money on welfare is justified by saying money needs spent on education, despite the education department spending being under-budget.

- $\mathcal{K}_p = \{\text{have\_money}, \text{education\_under\_budget}, \text{war}\}$
- $\mathcal{R}_d = \begin{cases} r1 : \text{war}, \text{have\_money} \Rightarrow \text{pay\_for\_war} \\ r2 : \text{have\_money} \Rightarrow \text{pay\_for\_education} \\ r3 : \text{have\_money} \Rightarrow \text{pay\_for\_welfare} \\ r4 : \text{education\_under\_budget} \Rightarrow \neg\text{pay\_for\_education} \end{cases}$
- $P_{\mathscr{L}} = \{\text{pay\_for\_war}\}$

And with:

- $r2 < r1, r3 < r1, r2 < r4, r3 < r2$
- $\overline{\text{pay\_for\_war}} = \{\text{pay\_for\_education}, \text{pay\_for\_welfare}\}$
- $\overline{\text{pay\_for\_education}} = \{\text{pay\_for\_war}, \text{pay\_for\_welfare}\}$
- $\overline{\text{pay\_for\_welfare}} = \{\text{pay\_for\_war}, \text{pay\_for\_education}\}$

The arguments in the system are:

$\mathscr{A}_1$: have_money
$\mathscr{A}_2$: education_under_budget
$\mathscr{A}_3$: war
$\mathscr{A}_4$: $\mathscr{A}_3, \mathscr{A}_1 \Rightarrow_{r1}$ pay_for_war
$\mathscr{A}_5$: $\mathscr{A}_1 \Rightarrow_{r2}$ pay_for_education
$\mathscr{A}_6$: $\mathscr{A}_1 \Rightarrow_{r3}$ pay_for_welfare
$\mathscr{A}_7$: $\mathscr{A}_2 \Rightarrow_{r4} \neg$pay_for_education

Using the last-link principle[3], argument preferences are $\mathscr{A}_5 \prec \mathscr{A}_4$; $\mathscr{A}_6 \prec \mathscr{A}_4$; $\mathscr{A}_6 \prec \mathscr{A}_5$; $\mathscr{A}_5 \prec \mathscr{A}_7$. The set of private arguments is $P_{\mathscr{A}rgs} = \{\mathscr{A}_3, \mathscr{A}_4\}$.

The resultant framework, evaluated under grounded semantics[4], is shown in Figure 2, with islands (those arguments with no attack interactions) being omitted for clarity. Acceptable arguments are marked with a tick ($\checkmark$), with private arguments surrounded by a dashed box.

The politician's beliefs are, therefore,

$$B = \{\text{money}, \text{war}, \text{education\_under\_budget}, \text{pay\_for\_war}\}.$$

Since $\mathscr{A}_3$ is a private argument, the only remaining defeater of $\mathscr{A}_6$ is $\mathscr{A}_5$. However, $\mathscr{A}_5$ is not acceptable because it is defeated by $\mathscr{A}_7$; thus, no honest defeater of $\mathscr{A}_6$ can be offered, without sharing a private argument.

4 Argument revision and lying

In this section, we show how Argument Revision techniques can be used to support lying.

---

[3] The last-link principle prefers an argument $\mathscr{A}$ over another argument $\mathscr{B}$ if the last defeasible rules used in $\mathscr{B}$ are less preferred than the last defeasible rules in $\mathscr{A}$ or, in case both arguments are strict, if the premises of $\mathscr{B}$ are less preferred than the premises of $\mathscr{A}$. A formal definition is provided in [Prakken, 2010].

[4] See [Dung, 1995, pp. 327–329] for the definition of grounded semantics

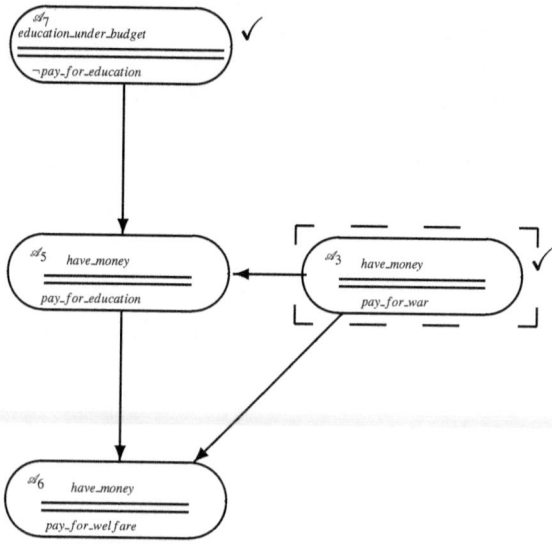

Fig. 2 Argumentation Framework for the politician example

Influenced by belief revision, the main principle in Argument Revision is minimal change, measured in terms of the loss and gain of arguments and argument acceptability [Snaith and Reed, 2012]. This model of Argument Revision was applied in a dialogical context to assist an agent in selecting a dialogue move and its content, especially when faced with needing to retract. The same principle can be extended to agents that are open to being dishonest, with Argument Revision being applied in the selection of a lie — chose the lie that has the smallest cost, and thus minimal maintenance.

Before describing the process of lying, and the use of Argument Revision to it, we first provide a formal definition of a lie in terms of the ASPIC$^+$ framework. Based on the intuition of Sakama et al. [2010], a lie, in response to an utterance, is the utterance of the conclusion of an unacceptable defeater of the argument which formed the basis of the utterance being replied to.

Definition 13. (Lie)
A locution $utter_\alpha(\phi)$ is a lie if $\phi \notin B_\alpha$ and $\exists \psi \in B_\alpha$ such that $\psi \in \overline{\phi}$ in $\mathscr{PAT}_\alpha$

Remark 5. We assume that α's intention for their opponent to believe ϕ is implicit in the utterance.

An argument is unacceptable when it is defeated by one or more acceptable arguments. Therefore, when lying, a participant needs to ensure it does not outwardly appear either to find those defeaters acceptable, or possess them at all. This is analogous to an Argument Revision process — the removal of or making unacceptable the defeaters will require either the addition or removal of elements of the knowledge base.

Four functions are defined in Snaith and Reed [2012] for identifying the effects of adding and removing formulae in an argumentation system's knowledge base; the outputs of these functions are sets of arguments that are affected. Using these functions will thus not only allow the minimal lie to be identified, but also show how it can be maintained.

4.1 The process of lying

Before uttering a lie, a participant first decides to lie, then, if more than one possible lie exists, chooses which one. Whether or not they do lie, the first stage is identifying what, if any, arguments (acceptable or otherwise, but not private) defeat the argument presented by their opponent. This is done through the *defeater identification function*:

Definition 14. (Defeater identification function)

$Def: \mathscr{A}rgs(\mathscr{AT}) \to 2^{\mathscr{A}rgs(\mathscr{AT})}$
$Def(\mathscr{A}) = \{\mathscr{B} \mid \mathscr{B} \text{ defeats } \mathscr{A}\}$

The set $Def(\mathscr{A})$ bifurcates into two disjoint subsets: $Def_a(\mathscr{A})$, those defeaters that are acceptable (in an extension $e(\mathscr{AT})$) and $Def_u(\mathscr{A})$, those defeaters that are not:

- $Def_a(\mathscr{A}) = e(\mathscr{AT}) \cap Def(\mathscr{A})$
- $Def_u(\mathscr{A}) = Def(\mathscr{AT}) \setminus e(\mathscr{AT})$

If $Def_a(\mathscr{A}) \setminus P_{\mathscr{A}rgs} = \emptyset$ (where $P_{\mathscr{A}rgs}$ is the set of private arguments), then the participant cannot provide an honest defeater for $\mathscr{A}$, thus it will either have to concede $\mathscr{A}$, or lie. If it is the case that $Def_u(\mathscr{A}) \setminus P_{\mathscr{A}rgs} = \emptyset$, the participant cannot lie because the only defeaters of $\mathscr{A}$ are private. This is a scenario where a participant may resort to BS (the utterance of a statement not grounded in truth or falsity), however the present paper shall retain its focus on the single act of lying.

4.2 Choosing a lie

The account of lying presented by Sakama et al. [2010] stipulates that an agent should not tell an unnecessarily strong lie — that is, if you are telling a lie, ensure it is the easiest possible to maintain. This is analogous to the principle of minimal change in Argument Revision, where, when revising an argumentation system, it should be done with the minimal possible effect on remaining arguments. Furthermore, the functions specified in Snaith and Reed [2012] capture the actual arguments that are affected and thus can be used by the agent as a means of ensuring their future utterances in the dialogue are compatible with the original lie.

In the current context, when choosing to lie, a participant, for the purposes and extent of the dialogue in which the lie occurs, considers the argument at the source of the lie to be acceptable. This, in turn, would mean the defeater(s) of the argument are unacceptable. We use the defeater identification function to identify the acceptable defeaters, all of which the participant needs to publicly consider unacceptable. Thus, if a participant lies with $utter_\alpha(Conc(\mathscr{A}))$ then it cannot publicly consider the arguments in $Def_a(\mathscr{A})$ to be acceptable. α therefore can perform an Argument Revision process with respect to every argument in $Def_a(\mathscr{A})$ — that is, the goal of the process is $\mathscr{AT} \dot{-} Def_a(\mathscr{A})$. The minimal lie (that is, the one that requires the least effort to maintain) is the method of achieving this contraction with the smallest change path cost.

With lying now available as an alternative to conceding, we update the process of selecting a dialogue move, as shown in figure 3. The initial steps of the process have been omitted, but can be seen in figure 1. Note that this process shows only one possible way of incorporating dishonesty into dialogue selection — being dishonest only if it is less costly than conceding, and the agent has no non-private way of defending their position. Other move selection strategies exist that allow for dishonesty, but exploring them in detail is beyond the scope of this paper and so is left to future work.

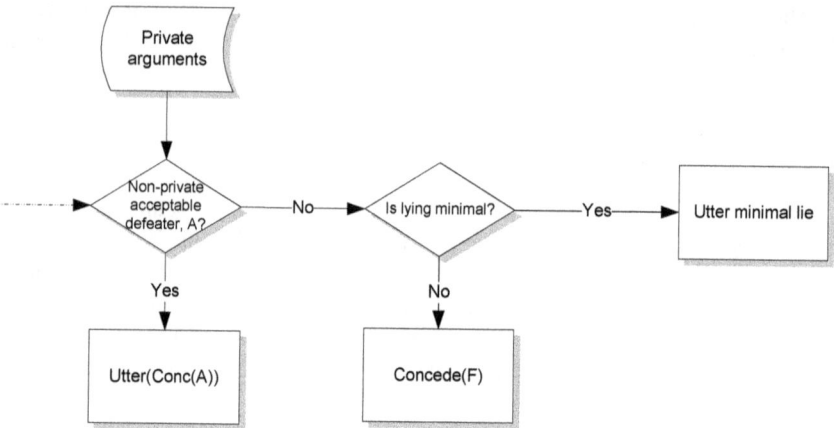

Fig. 3 Dialogue move selection including lying (initial steps omitted)

4.3 Maintaining a lie

Once a participant chooses to lie, it needs to maintain it. The choice of lie will have been made based on the minimal change required to its public opinions, but those opinions that have changed will need to be captured.

The four functions for Argument Revision identify drops and gains in terms of both arguments themselves and acceptability. If we consider drops, the participant will not be able to communicate arguments, or conclusions of arguments, in either drop set without appearing inconsistent. When an argument is dropped, it is because at least one of its sub-arguments has been dropped; when the acceptability of an argument is dropped, it's because at least one defeater has either gained acceptability, or been newly added to the system. In either case, the reason for the drop is the accommodation of the lie. However, the arguments have only been dropped in terms of the epistemic state the participant wishes to externally present. They still believe the conclusions of the dropped arguments, but can't communicate them — in other words, they have become private. Thus, the participant can simply add the content of the outputs of the drop functions to his store of private arguments. This allows the drops to be considered when contemplating future dialogue moves.

In terms of arguments and acceptability that are gained, there are two scenarios in which a participant needs to be aware of these; the first is when its opponent advances a conclusion of one of these arguments. The participant would need to concede that formula, or else risk exposing its lie, unless they construct a further lie that accommodates both rejecting their opponent's claim, and their original lie. The second situation is when the lie itself has been attacked and needs defended. To defend the lie would require the participant to use an argument whose conclusion they don't believe, but as a result of the lie must appear to, due to it either gaining acceptability, or being added to the argumentation system.

While arguments that are dropped or lose acceptability can be added to the set of private arguments, there is no current provision for a participant to keep a record of those arguments that are gained or gain acceptability. We therefore introduce a new set, $G_{\mathcal{A}rgs}$ such that for a formula removal or addition, $\mathcal{AT} \pm (\phi, \mathcal{AS})$, $\Gamma_A(\mathcal{AT}, \mathcal{AT} \pm (\phi, \mathcal{AS})) \cup \Gamma_S(\mathcal{AT}, \mathcal{AT} \pm (\phi, \mathcal{AS})) \subseteq G_{\mathcal{A}rgs}$.

5 Example

Consider a participant α with a personal argumentation theory $\mathcal{PAT}_\alpha$ containing a knowledge base and rule set thus:

$$\mathcal{K}_p = \begin{cases} a_1, a_2, b_1, \\ c_1, d_1, d_2, \\ e_1, e_2, f_1, \\ g_1, h \end{cases}$$

$$\mathcal{R}_d = \begin{cases} a_1, a_2 \Rightarrow_{r1} a, \ b_1 \Rightarrow_{r2} b, \ c_1 \Rightarrow_{r3} c \\ d_1, d_2 \Rightarrow_{r4} d, \ e_1 \Rightarrow_{r5} e, \ e_2 \Rightarrow_{r6} e \\ f_1 \Rightarrow_{r7} f, \quad g_1 \Rightarrow_{r8} g \end{cases}$$

And that:

- $b \in \overline{a}, c \in \overline{b}, d \in \overline{b}, e_2 \in \overline{h}, e \in \overline{d}, f \in \overline{b}, g \in \overline{f_1}$
- $P_{\mathscr{L}} = \{c\}$

The following arguments can be constructed:

$\mathscr{A}_1 : a_1$ $\mathscr{A}_2 : a_2$
$\mathscr{A}_3 : b_1$ $\mathscr{A}_4 : c_1$
$\mathscr{A}_5 : d_1$ $\mathscr{A}_6 : e_1$
$\mathscr{A}_7 : e_2$ $\mathscr{A}_8 : e_3$
$\mathscr{A}_9 : f_1$ $\mathscr{A}_{10} : g_1$
$\mathscr{A}_{11} : h$ $\mathscr{A}_{12} : \mathscr{A}_1, \mathscr{A}_2 \Rightarrow_{r1} a$
$\mathscr{A}_{13} : \mathscr{A}_3 \Rightarrow_{r2} b$ $\mathscr{A}_{14} : \mathscr{A}_4 \Rightarrow_{r3} c$
$\mathscr{A}_{15} : \mathscr{A}_5 \Rightarrow_{r4} d$ $\mathscr{A}_{16} : \mathscr{A}_6, \mathscr{A}_7 \Rightarrow_{r5} e$
$\mathscr{A}_{17} : \mathscr{A}_8 \Rightarrow_{r6} e$ $\mathscr{A}_{18} : \mathscr{A}_9 \Rightarrow_{r7} f$
$\mathscr{A}_{19} : \mathscr{A}_{10} \Rightarrow_{r8} g$

The abstract framework derived from $\mathscr{PAT}_\alpha$ is shown in figure 5, with islands (those arguments with no attack interactions) removed for clarity. Arguments marked with a tick (✓) are acceptable, while those surrounded with a dashed box are private. The grounded extension, again with islands removed, is $\{\mathscr{A}_{12}, \mathscr{A}_{14}, \mathscr{A}_{16}, \mathscr{A}_{17}, \mathscr{A}_{19}\}$ and the private arguments are $\{\mathscr{A}_4, \mathscr{A}_{14}\}$.

Assume that in a dialogue, α needs to advance an argument to defend $\mathscr{A}_{12}$ from $\mathscr{A}_{13}$. α has one acceptable defeater, $\mathscr{A}_{14}$, however due to c_1 being a private formula, $\mathscr{A}_{14}$ is a private argument. The only other defeaters of $\mathscr{A}_{13}$ are $\mathscr{A}_{15}$ and $\mathscr{A}_{18}$, but neither are acceptable. Thus if α advances the conclusion of either argument, they will be uttering a lie.

In order to determine the minimal lie, and assess its impact, we first use the defeater identification function for both $\mathscr{A}_{15}$ and $\mathscr{A}_{18}$:

- $Def(\mathscr{A}_{15}) = \{\mathscr{A}_{16}, \mathscr{A}_{17}\}$
- $Def(\mathscr{A}_{18}) = \{\mathscr{A}_{19}\}$

α therefore needs to contract $\mathscr{PAT}_\alpha$ with respect to either $\mathscr{A}_{16}$ and $\mathscr{A}_{17}$, or $\mathscr{A}_{19}$.

There are two possible methods of achieving $\mathscr{PAT}_\alpha \dot{-} \{\mathscr{A}_{16}, \mathscr{A}_{17}\}$: both involve removing e_3, since it is the only premise of $\mathscr{A}_{17}$; then, we can either remove e_1 or

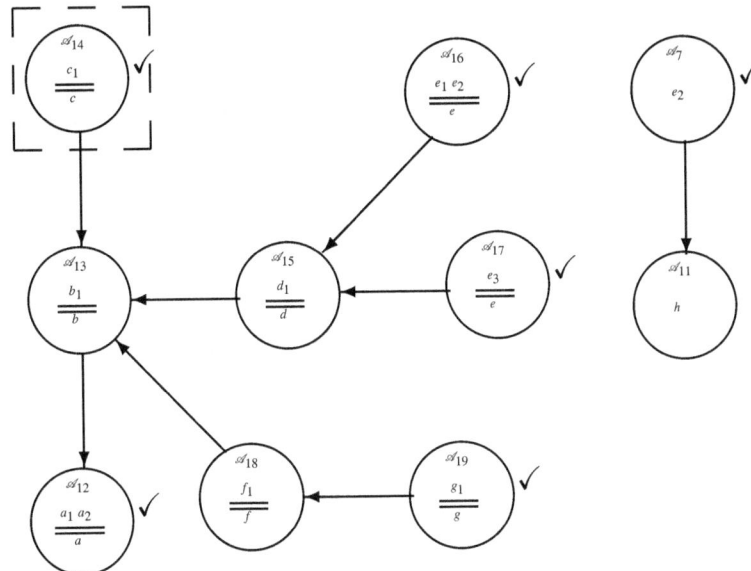

Fig. 4 Argumentation Framework from $\mathscr{AT}_{PAS\alpha}$

e_2, thus:

$\Delta_A(\mathscr{PAT}_\alpha, \mathscr{PAT}_\alpha - (e_3, \mathscr{AS}) - (e_1, \mathscr{AS})) = \{\mathscr{A}_6, \mathscr{A}_8, \mathscr{A}_{16}, \mathscr{A}_{17}\}$
$\Delta_S(\mathscr{PAT}_\alpha, \mathscr{PAT}_\alpha - (e_3, \mathscr{AS}) - (e_1, \mathscr{AS})) = \{\}$
$\Gamma_A(\mathscr{PAT}_\alpha, \mathscr{PAT}_\alpha - (e_3, \mathscr{AS}) - (e_1, \mathscr{AS})) = \{\}$
$\Gamma_S(\mathscr{PAT}_\alpha, \mathscr{PAT}_\alpha - (e_3, \mathscr{AS}) - (e_1, \mathscr{AS})) = \{\mathscr{A}_{15}\}$

$\Delta_A(\mathscr{PAT}_\alpha, \mathscr{PAT}_\alpha - (e_3, \mathscr{AS}) - (e_2, \mathscr{AS})) = \{\mathscr{A}_7, \mathscr{A}_8, \mathscr{A}_{16}, \mathscr{A}_{17}\}$
$\Delta_S(\mathscr{PAT}_\alpha, \mathscr{PAT}_\alpha - (e_3, \mathscr{AS}) - (e_2, \mathscr{AS})) = \{\}$
$\Gamma_A(\mathscr{PAT}_\alpha, \mathscr{PAT}_\alpha - (e_3, \mathscr{AS}) - (e_2, \mathscr{AS})) = \{\}$
$\Gamma_S(\mathscr{PAT}_\alpha, \mathscr{PAT}_\alpha - (e_3, \mathscr{AS}) - (e_2, \mathscr{AS})) = \{\mathscr{A}_{11}, \mathscr{A}_{15}\}$

The change graph for $\mathscr{PAT}_\alpha \dot{-} \{\mathscr{A}_{16}, \mathscr{A}_{17}\}$ is shown in figure 5.
The path cost for $\mathscr{PAT}_\alpha \dot{-} \{\mathscr{A}_{16}, \mathscr{A}_{17}\}$ are:

$V(\mathscr{PAT}_\alpha, \mathscr{PAT}_\alpha - (e_3, \mathscr{AS}) - (e_1, \mathscr{AS})) =$
$|\{\mathscr{A}_6, \mathscr{A}_8, \mathscr{A}_{16}, \mathscr{A}_{17}\} \cup \{\} \cup \{\} \cup \{\mathscr{A}_{15}\}| = 5$

$V(\mathscr{PAT}_\alpha, \mathscr{PAT}_\alpha - (e_3, \mathscr{AS}) - (e_2, \mathscr{AS})) =$
$|\{\mathscr{A}_7, \mathscr{A}_8, \mathscr{A}_{16}, \mathscr{A}_{17}\} \cup \{\} \cup \{\} \cup \{\mathscr{A}_{11}, \mathscr{A}_{15}\}| = 6$

There is only one way of performing $\mathscr{PAT}_\alpha \dot{-} \{A_{19}\}$ — the removal of g_1 from the knowledge base:

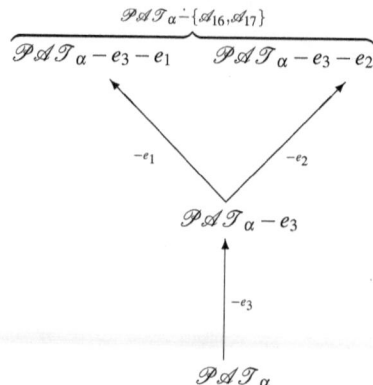

Fig. 5 Change graph showing $\mathcal{AT}\dot{-}\{\mathcal{A}_{16},\mathcal{A}_{17}\}$.

$\Delta_A(\mathcal{PAT}_\alpha, \mathcal{PAT}_\alpha - (g_1, \mathcal{AS})) = \{\mathcal{A}_{10}, \mathcal{A}_{19}\}$
$\Gamma_A(\mathcal{PAT}_\alpha, \mathcal{PAT}_\alpha - (g_1, \mathcal{AS})) = \{\}$
$\Delta_S(\mathcal{PAT}_\alpha, \mathcal{PAT}_\alpha - (g_1, \mathcal{AS})) = \{\}$
$\Gamma_S(\mathcal{PAT}_\alpha, \mathcal{PAT}_\alpha - (g_1, \mathcal{AS})) = \{\mathcal{A}_{18}\}$

The change graph for $\mathcal{PAT}_\alpha \dot{-} \{\mathcal{A}_{19}\}$ has only one path, whose cost is:

$$V(\mathcal{PAT}_\alpha, \mathcal{PAT}_\alpha - (g_1, \mathcal{AS})) = |\{\mathcal{A}_{10}, \mathcal{A}_{19}\} \cup \{\mathcal{A}_{18}\}| = 3$$

The minimal lie is, therefore, $utter_\alpha(f)$, since it has the lowest overall cost. This, in turn, results in $\mathcal{A}_{10}, \mathcal{A}_{19} \in P_{\mathcal{A}rgs}$ (the arguments for g_1 and g respectively) and $\mathcal{A}_{18} \in G_{\mathcal{A}rgs}$ (the argument for f).

6 Conclusions

We have in this paper shown how Argument Revision techniques can be used to support a participant in a dialogue to be dishonest. We first presented dishonesty as a form of belief revision and, by extension, Argument Revision, because to be dishonest requires a participant to externally present an epistemic state that is different from their actual beliefs, with the intention that their opponent(s) believe the dishonest remarks. It is in this intention a connection to belief (and argument) revision is found — the participant revises their beliefs or arguments before presenting them.

Inspired by the logical account of lying specified by Sakama et al. [2010], we formally defined what constitutes a lie in terms of the ASPIC⁺ framework for argumentation, which we then used to illustrate how Argument Revision can assist not only the selection of a dishonest act, but also maintaining it. By using four measures of minimal change, for a system of structured argumentation, specified in Snaith and Reed [2012], a participant can first determine a minimal lie (with respect to the number of changes to an argumentation system), then use the arguments in the outputs of those measures to maintain the lie.

Future work will focus on several areas. First, we intend to explore whether or not it is possible for a lie to never be discovered when using Argument Revision; conversely, we will also investigate if Argument Revision can be used to detect dishonesty. Secondly, dishonesty has different costs and benefits for different types of dialogue — for instance, being dishonest in a persuasion dialogue can increase the chances of beating your opponent, while in a negotiation it can improve your pay-off. It is our intention to examine these different application areas. Finally, the work presented here has focused on the single act of lying characterised by Sakama et al. [2010] (which is similar to Castelfranchi [2000]'s strong deception). Extending this work to account for other types of dishonesty, such as BS (similar to weak deception) and deceit will also be explored.

What we have demonstrated with this work is that methods for determining minimal change in the revision of an argumentation system can be used to support dishonesty, both in terms of identifying a minimal lie and then maintaining it.

Acknowledgements

This work was supported by the Engineering and Physical Sciences Research Council (EPSRC) of the UK government under grant number EP/G060347/1.

References

C.E. Alchourrón, P. Gärdenfors, and D. Makinson. On the logic of theory change: Partial meet contraction and revision functions. *The Journal of Symbolic Logic*, 50(2):510–530, 1985.

L. Amgoud, L. Bodenstaff, M. Caminada, P. McBurney, S. Parsons, H. Prakken, J. van Veenen, and G.A.W. Vreeswijk. Final review and report on formal argumentation system. Deliverable D2.6, ASPIC IST-FP6-002307, 2006.

C. Castelfranchi. Artificial liars: Why computers will (necessarily) deceive us and each other. *Ethics and Information Technology*, 2:113–119, 2000.

P. M. Dung. On the acceptability of arguments and its fundamental role in nonmonotonic reasoning, logic programming and n-person games. *Artificial Intelligence*, 77:321–357, 1995.

M. A. Falappa, G. Kern-Isberner, and G. R. Simari. Belief revision and argumentation theory. In I. Rahwan and G.R. Simari, editors, *Argumentation in Artificial Intelligence*, pages 314–360. Springer, 2009.

M. A. Falappa, A. J. Garcia, G. Kern-Isberner, and G. R. Simari. On the evolving relation between belief revision and argumentation. *The Knowledge Engineering Review*, 26(1):35–43, 2011.

P. Gärdenfors. *Knowledge in Flux*. MIT Press, 1988.

P. Gärdenfors. Belief revision: An introduction. In P. Gärdenfors, editor, *Belief Revision*, pages 1–27. Cambridge University Press, 1992.

J.L. Pollock. Defeasible reasoning. *Cognitive Science*, 11:481–518, 1987.

H. Prakken. An abstract framework for argumentation with structured arguments. *Argument and Computation*, 1(2):93–124, 2010.

C. Sakama, M. Caminada, and A. Herzig. A logical account of lying. In M. Fisher, W. van der Hoek, B. Konev, and A. Lisitsa, editors, *Logics in Artifical Intelligence*, pages 286–299. Springer, 2010.

M. Snaith and C. Reed. Justified argument revision in agent dialogue. In *Proceedings of the Ninth International Workshop on Argumentation in Mult-Agent Systems (ArgMAS 2012)*, 2012.

M. I. Snaith. *Argument Revision and its Role in Dialogue*. PhD thesis, University of Dundee, 2012.

D.N. Walton and E.C.W. Krabbe. *Commitment in Dialogue: Basic Concepts of Interpersonal Reasoning*. State University of New York Press, New York, 1995.

Layered Argumentation Frameworks with subargument relation

Beishui Liao

Abstract For an argumentation framework with subargument relation (AFwS), those arguments whose status can be indirectly determined by their proper subarguments as well as those arguments that are unrelated to any attacks may not be involved in the process of status evaluation of arguments. In this chapter, we propose a layered AFwS in which arguments are classified into two subsets: conflict-handling arguments and status-dependent arguments, such that the conflicts among arguments are handled centrally, within a sub-framework induced by the set of conflict-handling arguments. Based on the notion and the semantics of a layered AFwS, some basic notions and properties of updating a layered AFwS are presented, which are expected to better support the application of argumentation to belief revision.

1 Introduction

In some other chapters of this book, argumentation and belief revision have been shown as two general patterns of reasoning that might benefit each other. On the one hand, argumentation is suitable for dealing with (potentially) incomplete and/or conflicting information [Dung, 1995], aiming at justifying why specific beliefs (propositions) hold. On the other hand, belief revision focuses on determining how the existing beliefs (propositions) should be changed or adjusted upon the coming of new information. A possible way to combine these two reasoning formalisms is as follows:

Put an argumentation component before a belief revision component: The former is regarded as a "filter"or a "guard"of the latter. When receiving a new piece of infor-

---

Beishui Liao
Center for the Study of Language and Cognition, Dept. of Philosophy, Zhejiang University, China, e-mail: baiseliao@zju.edu.cn

mation, a set of accepted arguments (and therefore, their conclusions) are produced by means of argumentation. Then, the existing beliefs are revised accordingly.

To achieve this vision, a vital problem is to decide which arguments (and the beliefs they support) should be evaluated and how they are evaluated. Due to the incompleteness and inconsistency of knowledge, the reasoning performed by argumentation is nonmonotonic. As a result, when new information comes, all the existing arguments and the arguments newly constructed should be considered. This is obviously not a rational (or an efficient) approach. In order to cope with this problem, we may exploit some existing approaches introduced very recently, such as [Liao et al., 2011] and [Baumann, 2011]. In [Liao et al., 2011], we presented a method to handle the dynamics of an abstract argumentation framework: When an argumentation framework is updated, the status of unaffected arguments remains unchanged, while only the status of affected arguments is reevaluated. In [Baumann, 2011] and [Baumann et al., 2012], Baumann proposed a similar approach that is based on Lifschitz and Turner's splitting results for logic programs [Lifschitz and Turner, 1994]. Since these approaches are oriented to abstract argumentation frameworks, they pay no attention to an important relation between structured arguments (i.e., subargument relation). This gives rise to the following problems:

- *There is a gap between the dynamics of an abstract argumentation and that of underlying beliefs.* To solve this problem, we may resort to concrete (instantiated) argumentation systems. However, although in various argumentation systems constructed from underlying knowledge [García and Simari, 2004, Governatori et al., 2004, Gorogiannis and Hunter, 2011], there is a connection between the underlying knowledge and the constructed argumentation framework, the link between the dynamics of argumentation and that of underlying knowledge is still unclear.
- *For a given argumentation framework, all conflicting arguments are involved in the process of conflict handling.* Since many natural problems regarding argument acceptability are computational intractable [Dunne, 2007], letting all possible arguments in the computation might be inefficient.
- *When an argumentation framework is updated, attack relation is the only relation that is exploited to identify and evaluate all affected arguments.* Since the status of those arguments without direct attackers is determined by their proper subarguments, some affected arguments could be identified in terms of (proper) subargument relation, and their status could be directly obtained according to the status of their proper subarguments. Due to this reason, the above-mentioned approaches oriented to abstract argumentation frameworks could be improved.

Motivated by these problems, in this chapter, we will study how the status and the status evolution of arguments in an *argumentation framework with subargument relation* (AFwS) could be evaluated, according to both attack relation and subargument relation. By exploiting the subargument relation, the above three problems could be resolved to some extent. First, when an argumentation framework is characterized with subargument relation, it becomes closer to concrete argumentation systems, and so to beliefs. Second, for an AFwS, the evaluation of the status of some arguments is determined by that of its subarguments. In this way, only a part of arguments should be involved in the process of handling the conflicts between arguments according to attack relation. As a result, the evaluation of the status of arguments as a whole might be made easier. Third, since the arguments involving in the process of conflict handling (called the *conflict-handling arguments*) are differentiated from the arguments whose status is simply determined by its subarguments (called the *status-dependent arguments*), for the dynamics of an AFwS, it is possible to develop more sophisticated methods to identify which arguments should be evaluated. As a result, the evaluation process might become more simple.

Since we do not consider the internal structure of arguments, in this chapter, we still deal with abstract argumentation frameworks, but a notion of (abstract) subargument relation is integrated.

The remainder of this chapter is organized as follows. Firstly, we introduce some motivating examples, which illustrate the basic idea of this chapter. Then, we present the notion of abstract argumentation frameworks and their semantics, to make this chapter self-contained. And then, we formulate the definition, the semantics and the dynamics of a layered AFwS, respectively[1]. Finally, we offer some concluding remarks.

2 Motivating examples

For an AFwS [García and Simari, 2004, Gorogiannis and Hunter, 2011, Vreeswijk, 1997], the status of some structured arguments depends on that of their proper subarguments[2]. Hence, it might be unnecessary to involve all arguments in the process of acceptability evaluation of arguments. Let us consider the following example.

Example 1. Given a set of rules and propositions {"Quakers are pacifists", "pacifists follow principles of nonviolence", "Republicans are not pacifists", "Nixon is a Quaker", "Nixon is a Republican"}, informally, we may construct a set of arguments as follows:

a_1: [Nixon is a Quaker]
a_2: [Since Nixon is a Quaker, and Quakers are pacifists, Nixon is a pacifist]
a_3: [Since Nixon is a pacifist, and pacifists follow principles of nonviolence, Nixon follows principles of nonviolence]
a_4: [Nixon is a Republican]
a_5: [Since Nixon is a Republican, and Republicans are not pacifists, Nixon is not a pacifist]

---

[1] The notion and the semantics of a layered AFwS were originally presented in [Liao, 2012].
[2] Informally, a structured argument is a set of premises supporting a specific conclusion. Given a structured argument α, its subarguments are regarded as the structures that support intermediate conclusions, while its proper subarguments are the ones except α.

Among these arguments, a_1 and a_4 have no attackers and do not attack any arguments; a_2 directly attacks a_5, and vice versa; a_3 is indirectly attacked by a_5[3]; $\{\alpha_1\}$, $\{\alpha_1, \alpha_2\}$, and $\{\alpha_4\}$ are sets of proper subarguments of a_2, a_3 and a_5, respectively. Let $\rightarrow$ denote a direct attack, $\dashrightarrow$ an indirect attack, and $\multimap$ a proper subargument relation which is transitive. The above arguments, attacks and subargument relation constitute an AFwS, denoted as F_1^{SUB} (Figure 1a).

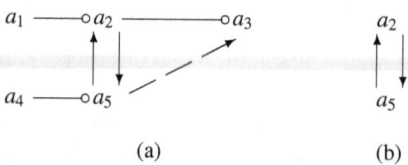

Fig. 1 An AFwS F_1^{SUB}, and the sub-framework for handling conflicts

In this example, a_3 has no direct attackers and does not attack any arguments (called a *status-dependent argument*). Its status depends on the status of its proper subargument a_2 that is related to direct attacks (called a *conflict-handling argument*). Intuitively, given a status-dependent argument α, for every argument β, if α is indirectly attacked by β, then there exists γ, a proper subargument of α, such that γ is directly attacked by β. As a result, the status of α can be determined by the status of its proper subarguments, i.e., if all of its proper subarguments are acceptable, then it is acceptable. In this example, a_2 (a proper subargument of a_3) is directly attacked by a_5. Since the status of a_2 can be independently evaluated in a sub-framework that is induced by the set of conflict-handling arguments (Figure 1b), the status of a_3 can then be assigned accordingly. In addition, if an argument has no attackers and does not attack any arguments (for instance, a_1 and a_4 in this example), then it is called a *trivial status-dependent argument*.

According to the above observations, for a given AFwS, it is possible to handle the conflicts of arguments within a sub-framework induced by the set of conflict-handling arguments, while the status of status-dependent arguments can be simply assigned according to the status of their proper subarguments.

Meanwhile, when an AFwS is updated, the subargument relation can also be exploited. Let us consider the following example.

Example 2. Continue Example 1. Suppose that at some time point, a new rule "if someone follows principles of nonviolence then he/she believs that nonviolent action is morally superior" is added to the system. With respect to this rule, we may construct a new argument a_6 as follows.

a_3: [Since Nixon is a pacifist, pacifists follow principles of

---

[3] Given two structured arguments α and β, α is regarded as a direct attacker (defeater) of β, if the conclusion of α and that of β are contradictory, and it is not the case that β is preferred over α; α is an indirect attacker of β, if α directly attacks a proper subargument of β).

nonviolence, and if someone follows principles of nonviolence then] he/she believs that nonviolent action is morally superior, Nixon believs that nonviolent action is morally superior]

After a_6 is added to F_1^{SUB}, we get an updated AFwS, denoted as F_2^{SUB} (Figure 2). In this case, no existing argument is affected, while the status of newly added argument a_6 could be directly obtained according to the status of a_3 (the only proper subargument of a_6).

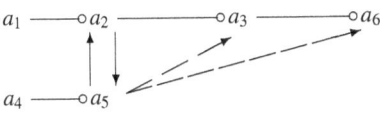

Fig. 2 An updated AFwS F_2^{SUB}

Example 2 shows that when an AFwS is updated, we may exploit the subargument relation to determine the status of some arguments. Besides this aspect, as explained by the following example, the subargument relation can also be used to identify some arguments, which are affected by an update, but not related to any direct attacks.

Example 3. To exemplify the above-mentioned property of subargument relation, let us suppose that the proposition "Nixon is a Republican" is not a fact, but an assumption, and therefore a_4 could be attacked by some other arguments. Then, we assume that at another time point, someone argues that "there is an evidence showing that Nixon is probably not a Republican". Accordingly, we may construct an argument a_7 as follows.

a_7: [Since there is an evidence shows that Nixon is probably not a Republican, Nixon is not a Republican]

After a_7 is added to F_2^{SUB}, we get an updated AFwS, denoted as F_3^{SUB} (Figure 3). In F_3^{SUB}, a_2, a_4, a_5 and a_7 are conflicting-handling arguments, while other arguments are status-dependent arguments. Note that a_4 changes from a status-dependent to a conflict-handling argument. Arguments a_2, a_3, a_4, a_5 and a_6 are affected arguments, in which a_2, a_4 and a_5 are identified and evaluated according to attack relation, while a_3 and a_6 are identified and evaluated according to subargument relation.

3 Dung-style argumentation framework

Since this chapter is focused on abstract argumentation frameworks with subargument relation (AFwSs), before presenting the nation of an AFwS as well as the

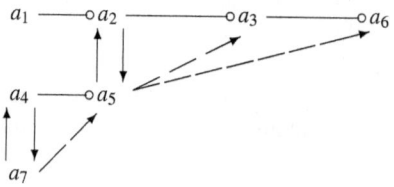

Fig. 3 An updated AFwS F_3^{SUB}

mechanisms to handle the semantics and the dynamics of an AFwS, let us first introduce the notion and the (extension-based) semantics of an abstract argumentation framework [Dung, 1995], called *Dung-style argumentation framework*.

Definition 1. An abstract argumentation framework (or simply, argumentation framework) is a tuple $F = \langle A, R \rangle$, where A is a set of arguments, and $R \subseteq A \times A$ is a set of attacks.

In Definition 1, arguments and attacks between them are abstract entities. We use $(\alpha, \beta) \in R$ to denote that α attacks β. Meanwhile, throughout this chapter, *we assume that A is generated by a reasoner at a given time point, and therefore is finite.*

Given an argumentation framework, a fundamental problem is to determine which arguments can be considered justified. According to [Dung, 1995], extension-based semantics is a formal way to answer this question. Here, an extension represents a set of arguments that are considered to be acceptable together, which is based on the following three important notions: *conflict-free sets, acceptability,* and *admissible sets*.

Definition 2. Let $F = \langle A, R \rangle$ be an argumentation framework.

- A set $E \subseteq A$ of arguments is *conflict-free* if and only if $\nexists \alpha, \beta \in E$, s.t. $(\alpha, \beta) \in R$.
- An argument $\alpha \in A$ is *acceptable* w.r.t. a set $E \subseteq A$ of arguments (also called α is defended by E), if and only if $\forall (\beta, \alpha) \in R, \exists \gamma \in E$, s.t. $(\gamma, \beta) \in R$.
- A conflict-free set of arguments $E \subseteq A$ is *admissible* if and only if each argument in E is acceptable w.r.t. E.

With the notion of admissible set, the extensions of an argumentation framework under some semantics (such as complete, preferred and grounded) can be defined as follows:

Definition 3. Let $F = \langle A, R \rangle$ be an argumentation framework, and $E \subseteq A$ be an admissible set of arguments.

- E is a complete extension if and only if each argument in A that is acceptable w.r.t. E is in E.
- E is a preferred extension if and only if E is a maximal (w.r.t. set-inclusion) complete extension.
- E is a grounded extension if and only if E is the minimal (w.r.t. set-inclusion) complete extension.

We use $\mathscr{E}_{adm}(F)$, $\mathscr{E}_{co}(F)$, $\mathscr{E}_{pr}(F)$, and $\mathscr{E}_{gr}(AF)$ to denote, respectively, the set of admissible extensions, complete extensions, preferred extensions, and the grounded extension of F. Meanwhile, for convenience, we use σ to denote a generic semantics, which could be one of the four semantics mentioned above, or other semantics (readers may refer to [Baroni et al., 2011] for a comprehensive introduction of argumentation semantics).

Example 4. According to Example 1, we may construct an argumentation framework $F_1 = (A_1, R_1)$, in which $A_1 = \{\alpha_1, \alpha_2, \alpha_3, \alpha_4, \alpha_5\}$ and $R_1 = \{(\alpha_2, \alpha_5), (\alpha_5, \alpha_2), (\alpha_5, \alpha_3)\}$. Then, $\{\alpha_1, \alpha_2, \alpha_3, \alpha_4\}$, $\{\alpha_1, \alpha_4, \alpha_5\}$, and $\{\alpha_1, \alpha_4\}$ are complete extensions.

Given $F = (A, R)$ and a subset $S \subseteq A$, let $F_S = (S, R_S)$ denote the sub-framework of F induced by S, in which $R_S = R \cap (S \times S)$. Let $S^- = \{\alpha \in A \setminus S \mid \exists \beta \in S, \text{ s.t. } (\alpha, \beta) \in R\}$. If $S^- = \emptyset$, i.e., the arguments in S are not attacked by the arguments outside S, then S is called an *unattacked set*. When $S \subseteq A$ is an unattacked set, under a semantics σ satisfying the *directionality* criterion [Baroni and Giacomin, 2007], it holds that:

$$\mathscr{E}_\sigma(F_S) = \{E \cap S \mid E \in \mathscr{E}_\sigma(F)\} \tag{1}$$

4 Argumentation framework with subargument relation (AFwS)

4.1 Subargument relation

The notion of subarguments mentioned above has been introduced in some work related to structured arguments [García and Simari, 2004, Gorogiannis and Hunter, 2011, Vreeswijk, 1997]. In this chapter, we do not treat with the internal structure of subarguments. So, given an argumentation framework $F = (A, R)$. If F is enriched with a binary relation (called *subargument relation*) over A, then conceptually, no difference is made between an *argument* and a *subargument*, in that any argument could be a subargument of another argument (and/or itself). Hence, a subargument relation is a binary relation between arguments. Formally, we have the following definition.

Definition 4. Let A be a set of (abstract) arguments. A subargument relation over A is a binary relation $\sqsubseteq$ between the arguments in A. For all $\alpha, \beta \in A$, we use $\alpha \sqsubseteq \beta$ (or $\beta \sqsupseteq \alpha$) to denote that α is a subargument of β, or β is a superargument of α.

When $\alpha \neq \beta$, we use $\alpha \sqsubset \beta$ (or $\beta \sqsupset \alpha$) to denote that α is a proper subargument of β, or β is a proper superargument of α.

Given a set of argument A and a subargument relation $\sqsubseteq$ over A, for all $\alpha \in A$, the set of proper subarguments (superarguments) of α, denoted as $sub^p(\alpha)$ (respectively, $sup^p(\alpha)$), is defined as follows:

$$sub^p(\alpha) = \{\beta \in A \mid \beta \sqsubset \alpha\} \qquad (2)$$
$$sup^p(\alpha) = \{\beta \in A \mid \beta \sqsupset \alpha\} \qquad (3)$$

According to Definition 4, every argument has at least one subargument (the argument itself), but may not have proper subargument. If an argument α has no proper subargument (i.e., $sub^p(\alpha) = \emptyset$), then it is called an *atomic argument*.

In addition, given an argument α, it can be a proper subargument of another argument β, which in turn can be a proper subargument of another argument γ, and so on. Since $\alpha \sqsubset \beta \sqsubset \gamma$, both α and β are proper subarguments of γ. When α is a proper subargument of γ, in order to decide whether there exists a third argument β such that α is a proper subargument of β and β is a proper subargument of γ, we introduce two notions: *direct proper subarguments* and *indirect proper subarguments*.

Definition 5. Let A be a set of arguments. For all $\alpha, \gamma \in A$, we say that α is a direct proper subargument of α if and only if $\alpha \sqsubset \gamma$ and $\nexists \beta \in A$ s.t. $\alpha \sqsubset \beta \sqsubset \gamma$. The set of direct proper subarguments of α is denoted as $sub^{dp}(\alpha)$.

Similarly, we may define *direct proper superarguments* and *indirect proper superarguments*. The set of direct proper superarguments of α is denoted as $sup^{dp}(\alpha)$.

4.2 Direct attacks and indirect attacks

When an argumentation framework $F = (A, R)$ is enriched with a subargument relation, the set of attacks R can be partitioned into two subsets: *direct attacks* and *indirect attacks*. We say that an argument α directly attacks another argument β, if there exists no proper subargument γ of β such that α attacks γ. Formally, we have the following definition.

Definition 6. Let $F = (A, R)$ be an argumentation framework, and $\sqsubseteq$ be the subargument relation over A. For all $(\alpha, \beta) \in R$, we say that α directly attacks β, if $\nexists \gamma \in A$ such that $\gamma \sqsubset \beta$ and $(\alpha, \gamma) \in R$; otherwise, say that α indirectly attacks β.

We use R^{dir} and R^{ind} to denote the set of direct attacks and indirect attacks, respectively. According to Definition 6, it holds that $R^{dir} \cup R^{ind} = R$.

Example 5. According to Example 1, α_2 directly attacks α_5 and vice versa, while α_3 is indirectly attacked by α_5. Let $R_1^{dir} = \{(\alpha_2, \alpha_5), (\alpha_5, \alpha_2)\}$, and $R_1^{ind} = \{(\alpha_5, \alpha_3)\}$. In terms of Example 4, it holds that $R_1 = R_1^{dir} \cup R_1^{ind}$.

According to Definition 6, we directly have the following proposition, which means that: if α attacks γ (directly or indirectly), then α indirectly attacks all proper superarguments of γ.

Proposition 1. *Let $F = (A,R)$ be an argumentation framework, and $\sqsubseteq$ the subargument relation over A. Let R^{dir} and R^{ind} be the set of direct attacks and the set of indirect attacks over A, respectively. For all $\alpha, \gamma \in A$, if $(\alpha, \gamma) \in R$, then $\forall \beta \in A$, if $\gamma \sqsubset \beta$ then $(\alpha, \beta) \in R^{ind}$.*

4.3 Formal definition of an AFwS

According to Proposition 1, indirect attacks can be represented by direct attacks and subargument relation. Hence, in an AFwS, indirect attacks may not be explicitly represented. Based on this consideration, we have the following definition.

Definition 7. Let A be a set of arguments, Sub^p be a set of proper subargument relations over A, and R^{dir} be a set of direct attacks over A. An argumentation framework with subargument relation (denoted as F^{SUB}) is defined as a triple as follows:

$$F^{SUB} = (A, R^{dir}, Sub^p) \qquad (4)$$

Example 6. Let $F_4^{SUB} = (A_4, R_4^{dir}, Sub_4^p)$ (Figure 4), where:

- $A_4 = \{a_1, \ldots, a_{10}\}$;
- $R_4^{dir} = \{(a_2, a_7), (a_7, a_2), (a_4, a_5), (a_5, a_4)\}$;
- $Sub_4^p = \{a_1 \sqsubset a_2, a_1 \sqsubset a_3, a_1 \sqsubset a_4, a_1 \sqsubset a_8, a_1 \sqsubset a_9, a_2 \sqsubset a_3, a_2 \sqsubset a_4, a_2 \sqsubset a_8, a_2 \sqsubset a_9, a_3 \sqsubset a_4, a_3 \sqsubset a_8, a_3 \sqsubset a_9, a_4 \sqsubset a_8, a_4 \sqsubset a_9, a_6 \sqsubset a_7, a_9 \sqsubset a_8, a_{10} \sqsubset a_8, a_{10} \sqsubset a_9\}$.

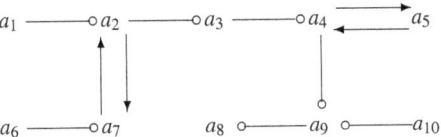

Fig. 4 An AFwS F_4^{SUB}

Given an AFwS $F^{SUB} = (A, R^{dir}, Sub^p)$, there exists a corresponding Dung-style argumentation framework $F = (A, R)$, in which $R = R^{dir} \cup R^{ind}$ where $R^{ind} = \{(\alpha, \beta) \mid \alpha, \beta \in A, \text{ and } \exists \gamma \in A \text{ s.t. } (\alpha, \gamma) \in R^{dir} \text{ and } \gamma \sqsubset \beta \in Sub^p\}$.

Example 7. Continue Example 6. We may construct a Dung-style argumentation framework $F_4 = (A_4, R_4)$ (Figure 5), in which $R_4 = R_4^{dir} \cup R_4^{ind}$ where $R^{ind} = \{(a_7, a_3), (a_7, a_4), (a_7, a_8), (a_7, a_9), (a_5, a_8), (a_5, a_9)\}$.

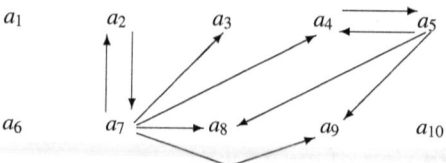

Fig. 5 The Dung-style argumentation framework of F_4^{SUB}

Note that when evaluating the status and the status evolution of the arguments in an AFwS, the Dung-style argumentation framework of the AFwS is not necessary to be constructed. A reference to the Dung-style argumentation framework of an AFwS is only used for the proofs of the properties of the AFwS.

5 Layered AFwS

Given an AFwS $F^{SUB} = (A, R^{dir}, Sub^p)$, we may evaluate the acceptability of arguments and handle the dynamics of argumentation semantics in the corresponding Dung-style argumentation framework. However, as mentioned in Sections 1 and 3, in an AFwS, not all arguments should participate in conflict resolving. Hence, before evaluating the status of arguments in A, we may partition A into two subsets: the set of conflict-handling arguments (denoted as A^{conf}) and the set of status-dependent arguments (denoted as A^{dep}), such that the status of the arguments in A^{conf} is evaluated in an independent sub-framework induced by A^{conf}, and the status of the arguments in A^{dep} is then directly assigned according to the subargument relation.

Now, let us formally define the notions of status-dependent arguments and conflict-handling arguments.

Definition 8. Let $F^{SUB} = (A, R^{dir}, Sub^p)$ be an AFwS. For all $\alpha \in A$, if $\nexists \beta \in A$, s.t. $(\beta, \alpha) \in R^{dir}$ or $(\alpha, \beta) \in R^{dir}$, then α is called a status-dependent argument. Otherwise, it is called a conflict-handling argument.

According to Definition 8, we have the following formulas.

$$A^{dep} = \{\alpha \in A \mid \nexists \beta \in A, \text{ s.t. } (\beta, \alpha) \notin R^{dir} \text{ or } (\alpha, \gamma) \notin R^{dir}\} \quad (5)$$
$$A^{conf} = A \setminus A^{dep} \quad (6)$$

Layered Argumentation Frameworks with subargument relation

Proposition 2. *It holds that: (i) A^{conf} is an unattacked set; and (ii) A^{dep} is conflict-free.*

Proof. First, assume that A^{conf} is not an unattacked set. Then, $\exists \alpha \in A \setminus A^{conf}$, $\exists \beta \in A^{conf}$, s.t. $(\alpha, \beta) \in R^{dir} \cup R^{ind}$. If $(\alpha, \beta) \in R^{dir}$, then $\alpha \in A^{conf}$, contradicting $\alpha \in A \setminus A^{conf}$. Else, if $(\alpha, \beta) \in R^{ind}$, then: according to Proposition 1, $\exists \gamma \sqsubseteq \beta$, s.t. $(\alpha, \gamma) \in R^{dir}$, and therefore $\alpha \in A^{conf}$, contradicting $\alpha \in A \setminus A^{conf}$.

Second, assume that A^{dep} is not conflict-free. Then, $\exists \alpha, \beta \in A^{dep}$, s.t. $(\alpha, \beta) \in R^{dir} \cup R^{ind}$. If $(\alpha, \beta) \in R^{dir}$, then $\alpha \in A^{conf}$, contradicting $\alpha \in A \setminus A^{conf}$. Else, if $(\alpha, \beta) \in R^{ind}$, then $\exists \gamma \sqsubseteq \beta$, s.t. $(\alpha, \gamma) \in R^{dir}$, and therefore $\alpha \in A^{conf}$, contradicting $\alpha \in A \setminus A^{conf}$.

Based on Proposition 2, since A^{conf} is an unattacked set, it is possible to first compute the status of arguments in A^{conf}, and then assign the status of arguments in A^{dep}. With this idea in mind, we propose a layered AFwS as follows.

Definition 9. Given $F^{SUB} = (A, R^{dir}, Sub^p)$, A^{conf} and A^{dep}, a layered AFwS is defined as follows:

$$F^{SUBL} = ((A^{conf}, R^{conf}), A^{dep}, Sub^p) \quad (7)$$

In formula 7, (A^{conf}, R^{conf}) is a Dung-style argumentation framework, called the *inner layer* of F^{SUBL}, in which $R^{conf} = R^{dir} \cup \{(\alpha, \gamma) \mid \alpha, \gamma \in A^{conf}, \text{ and } \exists \beta \in A^{conf}, \text{ s.t. } \beta \sqsubseteq \gamma \text{ and } (\alpha, \beta) \in R^{dir}\}$; A^{dep} is the set of status-dependent arguments, called the *outer layer* of F^{SUBL}.

Example 8. Continue Example 6. According to Definition 9, $F_4^{SUBL} = ((A_4^{conf}, R_4^{conf}), A_4^{dep}, Sub_4^p)$ (Figure 6), in which $A_4^{conf} = \{a_2, a_4, a_5, a_7\}$, $R_4^{conf} = \{(a_2, a_7), (a_7, a_2), (a_4, a_5), (a_5, a_4)\} \cup \{(a_7, a_4)\}$, $A^{dep} = \{a_3, a_8, a_9, a_1, a_6, a_{10}\}$.

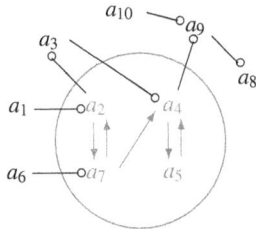

Fig. 6 F_4^{SUBL}

6 Semantics of a layered AFwS

Since status-dependent arguments can be evaluated in accordance with their proper subarguments, we may define the semantics of a layered AFwS with the help of status dependence relation (which is based on subargument relation) between different arguments.

6.1 Status dependence relation between arguments

For the status-dependent arguments defined in Definition 8, their status can be determined by that of their proper subarguments. The status dependence between an argument and its proper subarguments can be formalized in a tree, called a *status dependence tree*.

Definition 10. Let $F^{SUB} = (A, R^{dir}, \text{Sub}^p)$ be an AFwS, and $\alpha_0 \in A$ a status-dependent argument. The status dependence tree of α_0, denoted $\mathbb{T}_{\alpha_0}$, is defined as follows:

- The root of the tree is labeled with α_0.
- Let N be a node labeled with α_n. If $\text{sub}^p(\alpha_n) = \emptyset$, then α_n is a leaf; else, the arguments in $\text{sub}^{dp}(\alpha_n)$ label the children of N.

The set of arguments labeled the leaves of $\mathbb{T}_{\alpha_0}$ is denoted as $\text{Lvs}(\mathbb{T}_{\alpha_0})$. For all $\alpha_m \in \text{Lvs}(\mathbb{T}_{\alpha_0})$, let $[\alpha_0, \ldots, \alpha_m]$ be a path from α_0 to α_m. We use $\text{Lgth}(\alpha_0, \alpha_m) = m$ to denote the length of the path $[\alpha_0, \ldots, \alpha_m]$, and $\text{Hgt}(\mathbb{T}_{\alpha_0}) = \max_{\beta \in \text{Lvs}(\mathbb{T}_{\alpha_0})} \text{Lgth}(\alpha_0, \beta)$, to denote the height of $\mathbb{T}_{\alpha_0}$.

As shown in Figure 7, $\mathbb{T}_{\alpha_8}$ is the status dependence tree of α_8. It has two paths: $[\alpha_8, \alpha_9, \alpha_{10}]$ and $[\alpha_8, \alpha_9, \alpha_4, \alpha_3, \alpha_2, \alpha_1]$. The height of $\mathbb{T}_{\alpha_8}$ is 5.

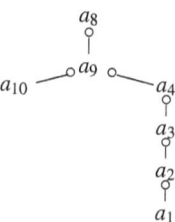

Fig. 7 A status dependence tree $\mathbb{T}_{\alpha_8}$

Note that for a *circular argument*, the height of its status dependence tree might be infinite. According to [Governatori et al., 2004], an argument is circular if a

Layered Argumentation Frameworks with subargument relation

conclusion depends on itself as a premise. Since circular arguments represent a very well known fallacy, they are very often not considered to be true arguments, and excluded from the set of arguments using syntactical definitions. In this chapter, we assume that there exists no circular argument in A, and A is finite. Based on this assumption, we have the following proposition.

Proposition 3. *Given* $F^{SUB} = (A, R^{dir}, Sub^p)$ *and* $F^{SUB_L} = ((A^{conf}, R^{conf}), A^{dep}, Sub^p)$, *if there exists no circular argument in A, and A is finite, then $\forall \alpha_0 \in A^{dep}$, $\mathrm{Hgt}(\mathbb{T}_{\alpha_0})$ is finite.*

Proof. Assume that $\mathrm{Hgt}(\mathbb{T}_{\alpha_0})$ is infinite, then there exists a path $[\alpha_0, \ldots, \alpha_m]$, s.t. m is infinite. Since the number of arguments in A is finite, $\exists \alpha_i, \alpha_j \in A$, s.t. $i < j$, and $\alpha_i = \alpha_j$. As a result, α_j is a proper subargument of α_i, and therefore α_j (α_i) is a circular argument. Contradiction.

On the other hand, as to the relation between the acceptability of a status-dependent argument and that of its proper subarguments, we have the following theorem.

Theorem 1. *Let* $F^{SUB} = (A, R^{dir}, Sub^p)$ *be an AFwS, and* $F = (A, R)$ *be the corresponding Dung-style argumentation framework, in which* $R = R^{dir} \cup R^{ind}$ *where* $R^{ind} = \{(\alpha, \beta) \mid \alpha, \beta \in A, \text{ and } \exists \gamma \in A \text{ s.t. } (\alpha, \gamma) \in R^{dir} \text{ and } \gamma \sqsubset \beta \in Sub^p\}$. *Given a subset $E \subseteq A$ and a status-dependent argument $\alpha \in A$, α is acceptable w.r.t. E, if and only if $\forall \beta \sqsubset \alpha$, β is acceptable w.r.t. E.*

Proof. ($\Rightarrow$) If α is acceptable w.r.t. E, then $\forall \beta \sqsubset \alpha$, β is acceptable w.r.t. E.
Assume the contrary, i.e., $\exists \beta \sqsubset \alpha$, s.t. β is not acceptable w.r.t. E. It follows that $\exists \gamma \in A$, s.t. $(\gamma, \beta) \in R$, and γ is not attacked by any argument in E. Since β is attacked by γ and $\beta \sqsubset \alpha$, according to Proposition 1, α is attacked by γ. As a result, α is not acceptable w.r.t. E. Contradiction.

($\Leftarrow$) If $\forall \beta \sqsubset \alpha$, β is acceptable w.r.t. E, then α is acceptable w.r.t. E.
Assume the contrary, i.e., α is not acceptable w.r.t. E. Since α receives no direct attack, $\exists \gamma \in A$, s.t. $(\gamma, \alpha) \in R^{ind}$, and γ is not attacked by any argument in E. According to Proposition 1, $\exists \beta \sqsubset \alpha$, s.t. $(\gamma, \beta) \in R^{dir}$. Thus, β is not acceptable w.r.t. E. Contradiction.

6.2 The defintion of the semantics of a layered AFwS babsed on an expansion function

According to Theorem 1, since the status of arguments only receiving indirect attacks depends on the status of their proper subarguments, when handling the conflicts between arguments, we may only consider the arguments that are related to direct attacks.

Given a layered AFwS $F^{SUBL} = ((A^{conf}, R^{conf}), A^{dep}, Sub^p)$, its semantics is defined according to the following two notions: (i) the extensions of (A^{conf}, R^{conf}), and (ii) the *expansion* of the extensions of (A^{conf}, R^{conf}) w.r.t. (A^{dep}, Sub^p).

Firstly, according to Proposition 2 and Formula 1, since A^{conf} is an unattacked set, under a semantics σ satisfying the directionality criterion, the extensions of (A^{conf}, R^{conf}) can be computed independently.

Example 9. Continue Example 8. The extensions of (A_4^{conf}, R_4^{conf}) in F_4^{SUBL} can be computed independently. For instance, under complete semantics, $\mathcal{E}_{co}((A^{conf}, R^{conf})) = \{E_1, E_2, E_3, E_4\}$, in which $E_1 = \{\alpha_2, \alpha_4\}$, $E_2 = \{\alpha_2, \alpha_5\}$, $E_3 = \{\alpha_5, \alpha_7\}$, and $E_4 = \emptyset$.

Then, the semantics of F^{SUBL} is obtained by expanding the extensions of (A^{conf}, R^{conf}) with the arguments in A^{dep}.

Definition 11. Given a layered AFwS $F^{SUBL} = ((A^{conf}, R^{conf}), A^{dep}, Sub^p)$, the expansion function of F^{SUBL} is defined as follows:
$\pi_{F^{SUBL}} : 2^{A^{conf} \cup A^{dep}} \to 2^{A^{conf} \cup A^{dep}}$,
$\pi_{F^{SUBL}}(E) = E \cup \{\alpha \in A^{dep} \setminus E \mid sub^p(\alpha) \subseteq E\}$.

It is obvious that if $sub^p(\alpha)$ is contained in E, then it is contained in any superset of E. Hence, we have the following proposition.

Proposition 4. $\pi_{F^{SUBL}}$ *is monotonic (w.r.t. set-inclusion).*

Thus, given $E \subseteq A^{conf} \cup A^{dep}$, there exists a unique fixed point of $\pi_{F^{SUBL}}$, denoted as E^*.

Example 10. According to Example 8, given $E_1 = \{\alpha_2, \alpha_4\}$, we have:
- $\pi_{F_2^{SUBL}}(E_1) = \{\alpha_2, \alpha_3, \alpha_4, \alpha_9\}$,
- $\pi^2_{F_2^{SUBL}}(E_1) = \{\alpha_2, \alpha_3, \alpha_4, \alpha_8, \alpha_9\}$,
- $\pi^3_{F_2^{SUBL}}(E_1) = \pi^2_{F_2^{SUBL}}(E_1)$.

So, $E_1^* = \pi^2_{F_2^{SUBL}}(E_1)$.

Now, let us present two basic properties of the expansion function $\pi_{F^{SUBL}}$ (for short, we write π instead of $\pi_{F^{SUBL}}$ if without confusion).

Theorem 2. *If $E \subseteq A$ is admissible, then $E^* = \pi^\infty(E)$ is admissible, i.e., E^* is conflict-free, and $\forall \alpha \in E^*$, α is acceptable w.r.t. E^*.*

Proof. Firstly, assume that $\pi^i(E)$, $i \geq 0$, is conflict-free, but $\pi^{i+1}(E)$ is not conflict-free (when $i = 0$, $\pi^i(E) = E$). According to Proposition 2, A^{dep} is conflict-free. Since both $\pi^i(E)$ and A^{dep} are conflict-free, but $\pi^{i+1}(E)$ is not conflict-free, $\exists \alpha \in \pi^i(E), \beta \in A^{dep} \cap \pi^{i+1}(E)$, s.t. $(\alpha, \beta) \in R^{ind}$. According to Proposition 1, $\exists \gamma \in sub^p(\beta)$, s.t. $(\alpha, \gamma) \in R^{dir}$. Since $\beta \in \pi^{i+1}(E)$, according to Definition 11,

$sub^p(\beta) \subseteq \pi^i(E)$. Thus, both α and γ belong to $\pi^i(E)$, but $(\alpha,\gamma) \in R^{dir}$, i.e., $\pi^i(E)$ is not conflict-free. Contradiction. Since E is conflict-free, and as long as $\pi^i(E)$ is conflict-free, $\pi^{i+1}(E)$ is conflict-free, we may infer that $E^* = \pi^\infty(E)$ is conflict-free.

Secondly, suppose that $\forall \eta \in \pi^i(E), i \geq 0$, η is acceptable w.r.t. $\pi^i(E)$. According to Definition 11, $\forall \alpha \in \pi^{i+1}(E)$, there are the following two possible cases:

- $\alpha \in \pi^i(E)$: In this case, since the arguments in A^{dep} do not attack any arguments in $A^{conf} \cup A^{dep}$, after the expansion of $\pi^i(E)$ with the arguments in A^{dep}, α is still acceptable w.r.t. $\pi^i(E) \subseteq \pi^{i+1}(E)$.
- $\alpha \in A^{dep} \cap \pi^{i+1}(E)$: In this case, according to Definition 11, $sub^p(\alpha) \subseteq \pi^i(E)$. Since the arguments in A^{dep} are only possibly attacked by the arguments in A^{conf}, $\forall \beta \in A^{conf}$, if β attacks α, then $(\beta,\alpha) \in R^{ind}$. According to Proposition 1, $\exists \gamma \in A^{conf}$, s.t. $\gamma \in sub^p(\alpha)$ and $(\beta,\gamma) \in R^{dir}$. Since $\gamma \in sub^p(\alpha) \subseteq \pi^i(E)$, $\gamma \in \pi^i(E)$, and therefore γ is acceptable w.r.t. $\pi^i(E)$. Since γ is acceptable w.r.t. $\pi^i(E)$ and γ is attacked by β, according to the definition of the acceptability of arguments, $\exists \xi \in \pi^i(E)$, s.t. $(\xi,\beta) \in R^{dir} \cup R^{ind}$. It turns out that $\forall \beta \in A^{conf}$, if $(\beta,\alpha) \in R^{ind}$, then $\exists \xi \in \pi^i(E) \subseteq \pi^{i+1}(E)$, s.t. $(\xi,\beta) \in R^{dir} \cup R^{ind}$. Thus, α is acceptable w.r.t. $\pi^{i+1}(E)$.

Since (i) $\forall \alpha \in E$, α is acceptable w.r.t. E, and (ii) $\forall i \geq 0$, if it holds that $\forall \alpha \in \pi^i(E)$, α is acceptable w.r.t. $\pi^i(E)$, then it holds that $\forall \alpha \in \pi^{i+1}(E)$, α is acceptable w.r.t. $\pi^{i+1}(E)$, we may infer that $\forall \alpha \in E^* = \pi^\infty(E)$, α is acceptable w.r.t. $E^* = \pi^\infty(E)$.

Theorem 3. *For all $\alpha_0 \in A^{dep}$ and $E \subseteq A^{conf}$, if $sub(\alpha_0) \cap A^{conf} \subseteq E$, then $\alpha_0 \in E^*$.*

Proof. According to Definition 10, for the status-dependent argument α_0, we may construct a status dependence tree $\mathbb{T}_{\alpha_0}$. Let $\text{Hgt}(\mathbb{T}_{\alpha_0})$ be the height of $\mathbb{T}_{\alpha_0}$. According to Proposition 3, $\mathbb{T}_{\alpha_0}$ is finite. Let $\mathbb{T}_{\alpha_0} = n$. It holds that $n \geq 1$ (according to Proposition 1, α_0 has at least a proper subargument). Then, $\mathbb{T}_{\alpha_0}$ is composed of $n+1$ levels, from the leaves to the root, such that the leaves are at level 0, the root is at level n, and if an argument β is at level i ($1 \leq i \leq n+1$), then all arguments in the set $sub^{dp}(\beta)$ are at level $i-1$. Now, let us check the arguments in $\mathbb{T}_{\alpha_0}$ from the leaves to the root.

First, for all γ at level 0, $\gamma \in A^{conf}$. Since $sub(\alpha_0) \cap A^{conf} \subseteq E$ and $\gamma \in sub(\alpha_0)$, $\gamma \in E$. So, at level 0, γ is in E.

Second, for all η at level 1, there are two possible cases: $\eta \in A^{conf}$, or $\eta \in A^{dep}$. When $\eta \in A^{conf}$, since $sub(\alpha_0) \cap A^{conf} \subseteq E$ and $\eta \in sub(\alpha_0)$, it holds that $\eta \in E \subseteq \pi(E)$. When $\eta \in A^{dep}$, since all arguments in $sub^p(\eta)$ are at level 0, and the arguments at level 0 are in E, we have $sub^p(\eta) \subseteq E$. According to Definition 11, $\eta \in \pi(E)$. So, at level 1, η is in $\pi(E)$.

Third, for all ξ at level j ($2 \leq j \leq n+1$), there are also two possible cases: $\xi \in A^{conf}$, or $\xi \in A^{dep}$. When $\xi \in A^{conf}$: since $sub(\alpha_0) \cap A^{conf} \subseteq E$ and $\xi \in sub(\alpha_0)$, it holds that $\xi \in E$. When $\xi \in A^{dep}$, suppose that all arguments at levels from 0 to $j-1$ are in $\pi^{j-1}(E)$. Since all arguments in $sub^p(\xi)$ are at levels from 0 to $j-1$, it hold that $sub^p(\xi) \subseteq \pi^{j-1}(E)$. According to Definition 11, $\eta \in \pi^j(E)$.

Accordingly, since $\alpha_0 \in A^{dep}$ is at level n, it holds that $\alpha_0 \in \pi^n(E) \subseteq E^*$.

Based on the above properties, the extensions of a layered AFwS can be formulated by means of expanding the extensions of (A^{conf}, R^{conf}). In this chapter, we consider the cases under some typical semantics, such as complete, preferred, and grounded, while the ones under other semantics are left to future work.

Theorem 4. $\mathcal{E}_\sigma(F) = \{E^* \mid E \in \mathcal{E}_\sigma((A^{conf}, R^{conf}))\}$, *for all* $\sigma \in \{co, pr, gr\}$.

Proof. $\Leftarrow$ (Soundness):

- Under complete semantics, let us prove that $\forall E \in \mathcal{E}_{co}((A^{conf}, R^{conf}))$, $E^* \in \mathcal{E}_{co}(F)$. Since E is admissible, according to Theorem 2, E^* is also admissible. Now, let us prove that $\forall \alpha \in A$, if α is acceptable w.r.t. E^*, then $\alpha \in E^*$. There are two possible cases. First, if $\alpha \in A^{conf}$, then: since α is acceptable w.r.t. E^*, and α is only possibly attacked and defended by the arguments in A^{conf} (arguments in A^{dep} do not attack any arguments), α is acceptable w.r.t. $E^* \cap A^{conf} = E$; since E is a complete extension, every argument acceptable w.r.t. E is in E; thus, $\alpha \in E$. Second, if $\alpha \in A^{dep}$, since $E \subseteq A^{conf}$, according to Theorem 3, in order to prove that $\alpha \in E^*$, we need only to prove that $sub(\alpha) \cap A^{conf} \subseteq E$. Assume the contrary. Then, $\exists \beta \in sub(\alpha) \cap A^{conf}$, s.t. $\beta \notin E$. Since E is a complete extension and $\beta \notin E$, β is not acceptable w.r.t. E. Hence, $\exists \gamma \in A^{conf}$, s.t. $(\gamma, \beta) \in R^{dir}$, and γ is not attacked by any argument in E. Since $(\gamma, \beta) \in R^{dir}$, $(\gamma, \alpha) \in R^{ind}$. As a result, α is not acceptable w.r.t. $E \subseteq E^*$. Contradiction.
- Under preferred semantics, $E \in \mathcal{E}_{pr}((A^{conf}, R^{conf}))$. we only need to prove that E^* is maximal (w.r.t. set-inclusion). Assume the contrary. Then, there exists $E' \in \mathcal{E}_{pr}(F)$, s.t. $E' \supset E^*$. Let $\Phi = E' \setminus E^*$. It holds that $\Phi \neq \emptyset$. Let $\Phi_1 = \Phi \cap A^{conf}$ and $\Phi_2 = \Phi \cap A^{dep}$. We have the following two possible cases:
    - When $\Phi_1 \neq \emptyset$, $E' \cap A^{conf} = (E^* \cup \Phi) \cap A^{conf} = (E^* \cap A^{conf}) \cup (\Phi \cap A^{conf}) = E \cup \Phi_1$. According to the directionality of argumentation, $E \cup \Phi_1$ is a preferred extension of (A^{conf}, R^{conf}). As a result, E is not a preferred extension of (A^{conf}, R^{conf}), contradicting $E \in \mathcal{E}_{pr}((A^{conf}, R^{conf}))$.
    - When $\Phi_1 = \emptyset$, it follows that $\Phi_2 \neq \emptyset$. For all $\alpha \in \Phi_2$, since $\alpha \notin E^*$ and E^* is a complete extension, it holds that α is not acceptable w.r.t. E^*. In other words, there exists $\beta \in A^{conf}$ s.t. $(\beta, \alpha) \in R^{ind}$, and β is not attacked by any argument in $E^* \cap A^{conf} = E$. Since $\Phi_1 = \emptyset$, it holds that $E' \cap A^{conf} = E^* \cap A^{conf} = E$. It follows that α is attacked by an argument β which is not attacked by any argument in E', and therefore α is not acceptable w.r.t. E', contradicting $\alpha \in E'$.

- Under grounded semantics, $E \in \mathcal{E}_{gr}((A^{conf}, R^{conf}))$. we only need to prove that E^* is minimal (w.r.t. set-inclusion). The proof is similar to that of the previous item (omitted).

$\Rightarrow$ (Completeness): Let us prove that $\forall E' \in \mathcal{E}_\sigma(F)$, $\exists E \in \mathcal{E}_\sigma((A^{conf}, R^{conf}))$, s.t. $E^* = E'$. Since A^{conf} is an unattacked set, $\mathcal{E}_\sigma((A^{conf}, R^{conf})) = \{E' \cap A^{conf} \mid E' \in \mathcal{E}_\sigma(F)\}$ (according to Proposition 2 and formula 1). Let $E = E' \cap A^{conf} \in \mathcal{E}_\sigma((A^{conf}, R^{conf}))$. According to the proof of the soundness of this theorem, $E^* \in \mathcal{E}_\sigma(F)$. According to Definition 11, given $E \in \mathcal{E}_\sigma((A^{conf}, R^{conf}))$, E^* is unique. In other words, given E', there exists a unique extension of F expanding from $E = E' \cap A^{conf}$. So, $E^* = E'$.

Example 11. Continue of Example 9. The fixed points of $\pi_{F_4^{SUBL}}$ w.r.t. E_1, E_2, E_3 and E_4, respectively, are $E_1^* = \{\alpha_1, \alpha_2, \alpha_3, \alpha_4, \alpha_6, \alpha_8, \alpha_9, \alpha_{10}\}$, $E_2^* = \{\alpha_1, \alpha_2, \alpha_3, \alpha_5, \alpha_6, \alpha_{10}\}$, $E_3^* = \{\alpha_1, \alpha_5, \alpha_6, \alpha_7, \alpha_{10}\}$, and $E_4^* = \{\alpha_1, \alpha_6, \alpha_{10}\}$. Let $F_4 = (A_4, R_4^{dir} \cup R_4^{ind})$. It is not difficult to verify that $\mathscr{E}_{co}(F_4) = \{E_1^*, E_2^*, E_3^*, E_4^*\}$.

7 Updating a layered AFwS

Since the status of status-dependent arguments can be determined by that of their proper subarguments, when a layered AFwS is updated, we may exploit subargument relation along with attack relation to identify and evaluate the status changing of arguments. In this section, let us first introduce some basic notions of removing a set of arguments from a layered AFwS.

When an argument is removed from a layered AFwS, we generally require that all of its (proper) superarguments should be removed. In this chapter, we say that removing a set of arguments B from a layered AFwS is *legal*, if for every argument of B, all of its (proper) superarguments are removed simultaneously, along with the attacks and subargument relations that are related to the arguments in B. Formally, we have the following definition.

Definition 12. Let $F^{SUBL} = ((A^{conf}, R^{conf}), A^{dep}, Sub^p)$ be a layered AFwS. Let $B \subseteq A$ be a set of arguments. Removing B from F^{SUBL} is legal, if:

(1) for all $\alpha \in B$, $sup^p(\alpha) \subset B$;
(2) for all $(\alpha, \beta) \in R^{conf}$, if $\alpha \in B$ or $\beta \in B$, then (α, β) is removed from R^{conf};
(3) for all $\alpha \in B$, for all $\beta \in sub^{dp}(\alpha)$, $\beta \sqsubset \alpha$ is removed from Sub^p.

Example 12. Consider F_4^{SUBL} in Example 8. Let $B_1 = \{a_4, a_8, a_9\}$. When we remove B_1 from F_4^{SUBL}, if $\{(a_4, a_5), (a_5, a_4), (a_7, a_4)\}$ is removed from R_4^{conf} and $\{a_3 \sqsubset a_4, a_4 \sqsubset a_9, a_9 \sqsubset a_8, a_{10} \sqsubset a_9\}$ is removed from Sub_4^p, then the removal is legal.

Let $B^{dep} = B \cap A^{dep}$ and $B^{conf} = B \cap A^{conf}$. It holds that $B^{dep} \cup B^{conf} = B$. We differentiate the following three cases.

First, when $B^{conf} = \emptyset$ (and so $B \subseteq A^{dep}$), all the removed arguments belong to the set of status-dependent arguments. In this case, no attack in R^{conf} is affected. So, all the arguments in A^{conf} are unaffected. Meanwhile, all arguments in $A^{dep} \setminus B$ are also unaffected.

Second, when $B^{dep} = \emptyset$ (and so $B \subseteq A^{conf}$), the status of a part of arguments in $A^{conf} \setminus B$ and A^{dep} might be affected. To figure out this kind of dynamics, we may follow the following two steps:

(1) Use the division-based method in [Liao et al., 2011] to identify and evaluate the set of arguments (denoted as Ψ) in $A^{conf} \setminus B$ that are affected by the removal of B (w.r.t. the attack relation R^{conf}).

(2) Identify and reevaluate the set of arguments in A^{dep} that are proper superarguments of the arguments in Ψ (w.r.t. the subargument relation Sub^p).

Example 13. Consider again $F_4^{SUB_L}$ in Example 8. Let $B_2 = \{a_5\}$. Let $B_2^{dep} = B_2 \cap A_4^{dep}$ and $B_2^{conf} = B_2 \cap A_4^{conf}$. It holds that $B_2^{dep} = \emptyset$ and $B_2^{conf} = \{a_5\}$. After removing B_2 along with the set of attacks $\{(a_4, a_5), (a_5, a_4)\}$ from $F_4^{SUB_L}$, according to [Liao et al., 2011], the set of arguments in $A_4^{conf} \setminus B_2$ that are affected by the removal of B_2 is $\{a_4\}$. After the status of the only argument a_4 in $\{a_4\}$ is recomputed according to the method inroduced in [Liao et al., 2011], according to subargument relation Sub_4^p, the set of arguments in A_4^{dep} that are proper superarguments of the arguments in $\{a_4\}$ is $\{a_8, a_9\}$, in which the status of each argument can be determined by the status of a_4.

Third, when $B^{dep} \neq \emptyset$ and $B^{conf} \neq \emptyset$, the dynamics of the system can be figured out by firstly applying the method for the first case, and then applying the method for the second case.

On the other hand, when adding a set of arguments to a layered AFwS, we generally require that for each argument of the set, all its (proper) subarguments should have been added in advance or to be added simultaneously. Compared to the definition of legally removing a set of arguments from a layered AFwS, the definition of legally adding a set of arguments to a layered AFwS is a little bit more complicated, since we have to consider the universe of arguments, as well as the universe of direct attacks, and of subargument relations, between the arguments. Instead of providing a formal definition, we give an example to explain some basic properties of adding a set of arguments to a layered AFwS.

Example 14. Consider Examples 1, 2 and 3 again. Corresponding to three AFwSs, we may construct three layered AFwSs: $F_1^{SUB_L}$, $F_2^{SUB_L}$ and $F_3^{SUB_L}$, as shown in Figure 8a, 8b and 8c, respectively. Let us observe the following two cases of changing.

First, after the set of arguments $\{a_6\}$ is added to $F_1^{SUB_L}$ along with the set of proper subargument relations $\{a_3 \sqsubset a_6\}$, we get $F_2^{SUB_L}$. In this case, no argument in $F_1^{SUB_L}$ is affected by the addition, while the status of newly added argument a_6 can be determined in terms of subargument relation.

Second, after the set of arguments $\{a_7\}$ is added to $F_2^{SUB_L}$ along with the set of direct attacks $\{(a_4, a_7), (a_7, a_4)\}$, we get $F_3^{SUB_L}$. In this case, two conflict-handling arguments a_2 and a_5 in $F_2^{SUB_L}$ are affected, while a status-dependent argument a_4 in $F_2^{SUB_L}$ becomes a conflict-handling argument in $F_3^{SUB_L}$. These affected arguments are identified in terms of attack relation, and their status could be evaluated by using the division-based method introduced in [Liao et al., 2011]. On the other hand, two status-dependent arguments a_3 and a_6 in $F_2^{SUB_L}$ are also affected. They can be identified and evaluated in terms of subargument relation.

From Example 14, we may observe that when a layered AFwS is updated by adding a set of arguments, both attack relation and subargument relation could be exploited to identify and evaluate the affected arguments.

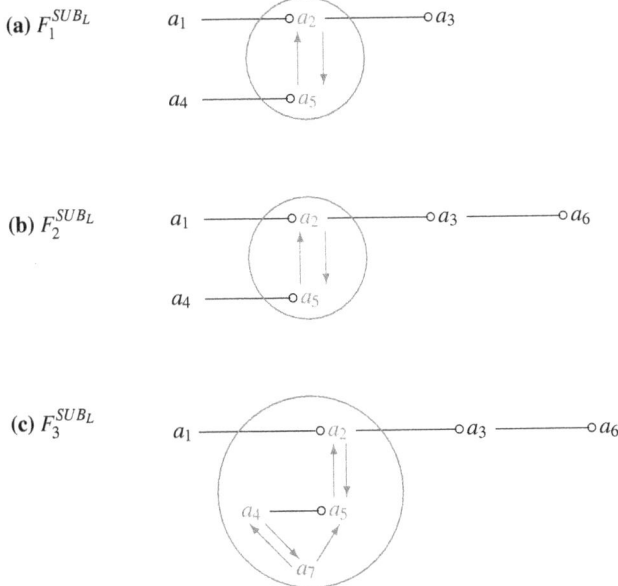

Fig. 8 Cases where a set of arguments are adding to a layered AFwS

8 Concluding Remarks

In this chapter, we have introduced the notion, the semantics and the dynamics of a layered AFwS.

The notion of subarguments is not new. It has been defined in various concrete argumentation systems [García and Simari, 2004, Gorogiannis and Hunter, 2011, Governatori et al., 2004, Vreeswijk, 1997, Prakken, 2012], and some abstract argumentation frameworks with subargument relation [Martínez et al., 2006, Baláž and Frtús, 2012]. Meanwhile, subargument relation (or more generally, support relations) has also been exploited in evaluating of the status of arguments: in [Martínez et al., 2007], to avoid the controversial situations in a dialectical process, Martínez et al proposed a notion of *progressive defeat paths*, in which subargument relation was exploited to identify controversial arguments; in [Cayrol and Lagasquie-Schiex, 2011, Cayrol and Lagasquie-Schiex, 2013], Cayrol and Lagasquie-Schiex introduced various kinds of support, which is one of the two basic relations between arguments in bipolar argumentation frameworks, and therefore plays a significant role in evaluating the status of arguments. However, how to integrate subargument relation into existing abstract argumentation theory and to handle the dynamics of argumentation systems is still an open problem.

A fundamental contribution of this work is the idea and the mechanism of handling the conflicts of arguments *centrally* within a sub-framework induced by the set of conflict-handling arguments. Conflict-handling is the core of reasoning under the context of disagreement. According to existing literature, in various argumentation systems, either abstract argumentation frameworks (with or without subargument relations) or concrete argumentation systems, all arguments are involved in conflict-handling, and it is not clear that the status of which kinds of arguments could be determined simply by subargument relation. The potential benefits of evaluating the status of arguments by exploiting the attack relation and the subargument relation respectively may include the following two aspects.

On the one hand, for a given AFwS, especially in the case where only a small part of arguments are related to conflict-handling, due to the decease of the size of arguments involved, the evaluation of the status of arguments might become easier.

On the other hand, when a layered AFwS is updated, we may develop more sophisticated and efficient approaches for evaluating the changes of the status of arguments. As presented in [Liao et al., 2011], for a Dung-style argumentation framework, given a set of arguments and/or a set of attacks to be added or removed, the only factor that determines which arguments are affected or unaffected is attack relation. On the contrary, for a layered AFwS, subargument relation and/or attack relation could be exploited to handle the dynamics of argumentation.

Finally, as the work presented in this chapter is rather preliminary, future work may include the following aspects.

First, *further explore the link between the dynamics of argumentation and that of underlying knowledge*. A possible question in this direction is: Besides the subargument relation that could be regarded as a bridge between abstract argumentation and concrete underlying knowledge, what other factors are worth being considered? And, how they could be exploited?

Second, *further study the dynamics of a layered AFwS*. In this chapter, we only introduce some basic notions and properties of updating a layered AFwS. More detailed and deeper study is necessary.

Third, *conduct empirical studies on layered AFwSs and their dynamics*. Based on the theory introduced in this chapter, it is interesting to first develop algorithms for computing the semantics and the semantics evolution of a layered AFwS, and then test their practical performance.

Acknowledgment

We are grateful to the anonymous reviewers, whose constructive comments helped to improve our work.

The research reported in this paper was partially supported by the National Grand Fundamental Research 973 Program of China (No.2012CB316400), the National

Science Foundation of China (No.61175058), and Zhejiang Provincial Natural Science Foundation of China (No. Y1100036).

References

[Baláž and Frtús, 2012] Baláž, M. and Frtús, J. (2012). Abstract argumentation with structured arguments. In *Proceedings of the 14th International Workshop on Non-Monotonic Reasoning*, Rome, Italy.

[Baroni et al., 2011] Baroni, P., Caminada, M., and Giacomin, M. (2011). An introduction to argumentation semantics. *The Knowledge Engineering Review*, 26:365–410.

[Baroni and Giacomin, 2007] Baroni, P. and Giacomin, M. (2007). On principle-based evaluation of extension-based argumentation semantics. *Artificial Intelligence*, 171:675–700.

[Baumann, 2011] Baumann, R. (2011). Splitting an argumentation framework. In *Proceedings of the 11th International Conference on Logic Programming and Nonmonotonic Reasoning*, pages 40–53.

[Baumann et al., 2012] Baumann, R., Brewka, G., Dvork, W., and S.Woltran (2012). Parameterized splitting: A simple modification-based approach. In *Correct Reasoning*, pages 57–71. Springer.

[Cayrol and Lagasquie-Schiex, 2011] Cayrol, C. and Lagasquie-Schiex, M. (2011). Bipolarity in argumentation graphs: Towards a better understanding. In *Proceedings of the 5th International Conference on Scalable Uncertainty Management*, pages 137–148.

[Cayrol and Lagasquie-Schiex, 2013] Cayrol, C. and Lagasquie-Schiex, M. (2013). Bipolarity in argumentation graphs: Towards a better understanding. *International Journal of Approximate Reasoning*, 54(7):876–899.

[Dung, 1995] Dung, P. (1995). On the acceptability of arguments and its fundamental role in nonmonotonic reasoning, logic programming and n-person games. *Artificial Intelligence*, 77:321–357.

[Dunne, 2007] Dunne, P. (2007). Computational properties of argument systems satisfying graph-theoretic constraints. *Artificial Intelligence*, 171:701–729.

[García and Simari, 2004] García, A. and Simari, G. (2004). Defeasible logic programming: an argumentative approach. *Theory and Practice of Logic Programming*, 4:95–138.

[Gorogiannis and Hunter, 2011] Gorogiannis, N. and Hunter, A. (2011). Instantiating abstract argumentation with classical logic arguments: Postulates and properties. *Artificial Intelligence*, 175:1479–1497.

[Governatori et al., 2004] Governatori, G., Maher, M. J., Antoniou, G., and Billington, D. (2004). Argumentation semantics for defeasible logic. *Journal of Logic and Computation*, 14:675–702.

[Liao, 2012] Liao, B. (2012). Dynamics of argumentation frameworks with subargument relations. In *the Workshop on the Dynamics of Argumentation, Rules and Conditional*, University of Luxembourg.

[Liao et al., 2011] Liao, B., Jin, L., and Koons, R. (2011). Dynamics of argumentation systems: a division-based method. *Artificial Intelligence*, 175:1790–1814.

[Lifschitz and Turner, 1994] Lifschitz, V. and Turner, H. (1994). Splitting a logic program. In *Principles of Knowledge Representation*, pages 23–37. MIT Press.

[Martínez et al., 2006] Martínez, D., García, A. J., and Simari, G. R. (2006). On acceptability in abstract argumentation frameworks with an extended defeat relation. In *Proceedings of 1st International Conference on Computational Models of Argument*, pages 273–278. IOS Press.

[Martínez et al., 2007] Martínez, D., García, A. J., and Simari, G. R. (2007). Modelling well-structured argumentation lines. In *Proceedings of the 20th International Joint Conference on Artificial Intelligence*, pages 465–470.

[Prakken, 2012] Prakken, H. (2012). An abstract framework for argumentation with structured arguments. *Argument & Computation*, 1:93–124.

[Vreeswijk, 1997] Vreeswijk, G. (1997). Abstract argumentation systems. *Artificial Intelligence*, 90:225–279.

Extending belief bases change to logic programs with ASP

Julien Hué, Odile Papini, and Éric Würbel

Abstract This chapter shows how ASP can be used as a unified formalism for both representing and implementing change operations for logic programs. In order to illustrate these two aspects, we present a framework, ΠRSF, which extends to logic programs the Removed Sets framework, proposed for revising and merging propositional belief bases. Properties of ΠRSF operators are studied and an implemementation of these operators is provided that can be performed with any ASP solver.

1 Introduction

Logic programs have been proved to be efficient formalisms for problems solving and for knowledge representation and reasoning. When logic programs represent the beliefs of a rational agent, belief change in this context has to be addressed. When dealing with belief change, the nature of information subject to change (beliefs, generic knowledge, preferences, goals), has to be taken into account as well as the representation of the agent's beliefs, belief base (finite set of formulae) or belief set (deductively closed set of formulae). When information subject to change is expert knowledge or generic knowledge the expectations are different from the case of belief change by new evidence incorporation. In this case, using logic programs

---

Julien Hué
University of Freiburg, Germany, e-mail: hue@informatik.uni-freiburg.de

Odile Papini
Aix-Marseille Université, CNRS, LSIS, UMR 7296, 13288 Marseille, France, e-mail: odile.papini@univ-amu.fr

Éric Würbel
Université de Toulon, CNRS, LSIS, UMR 7296, 83957 La Garde, Francee-mail: wurbel@univ-tln.fr

is quite suitable as unified formalism for both representing generic knowledge and reasoning from it.

In this paper, we consider two belief change operations: merging and revision. Merging gives a global logic program from the logic programs provided by the sources, exploiting their complementarity, solving possible inconsistencies, reducing imprecision and uncertainty, as well as removing redundancies coming from independent sources. Revision modifies a logic program in presence of a new rule, preserving consistency, minimally changing the logic program while keeping the new rule. In [Delgrande et al., 2006] three kinds of revision have been defined according to the nature of information. *Belief revision as defeasible inference* amounts to find the most plausible interpretations satisfying new information. *Belief revision as incorporating evidence* leads to change the order of plausibility in presence of new information. *Generic knowledge revision*: when in the propositional case, generic knowledge is described by a plausibility order on interpretations. Logic programs are suitable for both representing beliefs and generic knowledge, and seem natural for the third kind of revision.

Logic programs in a dynamic setting have been studied in [Przymusinski and Turner, 1997], [Zhang and Foo, 1998], [Alferes et al., 2000], [Sakama and Inoue, 1999], [Eiter et al., 2002], [Leite, 2003], [Delgrande et al., 2007] for the problem of updating knowledge bases represented by logic programs.

Answer Set Programming (or ASP) [Baral, 2003, Lifschitz, 2008b] now is a well established knowledge representation formalism as well as an efficient computation paradigm and a very active research area. ASP has an elegant and conceptually simple theoretical foundation and several definitions of an answer set have been proposed in the literature [Lifschitz, 2008a] bridging ASP with different knowledge representation domains [Schaub, 2008]. In the domain of belief change, ASP has been successfully used for implementing propositional belief bases change operations [Würbel et al., 2000, Hué et al., 2008, Benferhat et al., 2010]. Hence, ASP bridges logic programming and nonmonotonic reasoning and provides a representation formalism as well as an implementation tool. Due to the nonmonotonic nature of logic programs under answer set semantics, the problem of change in logic programs is different and more difficult than in the propositional case.

Belief change in logic programs under answer set semantics has been reformulated in [Delgrande et al., 2008, Delgrande et al., 2009] in a similar way than belief sets change in propositional logic. A model theory of logic programs stemming from SE-interpretations is used. A SE-interpretation is a model of a logic program in the same way that an interpretation is a model of a propositional formula. Belief sets change operations (revision and merging) based on distance between interpretations have been extended to logic programs. Belief sets update has been addressed in the same spirit by Slota and Leite [Slota and Leite, 2010]. However, they noted that belief sets change operations for logic program under answer set semantics have drawbacks and show undesired results from the ASP point of view.

We argue that extending belief sets change operations to logic programs seems problematic since belief sets are infinite structures while logic programs under answer set semantics are finite sets of rules with finite answer sets. Moreover, deductive closure is not defined for logic programs. Since ASP is a syntax-based approach, it seems natural to extend belief bases change operations to logic programs under answer set semantics. Hué and al. [Hué et al., 2009] proposed an extension of belief bases merging to logic programs under answer set semantics.

This chapter presents a more general framework ΠRSF for revising and merging belief bases represented by logic programs. The core idea, in case of inconsistency, is to pinpoint and remove from the union of logic programs, rules involved within the inconsistency. The most straightforward definition of inconsistency of logic programs is to consider as consistent a logic program which has at least one answer set. This first definition of consistency leads to a family of merging operations called strong Removed Sets Fusion. Nevertheless, this definition has numerous drawbacks starting by the impossibility to define a consequence relation thanks to it. This is the reason why we consider another definition of consistent logic programs based on the SE-models of the logic of Here-and-There [Pearce, 1996]. This definition leads to another family of merging operations, we call weak Removed Sets Fusion, in order to cope with these problems. It appears that weak Removed Sets Fusion keeps more beliefs than the strong Removed Sets Fusion operation.

This chapter extends the work presented in [Hué et al., 2009]. It presents within the same framework revision and merging of logic programs. It shows that the proposed operators satisfy most of the postulates defined for belief bases change. It provides an implementation of the proposed ΠRSF operators in ASP. The problem of logic programs merging is translated into a logic program under answer set semantics, a preference relation is then defined between answer sets according to the selected merging strategy. Then, a one-to-one correspondence between removed sets and preferred answer sets is established. The computation of answer sets is performed with any ASP solver supporting the *minimize* statement.

We show that ASP can be used as a unifying formalism for both representing and implementing belief change operations, extending belief bases change operations like RSR (Removed Set Revision) [Benferhat et al., 2010] and RSF (Removed Set Fusion) [Hué et al., 2008] initially proposed for revising and merging propositional belief bases respectively.

The paper is organized as follows. After a preliminary section that provides a reminder about belief change (revision and merging) and ASP, Section 3 shows how the Removed Set framework is extended to the revision and the merging of logic programs. A first extension, where the consistency of a logic program stems from the existence of answer sets, provides a family of strong operators. However due to the drawbacks of the obtained operators, a second extension is proposed, stemming from a different definition of the consistency of logic programs based on the existence of SE-models. It provides a more suitable family of weak operators. Section 4 presents an implementation of these merging operators using ASP. Finally Section 5 discusses related works before concluding.

Proofs of the results are gathered at the end of the paper.

2 Preliminaries

2.1 Notation

Throughout the paper we consider a propositional language $\mathscr{L}$ over a finite alphabet $\mathscr{P}$ of propositional variables (propositions). A literal is a proposition or the negation of a proposition. The usual propositional connectives are denoted by $\neg, \wedge, \vee, \rightarrow, \leftrightarrow$ and the constants True and False by $\top$ and $\bot$, respectively. Lowercase, greek lowercase and uppercase letters respectively denote propositions, formulae and sets of formulae. A *belief base* K is a finite set of propositional formulae over $\mathscr{L}$. We denote by IC the belief base representing the integrity constraints, i.e., general laws of the real world. We denote by $\mathscr{W}$ the set of interpretations of $\mathscr{L}$ and an interpretation is represented by a set of the propositional variables. $Mod(\alpha)$ denotes the set of models of a formula α, that is $Mod(\alpha) = \{\omega \in \mathscr{W} : \omega \models \alpha\}$ where $\models$ denotes the inference relation used for drawing conclusions. By extension, Let F be a set of formulae, $Mod(F) = \{\omega \in \mathscr{W} : \forall \alpha \in F, \omega \models \alpha\}$. Let F and G be two sets of formulae, $F \models G$ denotes $Mod(F) \subseteq Mod(G)$, moreover $F \equiv G$ if and only if $Mod(F) = Mod(G)$. We denote by $Cn(K)$ the set of logical consequences of the belief base K, more formally, $Cn(K) = \{\alpha : K \models \alpha\}$.

A belief profile E is a multi-set, $E = \{K_1, \ldots, K_n\}$, where K_i, $1 \leq i \leq n$ is a belief base. A belief profile denotes different sources of information. Let $E_1 = \{K_{1_1}, \ldots, K_{1_n}\}$ and $E_2 = \{K_{2_1}, \ldots, K_{2_n}\}$ be two belief profiles, E_1 and E_2 are equivalent, denoted by $E_1 \equiv E_2$ if and only if there exists a bijection f from E_1 to E_2 such that $\forall K \in E_1, f(K) \equiv K$.

In the following Π_i, $1 \leq i \leq n$ denotes a logic program, i.e. a finite set of rules. The notion of profile is extended to logic programs and is denoted by $E = \{\Pi_1, \ldots, \Pi_n\}$. When dealing with logic programs IC denotes a logic program consisting of the integrity constraints, and for $\bigcup E$ denotes $\Pi_1 \cup \ldots \cup \Pi_n$ and $E \cup IC$ denotes $(\Pi_1 \cup \ldots \cup \Pi_n) \cup IC$. Moreover $Atoms(\Pi_i)$ denotes the set of atoms occuring in Π_i. $Atoms(E), Atoms(E \cup IC)$ denotes the set of atoms occuring in $\Pi_1 \cup \ldots \cup \Pi_n$, $(\Pi_1 \cup \ldots \cup \Pi_n) \cup IC$ respectively. Similarly, $Lits(\Pi_i)$ denotes the set of literals occuring in Π_i. $Lits(E), Lits(E \cup IC)$, denotes the set of literals occuring in $\Pi_1 \cup \ldots \cup \Pi_n$, $(\Pi_1 \cup \ldots \cup \Pi_n) \cup IC$ respectively.

A preorder on a set A is a reflexive and transitive binary relation. A total preorder, denoted by $\leq$, is a preorder such that $\forall x, y \in A$ either $x \leq y$ or $y \leq x$ holds. The equivalence is defined by $x \simeq y$ if and only if $x \leq y$ and $y \leq x$. The corresponding strict total preorder, denoted by $<$, is the relation defined by $x < y$ if and only if $x \leq y$ holds but $x \simeq y$ does not hold. Let M be a subset of A, the set of minimal elements of M with respect to $\leq$, denoted by $min(M, \leq)$, is defined as: $min(M, \leq) = \{x \in M, \nexists y \in M : y < x\}$. The lexicographic total preorder is represented by the symbol $\leq_{lex}$.

2.2 Belief Change

2.2.1 Belief bases revision

Belief revision has been studied extensively in the field of Knowledge Representation and Reasoning in Artificial Intelligence. Characterization of the revision of theories or belief sets (deductively closed sets of formulae) has been proposed by Alchourròn, Gärdenfors and Makinson [Alchourrón et al., 1985] in terms of postulates that any revision operation should satisfy. When a theory is represented by a propositional formula ψ, Katsuno and Mendelzon [Katsuno and Mendelzon, 1991] have shown that any revision operator satisfying the AGM postulates amounts to associate to each formula ψ a total preorder $\leq_\psi$, which represents the plausible interpretation for each agent. $\omega \leq_\psi \omega'$ means that the interpretation ω is at least as plausible (closer to current beliefs ψ than ω'). All AGM revision operators can then be defined as functions selecting the most plausible interpretations when taking into account the new information μ: the models of new beliefs are then defined as $Mod(\psi \circ \mu) = min(Mod(\mu), \leq_\psi)$.

Revision has been extensively studied within the framework of belief bases i.e. finite sets of formulae [Hansson, 1999]. Let K be a belief base, μ be a formula and $\star$ be a base revision operator. $K \star \mu$ denotes the revised belief base. The symbol $+$ denotes the expansion operator (a non-closing operator) defined such that $K + \mu = K \cup \{\mu\}$. The basic postulates have been established as follows.

Success $\mu \in K \star \mu$.
Inclusion $K \star \mu \subseteq K + \mu$.
Vacuity   If $K \cup \mu$ is consistent then $K + \mu \subseteq K \star \mu$.
Consistency If μ is consistent so is $K \star \mu$.
Relevance   If $\alpha \in (K \cup \mu) \setminus (K \star \mu)$ then there is a set C such that
                $K \star \mu \subseteq C \subseteq (K \cup \mu)$, C is consistent
                but $C \cup \{\alpha\}$ is inconsistent.
Uniformity If for all $K' \subseteq K$, $K' \cup \alpha$ is inconsistent iff
                $K' \cup \beta$ is inconsistent, then $K \cap (K \star \alpha) = K \cap (K \star \beta)$

The *Success* postulate states that the new information is accepted. *Inclusion* says that revision does not introduce more information than expansion. *Vacuity* expresses that if the new information is consistent with the belief base then no information is removed. *Consistency* ensures that no inconsistency is introduced during revision. *Relevance* states that any removed item of information leads to inconsistency in a super set of the revision result. *Uniformity* says that the effect of revision is determined by subsets which are inconsistent with the new information. Several belief base revision operators have been proposed, some stemming from looking for maximal consistent subsets [Nebel, 1991, Benferhat et al., 1993, de Kleer, 1990, Lehmann, 1995, Fuhrmann, 1997], other from removing formulae in order to restore consistency [Papini, 1992, Hansson, 1997, Würbel et al., 2000, Falappa et al., 2012].

2.2.2 Belief bases merging

The aim of belief base merging is to obtain a global point of view, by exploiting the complementarity between belief sources, solving different existing conflicts, reducing imprecision and uncertainty, as well as removing redundancies coming from independent sources.

In the same spirit of AGM postulates for revision [Alchourrón et al., 1985] Konieczny and Pino Perez proposed postulates (KP postulates) [Konieczny and Pino Pérez, 1998] characterizing the rational behavior of merging operations which capture the following basic assumptions. The sources are mutually independent and no implicit link between the information from the different sources are assumed. All sources have the same level of importance and provide consistent belief bases. All information from a given source has the same level of reliability and priority.

In order to cope with the aim of merging and to obtain a global point of view, strategies are necessary. There are two straightforward strategies for defining the result of merging depending on whether the sources are conflicting. The classical conjunctive merging performs the conjunction of the belief bases and is suitable when the sources are not conflicting. The classical disjunctive merging performs the disjunction of the belief bases and is appropriate in case of conflicting sources. Because the two opposite cases are not satisfactory, several methods have been proposed for merging. In particular, the following classical merging strategies have been proposed: (i) the *Sum* strategy, denoted by Σ, [Lin and Mendelzon, 1998, Revesz, 1993] which follows the point of view of the majority of the belief bases; (ii) the *Cardinality* strategy, denoted by *Card*, [Baral et al., 1991] which is similar to Σ but without taking repetitions across the bases into account; (iii) the *Max* strategy [Revesz, 1995], which tries to best satisfy all the belief bases of E; (iv) the *Leximax* strategy, denoted by *GMax*, [Konieczny and Pino Pérez, 1998] which is the lexicographic refinement of *Max*.

Merging operators are broadly classified into two families: majority operators which follow the point of view of the majority of the belief bases and arbitration operators which try to best satisfy each belief base. The various works can be classified according to semantic and syntactic points of view. Semantic (or model-based) operators select interpretations that are the "closest" to the original belief bases [Baral et al., 1991, Revesz, 1993, Lin, 1996, Revesz, 1995, Cholvy, 1998, Konieczny and Pino Pérez, 2002, Konieczny et al., 2004, Bloch and Lang, 2002]. Many operators have been defined according to explicit or implicit priorities [Konieczny and Pino Pérez, 1998, Lafage and Lang, 2000, Konieczny, 2000, Delgrande et al., 2006, Fagin et al., 1986, Konieczny and Pino Pérez, 1999]. By contrast, syntactic approaches focus on explicitely given formulae. They select some formulae from the initial belief bases. More recently, within the context of syntax-based approaches, Falappa et al. [Falappa et al., 2012] proposed the following postulates in order to classify reasonable belief bases merging operators. These operators are considered as binary symmetric operators where two belief bases K and A

are symmetrically treated. Let K, H, A, B be belief bases, α be a formula and $\triangle$ be a merging operator.

Inclusion	$K \triangle A \subseteq K \cup A$.
Symmetry	$K \triangle A = A \triangle K$.
Strong Consistency	$K \triangle A$ is consistent.
Congruence	If $K \cup A = H \cup B$ then $K \triangle A = H \triangle B$.
Vacuity	If $K \cup A$ is consistent then $K \triangle A = K \cup A$.
Reversion	If $K \cup A$ and $H \cup B$ have the same minimal inconsistent subsets then $(K \cup A)\backslash(K \triangle A) = (H \cup B)\backslash(H \triangle B)$.
Global Relevance	If $\alpha \in (K \cup A)\backslash(K \triangle A)$ then there is a set C such that $K \triangle A \subseteq C \subseteq (K \cup A)$, C is consistent but $C \cup \{\alpha\}$ is inconsistent.
Global Core-retainment	If $\alpha \in (K \cup A)\backslash(K \triangle A)$ then there is a set C such that $C \subseteq (K \cup A)$, C is consistent but $C \cup \{\alpha\}$ is inconsistent.

Inclusion states that the union of the initial belief bases is the upper bound of any merging operation. *Symmetry* establishes that all the belief bases are considered of equal importance. *Strong consistency* requires the consistency of the result of merging. *Congruence* requires that the result of merging should not depend on syntactic properties of the belief bases. *Vacuity* says that if the union of the belief bases is consistent then the result of merging equals this union. *Reversion* says that if two sets of belief bases have the same minimally inconsistent subsets then the sentences removed in the respective sets are the same. *Global relevance* and *Global Core-retainment* express the intuition that nothing is removed from the original beliefs bases unless its removal helps to make the result consistent. Note that *Global relevance* entails *Global Core retainment* ([Falappa et al., 2012]). Inclusion, Symmetry, Congruence and Global Relevance have been proposed in [Fuhrmann, 1997] while *Reversion* and *Global Core-Retainment* have been proposed by Falappa and al. in [Falappa et al., 2002].

Combination operators have been first proposed in [Baral et al., 1991] and [Baral et al., 1992]. These operators perform the union of the bases and try to obtain a consistent result by the construction of maximal consistent subsets of the possibly inconsistent union. These operators have been reformulated in [Konieczny, 2000, Konieczny and Pino Pérez, 2011]. They select some maximal consistent sets of formulae where the notion of maximal consistent subset has been extended to belief profiles and the maximality is defined w.r.t. set inclusion or cardinality. However, as argued in [Konieczny, 2000, Konieczny and Pino Pérez, 2011] these operators are not satisfactory since they do not take into account the origin of the beliefs. In order to overcome this problem, three other syntax-based operators have been proposed by Konieczny in order to select maximal consistent subsets of formulae that best satisfy some merging strategies. The first one selects the maximal consistent subsets that are consistent with the greatest number of bases. The second one selects the maximal consistent subsets that have the smallest differences w.r.t. cardinality with the bases and the third one selects the maximal consistent subsets that have the biggest intersection w.r.t. cardinality with the bases.

Another approach stemming from maximal consistent subsets of formulae has been proposed in [Fuhrmann, 1997]. It extends partial meet revision of a belief base to merging and stems from the notion of remainder set, i.e., maximal subsets of the belief base that fail to entail the added formula [Alchourrón et al., 1985]. The partial meet merging operator intersects some maximal consistent subsets obtained from a general selection function [Falappa et al., 2012].

A dual point of view is to focus on minimal inconsistent subsets of formulae, and to remove at least one formula in each minimal inconsistent subset. Following this line Falappa and al. proposed a kernel merging operator in [Falappa et al., 2012]. The notion of α-kernel of K (minimal subset of K that entails α) was originally introduced by Hansson for the definition of kernel contraction [Hansson, 1994]. Kernel merging involves the definition of $\bot$-kernels (i.e. minimal inconsistent subsets) of the union of the original bases, and incision functions which select the formulae to remove in order to restore consistency. Another approach focusing on minimal inconsistent subsets of formulae is Removed Sets Fusion (RSF). It extends Removed Sets Revision (RSR) [Benferhat et al., 2010] and attempts to remove subsets of formulae according to different merging strategies (Sum, Card, Max, GMax) encoded by a preference relation between subsets of formulae [Hué et al., 2007, Hué et al., 2008].

2.3 Answer Set Programming

ASP has been successfully used for implementing nonmonotonic reasoning during the last decade. We briefly recall some essential notions.

2.3.1 Logic programs

An extended logic program (or ELP) is a set of rules of the form

$$r : l_0 \leftarrow l_1, \ldots, l_m, not\ l_{m+1}, \ldots, not\ l_n. \tag{1}$$

where l_i are literals, that is, atoms a or their *strong negation* $\neg a$. The set of all ASP atoms is denoted by $\mathscr{A}$. The keyword *not* denotes negation as failure. A literal of the form $not\ l_i$ is called a NAF-literal. A program (or a rule), is said to be *not-free* if it contains no NAF-literal. The literal l_0 is called the head of the rule and is denoted by $head(r)$. $l_i, i \in \{1,\ldots,n\}$, make up the body of the rule, which is denoted by $body(r)$. The body can be divided into a positive and negative body. Atoms l_i, $i \in \{1,\ldots,m\}$ represent the positive body, denoted by $body^+(r)$, and atoms $l_j, j \in \{m+1,\ldots,n\}$, represent the negative body, denoted by $body^-(r)$. Thus, $body(r) = body^+(r) \cup body^-(r)$. Intuitively, a rule can be understood as follows: if all the literals l_i of the positive body of the rule are true, and if none of the literals l_j of the negative body are true, then the head of the rule can be inferred. For a rule r of type (1), we denote by r^+ the rule $head(r) \leftarrow body^+(r)$. For any set of rules Π, $Lits(\Pi)$ denotes the literals appearing in Π, and $Atoms(\Pi)$ the set of atoms appearing in Π. By a slight abuse of notation, for a single rule r we write $Atoms(r)$ (respectively $Lits(r)$) to denote the set of atoms (respectively literals) appearing in r.

The usual semantics for extended programs is given by the *answer set semantics*. For any program Π, a subset $S \subseteq Lits(\Pi)$ *satisfies* a rule r of the form (1) if $body^+(r) \subseteq S$ and $body^-(r) \cap S = \emptyset$ implies $head(r) \in S$. For any not-free ELP Π, S is an *answer set* of Π if and only if it is a minimal subset of $Lits(\Pi)$ according to set inclusion which satisfies the following two conditions: (i) S satisfies all rules of Π ; (ii) if there exists $l \in S$ such that $\neg l \in S$, then $S = Lits(\Pi)$.

The Gelfond-Lifschitz reduct [Gelfond and Lifschitz, 1988], Π^X of an extended program Π by a set $X \subseteq Lits(\Pi)$ is defined as $\Pi^X = \{r^+ \,|\, r \in \Pi$ and $body^-(r) \cap X = \emptyset\}$. It allows for a simple definition of an answer set. Let X be a set of literals and Π a logic program. X is an answer set of Π if and only if X is an answer set of Π^X.

A program Π is *AS-consistent* if and only if it has an answer set, otherwise it is *AS-inconsistent*. The set of all answer sets of Π is denoted by $AS(\Pi)$.

Finally, for a program Π and a set of literals X, we denote by $X \models \Pi$ the fact that X satisfies Π in the sense of propositionnal logic, interpreting rules as formulae of propositional logic, and dealing with the *not* as negation in propositional logic.

2.3.2 SE-model semantics and properties of logic programs

Among the different ways to define an answer set [Lifschitz, 2010] we focus, in addition to the above definition, on the definition stemming from the *here-and-there logic* [1][Pearce, 1996]. In this context, a SE-interpretation is a pair of sets of atoms (X,Y) such that $X \subseteq Y$.

Let Π be a normal program and (X,Y) be a SE-interpretation. (X,Y) is a SE-model of Π if and only if $Y \models \Pi$ and $X \models \Pi^Y$ [Turner, 2003]. The set of SE-models of a program Π is denoted by $SE(\Pi)$. The terms SE-model and SE-interpretation were introduced by Turner [Turner, 2001]. It corresponds to a more direct characterization of respective notions of HT-interpretation and HT-model without the use of the *here-and-there* logic. The definition of an answer set from the notion of SE-model is as follows. Let Π be a normal program and let X be a set of atoms. X is an answer set of Π if and only if $(X,X) \in SE(\Pi)$ and there is no $(X',X) \in SE(\Pi)$ such that $X' \subset X$.

An extended logic program Π is *SE-consistent* if and only if $SE(\Pi) \neq \emptyset$. Two normal logic programs Π_1 and Π_2 are said to be strongly equivalent if and only if $SE(\Pi_1) = SE(\Pi_2)$, denoted by $\Pi_1 \equiv_s \Pi_2$.

---

[1] More precisely, the notion of answer set corresponds to the notion of model in the *equilibrium logic* [Pearce, 2006] which relies on the *here-and-there* logic equipped with the strong negation (i.e. classical).

Another meaning of this equivalence is that for any program Π, $AS(\Pi_1 \cup \Pi) = AS(\Pi_2 \cup \Pi)$. Similarly, we write $\Pi_1 \models_s \Pi_2$ for $SE(\Pi_1) \subseteq SE(\Pi_2)$.

2.4 ASP langage and solvers

The computation of the answer sets of logic programs is performed with programs called ASP solvers [Anger et al., 2002, Gebser et al., 2007, Giunchiglia et al., 2004, Leone et al., 2006, Lin and Zhao, 2004, Niemelä and Simons, 1997]. They provide several shorthands and statements in order to simplify the writing of collections of similar rules. A logic program is not usable in its primary form, and the computation of answer sets requires two separate steps: the logic program is first instantiated and transformed by a preprocessor then the translated program is provided to an ASP solver.

We present some of the statements and shorthands proposed by these solvers. All these statements are available in *lparse* [Syrjanen, 2002] and *Gringo* [Gebser et al., 2009] preprocessors.

2.4.1 Cardinality handling

A cardinality literal is of the form: $l\{a_1,\ldots,a_m, not\ b_1,\ldots, not\ b_n\}u$. This literal will be satisfied if at least l and at most u atoms among the set are satisfied. It is possible to omit one of the cardinality restrictions l or u.

It is also possible to assign a weight to each element of the set. In the following atom, the w_i are weights, expressed as integer quantities: $l[a_1 = w_1,\ldots, a_m = w_m, not\ b_1 = w_{m+1},\ldots, not\ b_n = w_{m+n}]u$. In this case, the weight literal is satisfied if the sum of the weights of satisfied literals in the rule body is between l and u (inclusive).

2.4.2 Optimization statements

Closely related to the cardinality handling is the question of the minimization (or maximization) of the number of satisfied atoms in an answer set model. For example, in a configuration problem, one would probably prefer to keep, among all solutions, the ones which require a minimal number of parts. Formally, a minimize statement is written: $minimize\{a_1,\ldots,a_m, not\ b_1,\ldots, not\ b_n\}$. The goal of this statement is to keep, among the set of answer sets of a program, those having the minimum number of satisfied literals from the set $\{a_1,\ldots,a_m, not\ b_1,\ldots, not\ b_n\}$. The dual statement is written as follows: $maximize\{a_1,\ldots,a_m, not\ b_1,\ldots, not\ b_n\}$, which keeps the answer sets containing the greatest number of satisfied literals from the set $\{a_1,\ldots,a_m, not\ b_1,\ldots, not\ b_n\}$. As for cardinality literals, it is possible to use weights:

$$minimize\ [a_1 = w_1, \ldots, a_m = w_m, not\ b_1 = w_{m+1}, \ldots, not\ b_n = w_{m+n}]$$

will minimize the accumulated weight of satisfied literals in the set.

2.4.3 Variables, domains of definition, intervals

It is possible to restrain the domain of a variable to a subset of the Herbrand universe of the program, thus controlling more finely the instantiation of the program. This declaration is specified using the directive #*domain a(X)*, where the variable X will be instantiated by the preprocessor to all values such that $a(X)$ is set as a fact. For example, the following declarations and facts: $\{\#domain\ a(X).\ a(1).\ a(2).\ f(X) \leftarrow not\ g(X).\}$ will lead to the following grounding: $\{a(1).\ a(2).\ f(1) \leftarrow not\ g(1).\ f(2) \leftarrow not\ g(2).\}$. Interval specifications for numerical values allows for even more compact representation: $\{\#domain\ a(X).\ a(1..3).\ f(X) \leftarrow not\ g(X).\}$ will lead to the following grounding:$\{a(1).\ a(2).\ a(3).\ f(1) \leftarrow not\ g(1).\ f(2) \leftarrow not\ g(2).\ f(3) \leftarrow not\ g(3).\}$.

3 Changing logic programs

When information subject to change is expert knowledge or generic knowledge the expectations are different from the case of belief change by new evidence incorporation. In this case, using logic programs under answer set semantics is quite suitable as unified formalism for both representing generic knowledge and reasoning with it.

For a more synthetic presentation, we introduce revision and merging in a common framework. ΠRSF denotes the extension of RSF to merging logic programs with constraints, the profile being denoted by $E = \{\Pi_1, \ldots, \Pi_n\}$ where Π_i, $1 \leq i \leq n$ are logic programs and *IC* is a logic program representing the constraints. In case of revision, ΠRSR denotes the extension of RSR to logic programs where the profile is replaced by a single logic program Π, the new information is the logic program *IC* representing the constraints to be included in the outcome of the revision and the strategy P is cardinality, denoted by *Card*.

The removed sets approach amounts to pinpoint a subset of rules from the union of logic programs in the profile and remove it according to a given strategy, in order to restore consistency and to satisfy the constraints. However, the notion of consistency of a logic program is not universally accepted. In a first approach, the notion of consistency of a logic program will rely on the existence of answer sets. As we shall see, this definition has several limitations, and this is why a second approach will be proposed, where the consistency of a logic program will be based on the existence of SE-models.

3.1 First extension of RSF to logic programs

For this first extension of RSF, we will rely on *AS*-consistency. We recall that a logic program Π_i is *AS*-consistent if and only if it has at least one answer set, a profile E is therefore consistent if and only if $\Pi_1 \cup \ldots \cup \Pi_n$ is consistent. The notion of strong potential removed set is then defined as follows:

Definition 1. Let $E = \{\Pi_1, \ldots, \Pi_n\}$ be a belief profile, *IC* be a program representing integrity constraints, and let X be a subset of rules from $\Pi_1 \cup \ldots \cup \Pi_n$. X is a strong potential removed set of E if and only if $((\Pi_1 \cup \ldots \cup \Pi_n) \backslash X) \cup IC$ is *AS*-consistent.

For any strategy P, a total preorder $\leq_P$ on the strong potential removed sets is defined. $X \leq_P Y$ means that X is preferred to Y according to the strategy P[2]. The merging strategies Σ, *Card*, *Max*, *GMax* are captured within the framework as follows. The Σ strategy minimizes the number of rules to remove from $\Pi_1 \cup \ldots \cup \Pi_n$, the total preorder $\leq_\Sigma$ is defined by $X \leq_\Sigma Y$ if and only if $\sum_{1 \leq i \leq n} |X \cap \Pi_i| \leq \sum_{1 \leq i \leq n} |Y \cap \Pi_i|$. The *Card* strategy attempts, just as Σ, to minimize the number of removed rules, but it does not take into account rules which are expressed several times. Let E be a profile, the total preorder $\leq_{Card}$ is defined by $X \leq_{Card} Y$ if and only if $|X \cap (\Pi_1 \cup \ldots \cup \Pi_n)| \leq |Y \cap (\Pi_1 \cup \ldots \cup \Pi_n)|$. The *Max* strategy tries to best satisfy all the logic programs. The total preorder $\leq_{Max}$ is defined by $X \leq_{Max} Y$ if and only if $\max_{1 \leq i \leq n} |X \cap \Pi_i| \leq \max_{1 \leq i \leq n} |Y \cap \Pi_i|$. The *GMax* strategy is a lexicographic refinement of the *Max* strategy, the total preorder $\leq_{GMax}$ is defined as follows. For every potential removed set X and every logic program Π_i, we define $p_X^{\Pi_i} = |X \cap \Pi_i|$. Let L_X^E be the sequence $(p_X^{\Pi_1}, \ldots, p_X^{\Pi_n})$ sorted by decreasing order. Let X and Y be two potential removed sets of E, $X \leq_{GMax} Y$ if and only if $L_X^E \leq_{lex} L_Y^E$. The notion of strong removed set is then defined according to P.

Definition 2. Let $E = \{\Pi_1, \ldots, \Pi_n\}$ be a belief profile, *IC* be a program representing integrity constraints, P be a strategy, and $X \subseteq \Pi_1 \cup \ldots \cup \Pi_n$. We say that X is a strong removed set of E according to P if and only if (1) X is a strong potential removed set of E; (2) There is no strong potential removed set Y of E such that $Y \subset X$. (3) There is no strong potential removed set Y of E such that $Y <_P X$.

The collection of strong removed sets of E according to the strategy P is denoted by $\mathcal{R}_P^1(E)$. The merging operation denoted by $\Delta_{P,IC}^{\Pi RSF,1}(E)$, is defined next.

Definition 3. Let $E = \{\Pi_1, \ldots, \Pi_n\}$ be a belief profile, *IC* be a program representing integrity constraints, P be a strategy, and f be a selection function from $\mathcal{R}_P^1(E)$ into $\mathcal{R}_P^1(E)$. Then,

$$\Delta_{P,IC}^{\Pi RSF,1}(E) = \{((\Pi_1 \cup \cdots \cup \Pi_n) \backslash f(\mathcal{R}_P^1(E))) \cup IC\}$$

The following example illustrates the first extension of RSF to logic programs.

---
[2] We define $<_P$ as the strict preorder associated to $\leq_P$ (i.e. $X <_P Y$ iff $X \leq_P Y$ and $Y \not\leq_P X$).

Example 1. Let $E = \{\Pi_1, \Pi_2\}$ be a profile such that

$$\Pi_1 = \left\{ \begin{array}{cc} a \leftarrow not\ b.\ c \leftarrow not\ a. \\ e. \quad d \leftarrow f. \end{array} \right\} \quad \Pi_2 = \left\{ \begin{array}{cc} b \leftarrow not\ c.\ \neg d. \\ d \leftarrow e. \quad f. \end{array} \right\}$$

and $IC = \{\top.\}$ be a logic program representing the constraints. The logic program $\Pi_1 \cup \Pi_2$ does not have any answer set since it leads to an inconsistent set of literals $\{d, \neg d, e, f\}$. The set of strong potential removed sets, which are minimal with respect to set inclusion, is $\{\{\neg d.\quad a \leftarrow not\ b.\}, \{\neg d.\quad b \leftarrow not\ c.\}, \{\neg d.\quad c \leftarrow not\ a.\}, \{e.\quad f.\quad a \leftarrow not\ b.\}, \{e.\quad f.\quad b \leftarrow not\ c.\},$
$\{e.\quad f.\quad c \leftarrow not\ a.\}, \{e.\quad d \leftarrow f.\quad a \leftarrow not\ b.\},$
$\{e.\quad d \leftarrow f.\quad b \leftarrow not\ c.\}, \{e.\quad d \leftarrow f.\quad c \leftarrow not\ a.\},$
$\{d \leftarrow e.\quad f.\quad a \leftarrow not\ b.\}, \{d \leftarrow e.\quad f.\quad b \leftarrow not\ c.\},$
$\{d \leftarrow e.\quad f.\quad c \leftarrow not\ a.\}, \{d \leftarrow e.\quad d \leftarrow f.\quad a \leftarrow not\ b.\},$
$\{d \leftarrow e.\quad d \leftarrow f.\quad b \leftarrow not\ c.\}, \{d \leftarrow e.\quad d \leftarrow f.\quad c \leftarrow not\ a.\}\}$

For the Σ and *Card* strategies, the set of strong removed sets is $\mathscr{R}_P^1(E) = \{\{\neg d.\quad a \leftarrow not\ b.\}, \{\neg d.\quad b \leftarrow not\ c.\}, \{\neg d.\quad c \leftarrow not\ a.\}\}$. The possible selection functions are $f_1(\mathscr{R}_P^1(E)) = \{\neg d.\quad a \leftarrow not\ b.\}$, $f_2(\mathscr{R}_P^1(E)) = \{\neg d.\quad b \leftarrow not\ c.\}$, $f_3(\mathscr{R}_P^1(E)) = \{\neg d.\quad c \leftarrow not\ a.\}$. Consequently,

$$\Delta_{\Sigma, IC}^{\Pi RSF,1}(E) = \Delta_{Card, IC}^{\Pi RSF,1}(E) = \left\{ \begin{array}{ll} c \leftarrow not\ a. & e. \quad d \leftarrow f. \\ b \leftarrow not\ c.\ d \leftarrow e. & f. \end{array} \right\}$$
$$\text{or} \left\{ \begin{array}{ll} a \leftarrow not\ b.\ c \leftarrow not\ a.\ e. \\ d \leftarrow f. \quad d \leftarrow e. \quad f. \end{array} \right\}$$
$$\text{or} \left\{ \begin{array}{ll} a \leftarrow not\ b. & e. \quad d \leftarrow f. \\ b \leftarrow not\ c.\ d \leftarrow e. & f. \end{array} \right\}$$

For the *Max* and *GMax* strategies, the set of strong removed sets is $\mathscr{R}_P^1(E) = \{\{\neg d.\quad a \leftarrow not\ b.\}, \{\neg d.\quad c \leftarrow not\ a.\}\}$. The possible selection fuctions are $f_1(\mathscr{R}_P^1(E)) = \{\neg d.\quad a \leftarrow not\ b.\}$, $f_2(\mathscr{R}_P^1(E)) = \{\neg d.\quad c \leftarrow not\ a.\}$. Consequently,

$$\Delta_{Max, IC}^{\Pi RSF,1}(E) = \Delta_{GMax, IC}^{\Pi RSF,1}(E) = \left\{ \begin{array}{ll} c \leftarrow not\ a. & e. \quad d \leftarrow f. \\ b \leftarrow not\ c.\ d \leftarrow e. & f. \end{array} \right\}$$
$$\text{or} \left\{ \begin{array}{ll} a \leftarrow not\ b. & e. \quad d \leftarrow f. \\ b \leftarrow not\ c.\ d \leftarrow e. & f. \end{array} \right\}$$

The first extension to logic programs has most of the desired properties.

Proposition 1. *For any strategy P the $\Delta_{P,IC}^{\Pi RSF,1}$ operator satisfies* Inclusion, Symmetry, Strong Consistency, Vacuity *and* Global Core-retainment. *Moreover, for the Σ and Card strategies it also satisfies* Congruence.

In case of revision, the profile is replaced by a single logic program Π, the new information is the logic program IC and the strategy is P. Every strategy $P \in \{Card, \Sigma, Max, GMax\}$ gives the same result, due to the fact that the profile is reduced to only one logic program. The revision operator is denoted by $\circ_{\Pi RSR,1}$.

However, this direct extension of the removed sets approach has a number of limitations. Defining a consequence relation between logic programs, is problematic if it only stems from the notion of answer set, as illustrated in the following example.

Example 2. Let $E = \{\{a.\}\}$ be a profile and $IC = \{c \leftarrow \text{not } a. \quad \neg b.\}$ be a logic program. The logic program $\Pi_1 \cup \ldots \cup \Pi_n \cup IC$ is AS-consistent (it has an answer set) and for any merging strategy P, the result of merging is $\Delta_{P,IC}^{\Pi RSF,1}(E) = \Pi_1 \cup \ldots \cup \Pi_n \cup IC$. Since the rules of IC belong to the result, this result should satisfy these rules. However, the only answer set of IC is $\{c, \neg b\}$ while the only answer set of $\Pi_1 \cup \ldots \cup \Pi_n \cup IC$ is $\{a, \neg b\}$. Although IC is included in the result of merging, $AS(\Pi_1 \cup \ldots \cup \Pi_n \cup IC) \not\subseteq AS(IC)$.

This problem is solved by choosing an alternative definition of the consequence relation in terms of SE-models introduced in 2.3. A logic program Π_1 is a consequence of logic program Π_2, denoted by $\Pi_1 \models_s \Pi_2$, if and only if $SE(\Pi_1) \subseteq SE(\Pi_2)$.

Example 3. Coming back to the previous example, the SE-models of IC are: $SE(IC) = \{(\{\neg b\}, \{\neg b, a\}), (\{\neg b, a\}, \{\neg b, a\}), (\{\neg b, c\}, \{\neg b, c\}),$ $(\{\neg b\}, \{\neg b, a, c\}), (\{\neg b, a\}, \{\neg b, a, c\}), (\{\neg b, c\}, \{\neg b, a, c\}),$ $(\{\neg b, a, c\}, \{\neg b, a, c\})\}$ and the SE-models of $E \cup IC$ are: $SE(E \cup IC) = \{(\{\neg b, a\}, \{\neg b, a\}), (\{\neg b, a\}, \{\neg b, a, c\}),$ $(\{\neg b, a, c\}, \{\neg b, a, c\})\}$. We have $SE(\Pi_1 \cup \ldots \cup \Pi_n \cup IC) \subseteq SE(IC)$ and therefore according to the definition of the consequence relation $(\Pi_1 \cup \ldots \cup \Pi_n \cup IC) \models_s IC$.

Another limitation of the direct extension of the removed sets approach is that it removes too much information, as illustrated in the following example.

Example 4. Among logic programs that do not have any answer set we can distinguish two kinds of programs. The first category is made up of programs that have as immediate consequence an inconsistent set of literals. For example, consider $\Pi = \{\neg a. \quad a \leftarrow b. \quad b.\}$. This program has no answer set because it leads to an inconsistent set of literals $\{a, \neg a, b\}$ and the only way to restore AS-consistency is the removal of rules. The second category is illustraded by the following example: consider the logic program $\Pi' = \{a \leftarrow \text{not } b. \quad b \leftarrow \text{not } c. \quad c \leftarrow \text{not } a.\}$. It does not have any answer set because it is not possible to derive a set of literals having a justification. However, this program does not lead to an inconsistent set of literals. It is possible to restore AS-consistency without deleting any rule but merely by adding justifications, for example by adding the rule $\{a.\}$. Thus the program $\Pi' \cup \{a.\}$ has $\{a, b\}$ as an answer set.

In the first extension of RSF to logic programs, the definition of strong removed set may result in unnecessary removal of rules, like rules with no justified consequences and whose justification can occur later. Indeed, the notion of AS-consistency of a logic program is only based on the existence of an answer set. Defining consistency of a logic program based on the existence of a SE-model (SE-consistency) allows us to solve this problem. For example, the program Π has no answer set but has the following SE-models: $SE(\Pi) = \{(\{b\}, \{b, a\}), (\{b, a\}, \{b, a\}), (\{c\}, \{b, c\}), (\{b, c\}, \{b, c\}),$

$(\{a\},\{a,c\}),(\{a,c\},\{a,c\}),(\{\emptyset\},\{a,b,c\}),(\{a\},\{a,b,c\}),(\{b\},\{a,b,c\}),$
$(\{c\},\{a,b,c\}),(\{a,b\},\{a,b,c\}),(\{b,c\},\{a,b,c\}),(\{a,c\},\{a,b,c\}),$
$(\{a,b,c\},\{a,b,c\})\}$. However, the program $\Pi' \cup \{a.\}$ has an answer set and its SE-models are:
$SE(\Pi' \cup \{a.\}) = \{(\{a,b\},\{a,b\}),(\{a\},\{a,b,c\}),(\{a,b\},\{a,b,c\}),$
$(\{a,c\},\{a,b,c\}),(\{a,b,c\},\{a,b,c\})\}$ where $\{a,b\}$ is an answer set.

It therefore seems natural to extend RSF to logic programs where consistency of logic programs is defined in terms of SE-models [Hué et al., 2009].

3.2 Second extension of RSF to logic programs

This extension relies on *SE*-consistency. We recall that a logic program Π_i is *SE*-consistent if and only if it has at least one SE-model. Therefore a profile is *SE*-consistent if and only if $\Pi_1 \cup \ldots \cup \Pi_n$ is *SE*-consistent. The notion of weak potential removed set is then defined as follows:

Definition 4. Let $E = \{\Pi_1, \ldots, \Pi_n\}$ be a belief profile, and *IC* be a logic program representing integrity constraints. Let X be a subset of $\Pi_1 \cup \ldots \cup \Pi_n$. We say that X is a weak potential removed set of E if and only if $((\Pi_1 \cup \ldots \cup \Pi_n) \setminus X) \cup IC$ is *SE*-consistent.

For any strategy P, a total preorder $\leq_P$ on the set of weak potential removed sets is defined. $X \leq_P Y$ means that X is preferred to Y according to the strategy P[3]. The merging strategies Σ, *Card*, *Max*, *GMax* are captured within the framework as detailed in the previous section. The notion of weak removed set according to P is then defined.

Definition 5. Let $E = \{\Pi_1, \ldots, \Pi_n\}$ be a belief profile, and *IC* be a logic program representing integrity constraints. Let P be a strategy, and $X \subseteq \Pi_1 \cup \ldots \cup \Pi_n$. We say that X is a weak removed set of E according to P if and only if (1) X is a weak potential removed set of E; (2) There is no weak potential removed set Y of E such that $Y \subset X$. (3) There is no weak potential removed set Y of E such that $Y <_P X$.

The collection of weak removed sets of E according to strategy P is denoted by $\mathscr{R}_P^2(E)$. The merging operation denoted by $\Delta_{P,IC}^{\Pi RSF,2}(E)$, is defined by:

Definition 6. Let $E = \{\Pi_1, \ldots, \Pi_n\}$ be a belief profile, *IC* be a logic program representing integrity constraints, P be a strategy, and f be a selection function defined on the set of weak removed set of E according to P.

$$\Delta_{P,IC}^{\Pi RSF,2}(E) = \{((\Pi_1 \cup \cdots \cup \Pi_n) \setminus f(\mathscr{R}_P^2(E))) \cup IC\}$$

The following example illustrates the second extension of RSF to logic programs.

---
[3] We define $<_P$ as the strict preorder associated to $\leq_P$ (i.e. $X <_P Y$ iff $X \leq_P Y$ and $Y \not\leq_P X$).

Example 5. We come back to example 1. The logic program $\Pi_1 \cup \Pi_2$ leads to an inconsistent set of literals $\{d, \neg d, e, f\}$. However, we note that the rules whose consequences are not justified are not involved in the deduction of this set, this is the set $\Pi'_i = \{a \leftarrow \text{not } b. \quad c \leftarrow \text{not } a. \quad b \leftarrow \text{not } c.\}$ and the set of rules leading to an inconsistent set of literals is $\Pi'_j = \{e. \quad d \leftarrow f. \quad \neg d. \quad d \leftarrow e. \quad f.\}$. The difference between the first and second extension of the removed set approach to logic programs is that the first extension will build strong removed sets drawing rules in Π'_i and Π'_j while the second extension will build weak removed sets by only drawing rules in Π'_j. More precisely, the set of weak potential removed sets, minimal with respect to set inclusion, is $\{\{\neg d.\}, \{e. \quad f.\}, \{e. \quad d \leftarrow f.\}, \{d \leftarrow e. \quad f.\}, \{d \leftarrow e. \quad d \leftarrow f.\}\}$. For the Σ and *Card* strategies, the set of weak removed sets is $\{\{\neg d.\}\}$. The selection function is $f(\mathcal{R}_P^2(E)) = \{\neg d.\}$. Consequently,

$$\Delta_{\Sigma,IC}^{\Pi RSF,2}(E) = \Delta_{Card,IC}^{\Pi RSF,2}(E) = \left\{ \begin{array}{l} a \leftarrow \text{not } b. \; c \leftarrow \text{not } a. \quad e. \\ d \leftarrow f. \quad b \leftarrow \text{not } c. \; d \leftarrow e. \\ f. \end{array} \right\}.$$

For the *Max* strategy, the set of removed sets is $\{\{f. \quad e.\}, \{d \leftarrow f. \quad d \leftarrow e.\}, \{\neg d.\}\}$. The possible selection functions are $f_1(\mathcal{R}_P^2(E)) = \{f. \quad e.\}$, $f_2(\mathcal{R}_P^2(E)) = \{d \leftarrow f. \quad d \leftarrow e.\}$, $f_3(\mathcal{R}_P^2(E)) = \{\neg d.\}$. Consequently,

$$\Delta_{Max,IC}^{\Pi RSF,2}(E) = \left\{ \begin{array}{ll} a \leftarrow \text{not } b. \; c \leftarrow \text{not } a. \; d \leftarrow f. \\ b \leftarrow \text{not } c. \quad \neg d. \quad d \leftarrow e. \end{array} \right\}$$
$$\text{or} \left\{ \begin{array}{ll} a \leftarrow \text{not } b. \; c \leftarrow \text{not } a. \; e. \\ b \leftarrow \text{not } c. \quad \neg d. \quad f. \end{array} \right\}$$
$$\text{or} \left\{ \begin{array}{l} a \leftarrow \text{not } b. \; c \leftarrow \text{not } a. \quad e. \\ d \leftarrow f. \quad b \leftarrow \text{not } c. \; d \leftarrow e. \\ f. \end{array} \right\}.$$

For the *GMax* strategy, the set of removed sets is $\{\{\neg d.\}\}$. The only selection function is $f_1(\mathcal{R}_P^2(E)) = \{\neg d.\}$. Consequently,

$$\Delta_{GMax,IC}^{\Pi RSF,2}(E) = \left\{ \begin{array}{l} a \leftarrow \text{not } b. \; c \leftarrow \text{not } a. \quad e. \\ d \leftarrow f. \quad b \leftarrow \text{not } c. \; d \leftarrow e. \\ f. \end{array} \right\}.$$

As illustrated in the example above, in the first approach which relies on *AS*-consistency, the consequence relation between logic programs cannot be defined adequately. Moreover, this approach removes more information than the second one relying on *SE*-consistency.

The second extension to logic programs has most of the desired properties.

Proposition 2. *For any strategy P the $\Delta_{P,IC}^{\Pi RSF,2}$ operator satisfies* Inclusion, Symmetry, Stong Consistency, Vacuity *and* Global Core-retainment. *Moreover, for the Σ and Card strategies it also satisfies* Congruence.

In the case of revision, the profile is replaced by a single logic program Π, the new information is the logic program *IC* and the strategy is *P*. Every strategy $P \in$

{*Card*, Σ, *Max*, *GMax*} gives the same result, due to the fact that the profile is reduced to only one logic program. The revision operator is denoted by $\circ_{\Pi RSR,2}$.

4 Implementation

The computation of strong and weak removed sets is performed by means of an ASP implementation. Concerning the strong version of ΠRSF, we build an extended logic program $PL_P^1(E,IC)$ such that the answer sets of this program correspond to the strong removed sets of E according to the strategy P. As for the weak version of ΠRSF, we build an extended logic program $PL_P^2(E,IC)$ such that the answer sets of this program correspond to the weak removed sets of E according to the strategy P.

More precisely each of the programs $PL_P^1(E,IC)$ and $PL_P^2(E,IC)$ contains two parts: one which is responsible for the computation of the strong (weak) potential removed sets named $PL^1(E,IC)$ ($PL^2(E,IC)$), respectively. Another part handles the strategy, that is, it keeps only the potential removed sets which are optimal according to the strategy, and thus are removed sets. This second part is common to both the weak and strong versions of ΠRSF, and is denoted by $\Pi^P(E,IC)$. In short, we have $PL_P^1(E,IC) = PL^1(E,IC) \cup \Pi^P(E,IC)$ and $PL_P^2(E,IC) = PL^2(E,IC) \cup \Pi^P(E,IC)$.

4.1 Computing strong potential removed sets

The implementation of this computation is largely inspired by the previous work on the computation of removed sets in the case of propositional belief bases [Hué et al., 2008]. Nevertheless, there are two major differences. First of all, in the propositional case, for each propositional variable a appearing in the belief profile, two rules $a \leftarrow not \neg a$ and $\neg a \leftarrow not\ a$ are added. The goal of these rules is to implement the closed world assumption. In the present case, the profile contains logic programs under answer set semantics, so this assumption is not valid, since an answer set can contain a, or $\neg a$, or none of the two literals. Moreover, in the propositional case, the literals present in a model must satisfy the formulae, while in the ASP framework, the literals present in an answer set must satisfy the program rules, but must also be justified regarding the rules. The construction of $PL^1(E,IC)$ is performed by the following steps:

1. Generate all the subsets of literals which can potentially lead to an answer set of $E \cup IC$.
2. Detect the rules which must be discarded.
3. Verify the integrity constraints.
4. Justify the presence of a literal in an answer set.

The first step generates all the subsets of literals which can lead to an answer set of $E \cup IC$. We recall that $Lits(E \cup IC)$ is the set of literals appearing in

$E \cup IC$. All these literals are added to the set of literals of $PL^1(E, IC)$, denoted by $Lits(PL^1(E, IC))$, and we add the following choice rule:

$$\{a_1, \neg a_1, \ldots, a_{|Atoms(E \cup IC)|}, \neg a_{|Atoms(E \cup IC)|}\}. \tag{2}$$

This rule states that any atom of the program or its strong negation can appear, or not, in an answer set. Depending on the solver and its handling of strong negation, it may be necessary to add for any atom $a \in Atoms(E \cup IC)$: $\leftarrow a, \neg a$.

The second step introduces in $Lits(E \cup IC)$, for each rule r of the profile E, an atom denoted by rm_r whose occurrence in an answer set corresponds to the presence of the corresponding rule in a strong potential removed set. This set of atoms is defined by $R^+ = \{rm_r \mid r \in \bigcup E\}$ and we call such atoms *rule atoms*. For all $S \subseteq Lits(PL^1(E, IC))$, we define $Rule(S) = \{r \mid r \in \bigcup E, rm_r \in S\}$. Then, a rule which does not appear in a strong potential removed set must be withdrawn if it is not satisfied by an answer set S. More formally, it means that $body^+(r) \subseteq S$, $body^-(r) \cap S = \emptyset$ and $head(r) \notin S$. Thus, for each atom $rm_r \in R^+$, the following rule is added:

$$rm_r \leftarrow not\ head(r), body(r). \tag{3}$$

The third step allows us to reject all subsets of literals which do not satisfy the integrity constraints in IC, and consequently cannot lead to an answer set. For each rule $r \in IC$, the following rule is introduced:

$$\bot \leftarrow not\ head(r), body(r) \tag{4}$$

The fourth step justifies the presence of a literal in an answer set. Every literal in an answer set must be either a fact, or inferred from the rules. For each $l \in Lits(E \cup IC)$, the atoms $auth(l)$ and $auth(\neg l)$ are added to $Lits(PL^1(E, IC))$. These atoms specify that l can be inferred. A literal can be inferred by a rule if and only if the body of the rule is satisfied. Thus, for each rule $r : c \leftarrow a_1, \ldots, a_n, not\ b_1, \ldots, not\ b_m$. the following rule is introduced:

$$auth(c) \leftarrow not\ rm_r, auth(a_1), \ldots, auth(a_n), \\ not\ auth(b_1), \ldots, not\ auth(b_m). \tag{5}$$

Next, we specify that a literal cannot be present in an answer set if it is not justified by adding the following rules for each $l \in Lits(E \cup IC)$:

$$\bot \leftarrow l, not\ auth(l). \tag{6}$$

Then, the correspondence between the answer sets of $PL^1(E, IC)$ and the strong potential removed sets of E is given by the following result.

Proposition 3. *S is an answer set of $PL^1(E, IC)$ if and only if $Rules(S)$ is a strong potential removed set of E.*

Example 6. We consider example 1 and detail its encoding.

Step 1. Generation of the subsets of literals which can lead to an answer set: $\{a,b,c,d,\neg d,e,f\}$. Depending on the solver, it may be necessary to add the rule $\bot \leftarrow d, \neg d$. Note that we only consider the strong negation of d. Indeed, the strong negation of the other atoms cannot appear in any answer set, because these negated atoms do not even belong to the vocabulary of the profile, and the strong negation in extended logic programs does not rely on closed world assumption.

Step 2. We then define the set $R^+ = \{rm_1, rm_2, rm_3, rm_4, rm_5, rm_6, rm_7, rm_8\}$, and we define the rules of type (3):

$$rm_1 \leftarrow not\ a, not\ b. \quad rm_2 \leftarrow not\ c, not\ a. \quad rm_3 \leftarrow not\ e.$$
$$rm_4 \leftarrow not\ d, f. \quad rm_5 \leftarrow not\ b, not\ c. \quad rm_6 \leftarrow not\ \neg d.$$
$$rm_7 \leftarrow not\ d, e. \quad\quad rm_8 \leftarrow not\ f.$$

Step 3. No step 3 in this case, as there are no constraints.

Step 4. Generation of the type (5) rules:

$$auth(a) \leftarrow not\ rm_1, not\ auth(b). \quad auth(c) \leftarrow not\ rm_2, no\ auth(a).$$
$$auth(e) \leftarrow not\ rm_3. \quad\quad auth(d) \leftarrow not\ rm_4, auth(f).$$
$$auth(b) \leftarrow not\ rm_5, not\ auth(c). \quad auth(\neg d) \leftarrow not\ rm_6.$$
$$auth(d) \leftarrow not\ rm_7, auth(e). \quad\quad auth(f) \leftarrow not\ rm_8.$$

Finally, we add the rules of type (6):

$$\bot \leftarrow a, not\ auth(a). \quad \bot \leftarrow b, not\ auth(b). \quad \bot \leftarrow c, not\ auth(c).$$
$$\bot \leftarrow d, not\ auth(d). \quad \bot \leftarrow \neg d, not\ auth(\neg d). \quad \bot \leftarrow e, not\ auth(e).$$
$$\bot \leftarrow f, not\ auth(f).$$

There are 44 answer sets for this program, which correspond to the 44 strong potential removed sets. For 15 of them, their intersection with R^+ is minimal according to set inclusion. These 15 answer sets are

$\{rm_5, rm_4, rm_7, a, e, \neg d, f\}$ $\{rm_1, rm_7, rm_4, e, c, \neg d, f\}$ $\{rm_1, rm_7, rm_8, e, \neg d, c\}$
$\{rm_1, rm_3, rm_4, c, dp, f\}$ $\{rm_1, rm_6, e, d, c, f\}$ $\{rm_2, rm_7, rm_4, e, f, \neg d, b\}$
$\{rm_1, rm_8, rm_3, c, \neg d\}$ $\{rm_8, rm_2, rm_3, \neg d, b\}$ $\{rm_5, rm_8, rm_3, a, \neg d\}$
$\{rm_6, rm_2, e, d, f, b\}$ $\{rm_8, rm_2, rm_7, e, \neg d, b\}$ $\{rm_8, rm_7, rm_5, a, \neg d, e\}$
$\{rm_3, rm_4, rm_5, a, \neg d, f\}$ $\{rm_2, rm_3, rm_4, f, \neg d, b\}$ $\{rm_6, rm_5, a, d, e, f\}$.

They correspond to the 15 strong potential removed sets which are minimal according to set inclusion, as presented in example 1.

4.2 Computing weak potential removed sets

In this section, we describe the construction of the program $PL^2(E, IC)$ which computes the weak potential removed sets of E. Intuitively, $PL^2(E, IC)$ tries to compute the SE-models of $E \cup IC$, while identifying the rules which prevent an SE-interpretation to be a SE-model. The construction of the program is performed in three steps:

1. generate all the subsets of atoms which can potentially lead to a SE-model of $E \cup IC$.
2. detect the rules which must be removed.
3. verify the integrity constraints.

In order to avoid a conflicting use of the strong negation in the original profile and in $PL^2(E,IC)$, we first translate the strong negation into another notation. Let $Atoms(E \cup IC)$ be the set of atoms appearing in $E \cup IC$ and let $Atoms'(E \cup IC)$ be the set of atoms a' representing $\neg a$ with $a \in Atoms(E \cup IC)$ and $\neg a$ appearing in a rule of $E \cup IC$. The first step generates all the subsets of atoms which can potentially lead to a SE-model. In order to do this, we introduce in the set of atoms of $PL^2(E,IC)$, denoted by $Atoms(PL^2(E,IC))$, new atoms $\{a_S, a_E, \tilde{a}_S, \tilde{a}_E \mid a \in Atoms(E \cup IC) \cup Atoms'(E \cup IC)\}$, where $\tilde{a}$ represents $not\ a$ and for all $a \in Atoms(E \cup IC) \cup Atoms'(E \cup IC)$, the four following rules are introduced:

$$a_S \leftarrow not\ \tilde{a}_S. \quad \tilde{a}_S \leftarrow not\ a_S. \quad a_E \leftarrow not\ \tilde{a}_E. \quad \tilde{a}_E \leftarrow not\ a_E. \tag{7}$$

The intuition behind a_S and a_E atoms is that they represent the S part and the E part of a SE-model (S,E). In order to avoid the presence of an inconsistent subset of literals in a SE-model, we introduce the following rules which forbid an atom and its strong negation to be present at the same time in an interpretation:

$$\bot \leftarrow a_S, a'_S. \quad \bot \leftarrow a_E, a'_E. \tag{8}$$

The following rules allows us for discarding the tuples (X,Y) of sets of atoms for which an atom a appears in X but not in Y (see definition of a SE-model in section 2.3.2).

$$\bot \leftarrow a_S, \tilde{a}_E. \tag{9}$$

For each rule $r \in \bigcup E$, step 2 introduces in $Atoms(PL^2(E,IC))$ an atom rm_r whose presence in an answer set of $PL^2(E,IC)$ reflects the presence of the corresponding rule in a weak potential removed set of E. This set of new atoms is denoted by $R^+ = \{rm_r \mid r \in \bigcup E\}$ and for all $S \subseteq Atoms(PL^2(E,IC))$, we define $Rules(S) = \{r \mid r \in \bigcup E, rm_r \in S\}$. Moreover, for any rule $r \in \bigcup E$, r_S (r_E) denotes the rule obtained by replacing each atom a appearing in r by a_S (a_E), respectively. The symbol $\tilde{r}$ denotes the rule r after the substitution of each atom a_S (a_E) by $\tilde{a}_S$ ($\tilde{a}_E$), respectively. For each rule $r \in \bigcup E$, the following rules are then introduced:

$$\begin{aligned} rm_r &\leftarrow head(\tilde{r}_E), body^+(r_E), body^-(\tilde{r}_E).\\ rm_r &\leftarrow head(\tilde{r}_S), body^+(r_S), body^-(\tilde{r}_E). \end{aligned} \tag{10}$$

The first rule allows us to discard any rule r which prevents a SE-interpretation (X,Y) from being a SE-model because this rule allows $Y \not\models r$, and thus $Y \not\models E \cup IC$. The second rule allows us to discard any rule r which prevents a SE-interpretation

Extending belief bases change to logic programs with ASP

(X,Y) from being a SE-model because this rule allows $X \not\models r^Y$, and thus $X \not\models (E \cup IC)^Y$.

The same notations are used in step three which discards SE-interpretations that violate the integrity constraints. For each rule $r \in IC$, the following rules are introduced:

$$\bot \leftarrow head(\tilde{r}_E), body^+(r_E), body^-(\tilde{r}_E).$$
$$\bot \leftarrow head(\tilde{r}_S), body^+(r_S), body^-(\tilde{r}_E).$$

This construction is inspired by the one appearing in [Delgrande *et al.*, 2009], but it differs from it in several aspects. In our case, the goal is not to compare SE-models, but to identify the rules which prevents an SE-interpretation to become a SE-model. The common point between the two approaches lies in the encoding, allowing for the generation of SE-models. Although our approach is restricted to extended programs, where the merging operators developed in [Delgrande *et al.*, 2009] can handle profiles made up of general programs (that is, programs accepting disjunction of literals and default negation in the head of the rules [Inoue and Sakama, 1998]). We establish the following result:

Proposition 4. *S is an answer set of $PL^2(E,IC)$ if and only if $Rules(S)$ is a weak potential removed set of E.*

Example 7. We present the encoding of example 1, and the results of the program. First we instantiate the rules of type (7):

$$a_S \leftarrow not\ \tilde{a}_S.\ \ \tilde{a}_S \leftarrow not\ a_S.\ \ a_E \leftarrow not\ \tilde{a}_E.\ \ \tilde{a}_E \leftarrow not\ a_E.$$
$$b_S \leftarrow not\ \tilde{b}_S.\ \ \tilde{b}_S \leftarrow not\ b_S.\ \ b_E \leftarrow not\ \tilde{b}_E.\ \ \tilde{b}_E \leftarrow not\ b_E.$$
$$c_S \leftarrow not\ \tilde{c}_S.\ \ \tilde{c}_S \leftarrow not\ c_S.\ \ c_E \leftarrow not\ \tilde{c}_E.\ \ \tilde{c}_E \leftarrow not\ c_E.$$
$$d_S \leftarrow not\ \tilde{d}_S.\ \ \tilde{d}_S \leftarrow not\ d_S.\ \ d_E \leftarrow not\ \tilde{d}_E.\ \ \tilde{d}_E \leftarrow not\ d_E.$$
$$d'_S \leftarrow not\ \tilde{d}'_S.\ \ \tilde{d}'_S \leftarrow not\ d'_S.\ \ d'_E \leftarrow not\ \tilde{d}'_E.\ \ \tilde{d}'_E \leftarrow not\ d'_E.$$
$$e_S \leftarrow not\ \tilde{e}_S.\ \ \tilde{e}_S \leftarrow not\ e_S.\ \ b_E \leftarrow not\ \tilde{e}_E.\ \ \tilde{e}_E \leftarrow not\ e_E.$$
$$f_S \leftarrow not\ \tilde{f}_S.\ \ \tilde{f}_S \leftarrow not\ f_S.\ \ f_E \leftarrow not\ \tilde{f}_E.\ \ \tilde{f}_E \leftarrow not\ f_E.$$

Next, we introduce the exclusion rules (8) for the strong negation of d:

$$\bot \leftarrow d_E, d'_E.\ \ \bot \leftarrow d_S, d'_S.$$

Then we add the rules of type (9):

$$\leftarrow a_S, \tilde{a}_E.\ \leftarrow b_S, \tilde{b}_E.\ \leftarrow c_S, \tilde{c}_E.\ \leftarrow d_S, \tilde{d}_E.\ \leftarrow d'_S, \tilde{d}'_E.\ \leftarrow e_S, \tilde{e}_E.\ \leftarrow f_S, \tilde{f}_E.$$

Finally we instantiate the rules of type (10):

$$rm_1 \leftarrow \tilde{a}_E, \tilde{b}_E.\ rm_1 \leftarrow \tilde{a}_S, \tilde{b}_E.\ rm_2 \leftarrow \tilde{c}_E, \tilde{a}_E.\ rm_2 \leftarrow \tilde{c}_S, \tilde{a}_E.$$
$$rm_3 \leftarrow \tilde{e}_E.\ \ \ \ \ rm_3 \leftarrow \tilde{e}_S.\ \ \ \ \ rm_4 \leftarrow \tilde{d}_E, f_E.\ rm_4 \leftarrow \tilde{d}_S, f_S.$$
$$rm_5 \leftarrow \tilde{b}_E, \tilde{c}_E.\ rm_5 \leftarrow \tilde{b}_S, \tilde{c}_E.\ rm_6 \leftarrow \tilde{d}'_E.\ \ \ \ \ rm_6 \leftarrow \tilde{d}'_S.$$
$$rm_7 \leftarrow \tilde{d}_E, e_E.\ rm_7 \leftarrow \tilde{d}_S, e_S.\ \ rm_8 \leftarrow \tilde{f}_E.\ \ \ \ \ rm_8 \leftarrow \tilde{f}_S.$$

The program has 1215 answer sets. For five of them their intersection with R^+ is minimal according to set inclusion:

$$S_1 = \{rm_7, rm_8, c_E, \tilde{d}_S, \tilde{a}_S, \tilde{b}_S, b_E, d'_S, \tilde{f}_E, e_E, e_S, d'_E, c_S, \tilde{a}_E, \tilde{d}_E, \tilde{f}_S\}$$
$$S_2 = \{rm_4, rm_3, c_E, \tilde{d}_S, \tilde{a}_S, \tilde{b}_S, f_E, d'_S, b_E, \tilde{e}_E, \tilde{e}_S, d'_E, c_S, \tilde{a}_E, \tilde{d}_E, f_S\}$$
$$S_3 = \{rm_6, c_E, b_E, \tilde{a}_S, d_E, f_E, \tilde{a}_E, e_E, \tilde{b}_S, \tilde{d}'_E, \tilde{d}'_S, c_S, d_S, e_S, f_S\}$$
$$S_4 = \{rm_3, rm_8, \tilde{d}_S, \tilde{a}_S, \tilde{b}_S, b_E, d'_S, \tilde{f}_E, \tilde{e}_E, \tilde{e}_S, d'_E, c_S, \tilde{a}_E, \tilde{d}_E, \tilde{f}_S\}$$
$$S_5 = \{rm_4, rm_7, \tilde{d}_S, \tilde{a}_S, \tilde{b}_S, f_E, d'_S, b_E, e_E, e_S, d'_E, c_S, \tilde{a}_E, \tilde{d}_E, f_S\}$$

Each of these answer sets corresponds to a weak potential removed set which is minimal according to set inclusion:

S_1 corresponds to $\{d \leftarrow e. \;\; f.\}$	S_2 corresponds to $\{e. \;\; d \leftarrow f.\}$
S_3 corresponds to $\{\neg d.\}$	S_4 corresponds to $\{e. \;\; f.\}$
S_5 corresponds to $\{d \leftarrow f. \;\; d \leftarrow e.\}$	

4.3 Encoding the strategy

We now explain how to build the set of rules $\Pi^P(E, IC)$ which encodes the chosen merging strategy P. This set is very similar to the one used in [Hué et al., 2008]. Note that this encoding is the same for the weak and strong version of the merging operators.

4.3.1 Σ strategy

The set of removed sets according to the Σ strategy is the set of potential removed sets of minimal cardinality, i.e., those containing less rules. Thus, the set of preferred answer sets of $PL^1(E, IC)$ according to the Σ strategy will be the set of answer sets of $PL^1(E, IC)$ containing the minimum number of rule atoms. So, all $\Pi^\Sigma(E, IC)$ only contains is a counting mechanism and a minimization statement which will only keep only the answer sets with a minimal value for the counter. This is achieved with a simple *minimize* statement: $\Pi^\Sigma(E, IC) = \{\#minimize \{rm_r^i \mid rm_r^i \in R^+\}\}$.

Proposition 5. *Let $E = \{\Pi_1, \ldots, \Pi_n\}$ be a belief profile and IC a set of integrity constraints. S is an answer set of $PL^1_\Sigma(E, IC) = PL^1(E, IC) \cup \Pi^\Sigma(E, IC)$ ($PL^2_\Sigma(E, IC) = PL^2(E, IC) \cup \Pi^\Sigma(E, IC)$) if and only if Rules(S) is a strong (weak) removed set of E according to Σ, respectively.*

Example 8. Going back to examples 6 and 7, the program encoding the merging strategy Σ in this case is defined by $\Pi^\Sigma(E, IC) = \{\#minimize\{ rm_1, rm_2, rm_3, rm_4, rm_5, rm_6, rm_7, rm_8 \}\}$.

The program $PL^1(E, IC) \cup \Pi^\Sigma(E, IC)$ has the following preferred answer sets:

$$\{a,d,e,f,rm_5,rm_6,auth(e),auth(f),auth(a),auth(d)\},$$
$$\{b,d,e,f,rm_2,rm_6,auth(e),auth(f),auth(b),auth(d)\},$$
$$\{c,d,e,f,rm_1,rm_6,auth(e),auth(f),auth(c),auth(d)\}$$

Each of these answer sets corresponds to a removed set. S_1 corresponds to $\{b \leftarrow not\ c. \neg d.\}$. S_2 corresponds to $\{c \leftarrow not\ a. \neg d.\}$. S_3 corresponds to $\{a \leftarrow not\ b. \neg d.\}$.

The program $PL^2(E,IC) \cup \Pi^\Sigma(E,IC)$ has 14 preferred answer sets, but they all contain the same R^+ atom, namely rm_6, which corresponds to a single answer set $\{\neg d.\}$.

4.3.2 Card strategy

The *Card* strategy is very similar to the Σ strategy. The difference lies in the fact that the *Card* strategy considers rules occuring several times as if they occur only once. The difference in the implementation is achieved by the mean of a preprocessing where the rules occuring more than once are deleted from the profile. We denote by $s(E)$ the belief profile E where the rules occuring several times are reduced to a singleton. We then define $\Pi^{Card}(E,IC) = \{\#minimize\ \{rm_r^i \mid rm_r^i \in R^+\}\}$.

Proposition 6. *Let $E = \{\Pi_1,\ldots,\Pi_n\}$ be a belief profile and IC a set of integrity constraints. S is an answer set of $PL^1_{Card}(E,IC) = PL^1(E,IC) \cup \Pi^{Card}(E,IC)$ ($PL^2_{Card}(E,IC) = PL^2(E,IC) \cup \Pi^{Card}(E,IC)$) iff Rules(S) is a strong (weak) removed set of E according to the Card strategy, respectively.*

Example 9. If we consider the previous example, the results are obviously the same as those for the Σ strategy, as there is no duplicate rules in the profile.

4.3.3 Max strategy

In the case of *Max* strategy, the set of preferred answer sets of $PL^1(E,IC)$ is the set of answer sets which minimize the number of rules to remove in the base where the greatest number of rules has to be removed. The practical implementation in ASP requires several steps, which we sketch quickly before explaining them in detail. First of all, we need to count the number of rule atoms present in a model for each logic program in the profile. This is done by the mean of a predicate $size(U)$. Then the greatest value of $size(U)$ is computed by the mean of a predicate $max(X)$. Finally, the smallest value of $max(X)$ is provided by a *#minimize* statement.

As said above, we first need to compute, for each answer set X, the value of $max_{1 \leq i \leq n} |\{rm_r^i \mid rm_r^i \in X\}|$. We describe this computation using two sets of rules. The first one computes the number of rules removed in each logic program for the given interpretation, namely $|\{rm_r^i \mid rm_r^i \in X\}|$:

$$\Pi^{Max,size}(E,IC) = \left\{ \begin{array}{l} \delta_0 : \#domain\ possible(U). \\ \delta_1 : possible(1..m). \\ \alpha_{i \in \{1...n\}} : size(U) \leftarrow U\{rm_r^i \mid r \in \Pi_i\}U. \end{array} \right.$$

where m is the number of rules of the largest logic program in the profile, and n is the number of logic programs in the belief profile. The meaning of the α_i rules is that if there exists a logic program Π_i in the profile for which the number of rule atoms rm_r^i true in X is equal to U, then $size(U)$ will be true in X.

The second set of rules allows for the computation of the greatest value of $size(U)$:

$$\Pi^{Max,bound}(E,IC) = \left\{ \begin{array}{l} \delta_2 \,:\, \#domain\ possible(W). \\ \beta_1 \,:\, negmax(W) \leftarrow size(U), U > W. \\ \beta_2 \,:\, max(U) \leftarrow size(U), not\ negmax(U). \end{array} \right\}$$

The rule β_1 determines all natural numbers W for which there exists $U > W$ such that $size(U)$ is true. Then, $max(U)$ is true for the greatest value of U such that $size(U)$ is true. Finally, in order to select the smallest value of $max(U)$, we add the following optimization statement:

$$\#minimize\,[max(1) = 1,\ldots,max(m) = m].$$

In order to sum up all the preceding ideas, the part of the program which handles the Max strategy is defined by

$$\Pi^{Max}(E,IC) = \Pi^{Max,size}(E,IC) \cup \Pi^{Max,bound}(E,IC)$$
$$\cup\,\{minimize\,[max(1) = 1,\ldots,max(m) = m]\},$$

and the full program whose answer sets are the preferred answer sets of $PL^1(E,IC)$ ($PL^2(E,IC)$) for the Max strategy is defined by $PL^1_{Max}(E,IC) = PL^1(E,IC) \cup \Pi^{Max}(E,IC)$ ($PL^2_{Max}(E,IC) = PL^2(E,IC) \cup \Pi^{Max}(E,IC)$) respectively.

Proposition 7. *Let $E = \{\Pi_1,\ldots,\Pi_n\}$ be a belief profile, IC be a logic program representing integrity constraints and S be set of atoms. S is an answer set of $PL^1_{Max}(E,IC) = PL^1(E,IC) \cup \Pi^{Max}(E,IC)$ ($PL^2_{Max}(E,IC) = PL^2(E,IC) \cup \Pi^{Max}(E,IC)$) such that $S \cap R^+$ is minimal according to set inclusion if and only if $Rules(S)$ is a strong (weak) removed set of E according to the Max strategy, respectively.*

Example 10. Going back to examples 6 and 7, the program encoding the merging strategy Max in this case is defined by

$$\Pi^{Card}(E,IC) = \left\{ \begin{array}{l} \#domain\ possible(U).\ \#domain\ possible(W).\ possible(1..4). \\ size(U) \leftarrow U\{rm_1,rm_2,rm_3,rm_4\}U. \\ size(U) \leftarrow U\{rm_5,rm_6,rm_7,rm_8\}U. \\ negmax(W) \leftarrow size(U), U > W. \\ max(U) \leftarrow size(U), not\ negmax(U). \\ \#minimize\,\{max(1) = 1, max(2) = 2, max(3) = 3, max(4) = 4\} \end{array} \right\}.$$

The program $PL^1(E,IC) \cup \Pi^{Max}(E,IC)$ has the following preferred answer sets:

$\{b,d,e,f,rm_2,rm_6,auth(e),auth(f),auth(b),auth(d)\}$,
$\{c,d,e,f,rm_1,rm_6,auth(e),auth(f),auth(c),auth(d)\}$

Each of these sets corresponds to a removed set. S_1 corresponds to the removed set $\{c \leftarrow not\ a.\ \neg d.\}$ and S_2 corresponds to the removed set $\{a \leftarrow not\ b.\ \neg d.\}$. The program $PL^2(E,IC) \cup \Pi^\Sigma(E,IC)$ has 76 preferred answer sets, but only three of them have their intersection with R^+ which is minimal according to set inclusion: $\{rm_6\}$, $\{rm_4,rm_7\}$, $\{rm_3,rm_8\}$, which corresponds to the answer sets $\{\neg d.\}$, $\{d \leftarrow e.d \leftarrow f\}$, $\{e.f.\}$.

4.3.4 GMax strategy

In order to implement the *GMax* strategy, we need to count rule atoms for each program. Then, we sort these sequences of numbers in decreasing order. A preferred answer set will be an answer set for which such a sequence is minimal according to the lexicographical order.

The *GMax* strategy compares the potential removed sets using the sequence of the $|X \cap R_i^+|$ values sorted in decreasing order. As for *Max*, we must know how many rules are removed from each logic program. This computation is represented by the following set of rules, where m represents the size of the logic program Π_V:

$$\Pi_{E,IC}^{GMax,size} = \left\{ \begin{array}{l} \delta_0 : \#domain\ possible(U). \\ \delta_1 : \#domain\ base(V). \\ \delta_2 : possible(1..m). \\ \delta_3 : base(1..n). \\ \gamma_1 : size(V,U) \leftarrow U\ \{rm_1^V,\ldots,rm_m^V\}\ U. \end{array} \right\}$$

If $size(V,U)$ is true, this means that in the logic program Π_V, the current answer set contains U rule atoms. Once these values are computed for all the logic programs Π_V, they are sorted by the rules $\Pi_E^{GMax,bound}$, where X_1 is the largest value of $size()$, and X_n the smallest one:

$$\Pi_{E,IC}^{GMax,bound} = \left\{ \begin{array}{l} \alpha_0 : max(X_1,\ldots,X_n) \leftarrow size(X_1,Y_1),\ldots,size(X_n,Y_n), \\ \quad X_1 \geq X_2,\ldots,X_{n-1} \geq X_n, neq(Y_1,\ldots,Y_n). \\ \alpha_1 : max_1(X_1) \leftarrow max(X_1,\ldots,X_n), \\ \quad X_1 \geq X_2,\ldots,X_{n-1} \geq X_n. \\ \alpha_i : \ldots \\ \alpha_n : max_n(X_n) \leftarrow max(X_1,\ldots,X_n), \\ \quad X_1 \geq X_2,\ldots,X_{n-1} \geq X_n. \end{array} \right\}$$

Finally, let X and X' be two sets of literals such that the sorted tuple of values $S_i = |X \cap R_i^+|$ is $(S_1,\ldots,S_n)$ and the sorted tuple of values $S_i' = |X' \cap R_i^+|$ is $(S_1',\ldots,S_n')$. Using ASP, the only way to select some answer sets among a set of answer sets is to translate the preference into a numerical value where the set of desired solutions will be the set of solutions whose numerical value is minimal (or maximal). Namely, we must translate $(S_1,\ldots,S_n)$ into an integer $Val(S_1,\ldots,S_n)$ such that $Val(S_1,\ldots,S_n) \leq Val(S_1',\ldots,S_n')$ if and only if $(S_1,\ldots,S_n) \leq_{lex} (S_1',\ldots,S_n')$.

In order to achieve this goal, we build a *minimize* statement on max_i predicates which are weighted in order to optimize the polynomial $S_n + S_{n-1} \times (m+1) + \ldots + S_1 \times (m+1)^{n-1}$. Indeed, each value of a tuple $(S_1, \ldots, S_n)$ can range from 1 to m, so we can represent this tuple by an integer with radix $m+1$.

$$\Pi_{E,IC}^{GMax} = \Pi_{E,IC} \cup \Pi_{E,IC}^{GMax,size} \cup \Pi_{E,IC}^{GMax,bound}$$
$$\cup minimize \begin{bmatrix} max_n(1) = 1, & \ldots, & max_n(m) = m, \\ \vdots & & \vdots \\ max_i(1) = 1 \times (m+1)^{n-i}, & \ldots & max_i(m) = m \times (m+1)^{n-i}, \\ \vdots & & \vdots \\ max_1(1) = 1 \times (m+1)^{n-1}, & \ldots & max_1(m) = m \times (m+1)^{n-1} \end{bmatrix}$$

The following proposition is valid:

Proposition 8. *Let $E = \{K_1, \ldots, K_n\}$ be a belief profile. The set of answer sets of $PL_{GMax}^1(E,IC) = PL^1(E,IC) \cup \Pi^{GMax}(E,IC)$ ($PL_{GMax}^2(E,IC) = PL^2(E,IC) \cup \Pi^{GMax}(E,IC)$) is the set of preferred answer sets of $PL^1(E,IC)$ ($PL^2(E,IC)$) according to the GMax strategy, respectively.*

Example 11. Example 6 provide the same results as for the *Max* strategy. However, for example 7 the program $PL^2(E,IC) \cup \Pi^{GMax}(E,IC)$ has 14 preferred answer sets, among which 13 have a minimal intersection with R^+ according to set inclusion. These 13 answer sets have the same intersection with R^+, namely $\{rm_6\}$, wihich corresponds to the weak removed set $\{\neg d\}$.

As a final remark about this implementation, it is worth noticing that the weak and strong approaches for the computation of removed sets have a similar computationnal complexity. Indeed, the encoding of the strategies is exactly the same for both operators, and concerning the encoding of the profile, the sizes of $PL^1(E,IC)$ and $PL^2(E,IC)$ are both polynomial in the number of rules in $E \cup IC$.

5 Related works

As mentioned in the introduction, previous works on changing logic programs focused on updating knowledge bases represented by logic programs [Przymusinski and Turner, 1997, Zhang and Foo, 1998, Alferes et al., 2000, Sakama and Inoue, 1999, Eiter et al., 2002, Leite, 2003, Delgrande et al., 2007].

Changing logic programs in the same spirit as changing beliefs sets in a propositional setting has first been addressed in [Eiter et al., 2002]. Belief sets for logic programs are defined as sets of rules satisfied by the interpretations corresponding to answer sets. However the proposed approach violates most of the AGM postulates for revision and update postulates. In [Delgrande et al., 2008, Delgrande et al., 2009], belief sets are then represented by SE-models and change operations (revision and merging) based on distance between interpretations have then been extended to logic programs with SE-model semantics. The proposed change operations stemming from a distance between SE-models have been implemented in ASP. Belief sets update has also been addressed in the same spirit by Slota and Leite [Slota and Leite, 2010].

A first extension of Belief Base Revision to logic programs has been proposed in [Kudo and Murai, 2004], a belief base for logic programs is defined as its set of rules. A method and an algorithm was provided stemming from state transition diagrams, but no formal results were given. More recently, Krümpelmann and Kern-Isberner addressed the problem of revising logic programs [Krümpelmann and Kern-Isberner, 2012] extending Belief Base Revision to logic programs. They first show that the direct transfer of the construction of belief base operators via Levi identity to logic programs does not work and they instead propose the construction of an ASP base revision. The base revision of an extended logic program P by an extended logic program Q involves removing rules from $P \cup Q$ such that the resulting program is consistent with Q. They follow the same line as ΠRSR, however concrete ASP Base revision operators stem from the extension of partial meet revision to logic programs while ΠRSF operators stem from the extension of removed sets approach to logic programs. Note that their extension deals with a definition of consistency stemming from the existence of answer sets as the first family of ΠRSF operators, while a second family of ΠRSF operators is proposed dealing with a definition of consistency based on SE-models.

More precisely, Krümpelmann and Kern-Isberner consider screened consolidation operator [Makinson, 1997], denoted by $!_R$, where R is a set of rules immune to change, defined from consolidation operator for semi-revision [Hansson, 1997]. Therefore the revision operator is defined by $P \star Q = (P \cup Q)!_Q$. In order to construct screened consolidation suitable for logic programs, using partial meet construction they extend the definition of remainder sets to logic programs. More formally, let R and P be sets of rules with $R \subseteq P$, the set of *screened remainder sets*, denoted by $P \perp_! R$, is such that for each $X \in P \perp_! R$: i) $R \subseteq M \subseteq P$, ii) M is consistent, iii) there is no M' such that $M \subset M' \subseteq P$ and M' is consistent. The links between remainder sets and potential removed sets is as follows. Let P be a logic program, a set of rules $M \subseteq P$ is a screened remainder set of $(P \cup Q)!Q$, if and only if $(P \cup Q)\backslash M$ is a potential removed set minimal w.r.t. set inclusion.

Due to the fact that the intersection of consistent logic programs may be inconsistent, partial meet revision amounts to only select one remainder set. Similarly, but according to a dual point of view, ΠRSF only selects one removed set. A global maxichoice selection function is defined such that if $P \perp_! R \neq \emptyset$ then $\gamma_P(P \perp_! R) = M$, for some $M \in P \perp_! R$, otherwise $\gamma_P(P \perp_! R) = P$. There are as many selection functions as remainder sets and there is no reason to choose one of them. On contrast, ΠRSF only provides one selection function stemming from a cardinality strategy that allows us to perform minimal change since it removes the smallest subsets of rules which intersect with the minimal inconsistent set of rules. Moreover, an implementation stemming from ASP is provided for ΠRSF.

6 Conclusion

This chapter presents a general framework, ΠRSF, for revising and merging belief bases represented by logic programs extending the RSR and RSF frameworks initially proposed for revising and merging propositional belief bases respectively. It provides two families of merging and revision operators for logic programs: Strong Removed Sets Fusion operators steming from answer set semantics, and weak Removed Sets Fusion operators steming from SE-model semantics. We show that the proposed operators satisfy most of the postulates for belief base merging. Moreover, it provides for each family of operators an implementation in ASP.

The extension of belief base change to logic programs change raises the question of the consistency of logic programs. In the first family of operators, consistency is based on the existence of answer sets, while in the second one consistency stems from the existence of SE-models, and allows us for defining change operators that keep more rules and therefore best satisfy the principle of minimum change. This question is related to the definition of equivalence between logic programs. Several definitions have been provided, stemming from the identity of sets of answer sets [Lifschitz et al., 2001], strong equivalence stemming from the identity of sets of SE-models [Turner, 2001], uniform equivalence [Eiter and Fink, 2003], syntactic identity of logic programs. Recently in [Krümpelmann and Kern-Isberner, 2012] new postulates for ASP revision parameterized with these different definitions of equivalence have been proposed. However they do not focus on the differences, in terms of minimal change, between the different definitions of equivalence, which seems to be an interesting question to investigate.

Concerning the belief change postulates, we used a direct translation of postulates provided for belief base merging, where some are not applicable since they refer to minimal inconsistent subsets. Compactness holds for classical logic, but does not hold for logic programs under answer set semantics since a logic program may be consistent but a sub-program can be inconsistent. Moreover, in case of revision of logic programs, the operators $\circ_{\Pi RSR,1}$ and $\circ_{\Pi RSR,2}$ satisfy the following postulates recently proposed for the revision of logic programs in [Krümpelmann and Kern-Isberner, 2012]. Namely, *Success, Inclusion, Vacuity, Consistency, NM-Consistency* and therefore, *Initialisation, Idempotence, Absorption*. An interesting question to investigate is the extension of belief base merging postulates to logic programs that captures the specificity of merging in the context of logic programs.

This chapter extends the RSF framework to logic programs, so it could be interesting to extend other propositional belief base merging operators to logic programs. In [Benferhat et al., 2010] RSR and RSF frameworks have been extended to prioritized belief bases and partially preordered belief bases. In future work, we plan to investigate the extension of ΠRSF to logic programs with preferences and partial preferences.

Finally we proposed an implementation in ASP. Experimentations has to be conducted, particularly with the benchmarks provided by Potassco[4].

Acknowledgements

This work has received support from the french Agence Nationale de la Recherche, ASPIQ project reference ANR-12-BS02-0003.

References

[Alchourrón et al., 1985] Carlos E. Alchourrón, Peter Gärdenfors, and David Makinson. On the logic of theory change : Partial meet contraction and revision functions. *Journal of Symbolic Logic*, 50(2):510–530, 1985.

[Alferes et al., 2000] José Júlio Alferes, João Alexandre Leite, Luís Moniz Pereira, Halina Przymusinska, and Teodor C. Przymusinski. Dynamic updates of non-monotonic knowledge bases. *Journal of Logic Programming*, 45(1-3):43–70, 2000.

[Anger et al., 2002] Christian Anger, Kathrin Konczak, and Thomas Linke. Nomore : Non-monotonic reasoning with logic programs. In Sergio Flesca, Sergio Greco, Nicola Leone, and Giovambattista Ianni, editors, *JELIA*, volume 2424 of *Lecture Notes in Computer Science*, pages 521–524. Springer, 2002.

[Baral et al., 1991] Chitta Baral, Sarit Kraus, Jack Minker, and V. S. Subrahmanian. Combining knowledge bases consisting of first order theories. In Zbigniew W. Ras and Maria Zemankova, editors, *ISMIS*, volume 542 of *Lecture Notes in Computer Science*, pages 92–101. Springer, 1991.

[Baral et al., 1992] Chitta Baral, Sarit Kraus, Jack Minker, and V. S. Subrahmanian. Combining knowledge bases consisting of first-order analysis. *Computational Intelligence*, 8:45–71, 1992.

[Baral, 2003] Chitta Baral. *Knowledge Representation, Reasoning and Declarative Problem Solving*. Cambridge University Press, 2003.

[Benferhat et al., 1993] Salem Benferhat, Claudette Cayrol, Didier Dubois, Jérôme Lang, and Henri Prade. Inconsistency management and prioritized syntax-based entailment. In *IJCAI*, pages 640–647, 1993.

[Benferhat et al., 2010] Salem Benferhat, Jonathan Bennaim, Odile Papini, and Éric Würbel. Answer set programming encoding of prioritized removed sets revision: Application to gis. *Applied Intelligence*, 32:60–87, 2010.

[Bloch and Lang, 2002] Isabelle Bloch and Jérôme Lang. *Towards mathematical morpho-logics*, pages 367–380. Physica-Verlag GmbH, Heidelberg, Germany, Germany, 2002.

[Cholvy, 1998] Laurence Cholvy. Reasoning about merging information. *Handbook of Defeasible Reasoning and Uncertainty Management Systems*, 3:233–263, 1998.

[Cohn et al., 2000] Anthony G. Cohn, Fausto Giunchiglia, and Bart Selman, editors. *Principles of Knowledge Representation and Reasoning : Proceedings of the Seventh International Conference, KR2000, Breckenridg, Colorado, April 12-15, 2000*. Morgan Kaufmann, 2000.

[de Kleer, 1990] Johan de Kleer. Using crude probability estimates to guide diagnosis. *Artificial Intelligence*, 45:381–392, 1990.

[Delgrande et al., 2006] James P. Delgrande, Didier Dubois, and Jerome Lang. Iterated revision as prioritized merging. In *Proceedings of KR*, pages 210–220, 2006.

---

[4] Potassco, the Potsdam Answer Set Solving Collection (http://potassco.sourceforge.net/)

[Delgrande et al., 2007] James P. Delgrande, Torsten Schaub, and Hans Tompits. A preference-based framework for updating logic programs. In *Proceedings of LPNMR*, pages 71–83, 2007.

[Delgrande et al., 2008] James P. Delgrande, Torsten Schaub, Hans Tompits, and Stefan Woltran. Belief revision of logic programs under answer set semantics. In Gerhard Brewka and Jérôme Lang, editors, *KR*, pages 411–421. AAAI Press, 2008.

[Delgrande et al., 2009] James P. Delgrande, Torsten Schaub, Hans Tompits, and Stefan Woltran. Merging logic programs under answer set semantics. In Patricia M. Hill and David Scott Warren, editors, *ICLP*, volume 5649 of *Lecture Notes in Computer Science*, pages 160–174. Springer, 2009.

[Eiter and Fink, 2003] Thomas Eiter and Michael Fink. Uniform equivalence of logic programs under the stable model semantics. In *ICLP*, pages 224–238, 2003.

[Eiter et al., 2002] Thomas Eiter, Michael Fink, Giuliana Sabbatini, and Hans Tompits. On properties of update sequences based on causal rejection. *Theory and Practice of Logic Programming*, 2(6):711–767, 2002.

[Fagin et al., 1986] Ronald Fagin, Gabriel M. Kuper, Jeffrey D. Ullman, and Moshe Y. Vardi. Updating logical databases. *Advances in Computing Research*, 3:1–18, 1986.

[Falappa et al., 2002] M. A. Falappa, G. Kern-Isberner, and G. R. Simari. Explanations, belief revision and defeasible reasoning. *Artif. Intell.*, 141(1/2):1–28, 2002.

[Falappa et al., 2012] M. A. Falappa, G. Kern-Isberner, M. D. L. Reis, and G. R. Simari. Prioritized and non-prioritized multiple change o, belief bases. *Journal of Philosophical Logic*, 41:77–113, 2012.

[Fuhrmann, 1997] A. Fuhrmann. *An essay on contraction.* CSLI Publications, Stanford. California, 1997.

[Gebser et al., 2007] Martin Gebser, Benjamin Kaufmann, Andre Neumann, and Torsten Schaub. Clasp: A conflict-driven answer set solver. In Chitta Baral, Gerhard Brewka, and John S. Schlipf, editors, *LPNMR*, volume 4483 of *Lecture Notes in Computer Science*, pages 260–265. Springer, 2007.

[Gebser et al., 2009] Martin Gebser, Max Ostrowski, Torsten Schaub, and Sven Thiele. On the input langage of ASP grounder gringo. In Esra Erdem, Fangzhen Lin, and Torsten Schaub, editors, *LPNMR*, volume 5753 of *Lecture Notes in Computer Science*, pages 502–508. Springer, 2009.

[Gelfond and Lifschitz, 1988] Michael Gelfond and Vladimir Lifschitz. The stable model semantics for logic programming. In *ICLP/SLP*, pages 1070–1080, 1988.

[Giunchiglia et al., 2004] Enrico Giunchiglia, Yuliya Lierler, and Marco Maratea. Sat-based answer set programming. In Deborah L. McGuinness and George Ferguson, editors, *AAAI*, pages 61–66. AAAI Press / The MIT Press, 2004.

[Hansson, 1994] Sven Ove Hansson. Kernel contraction. *Journal of Symbolic Logic*, 59:845–859, 1994.

[Hansson, 1997] Sven Ove Hansson. Semi-revision (invited paper). *Journal of Applied Non-Classical Logics*, 7(2):151–175, 1997.

[Hansson, 1999] Sven Ove Hansson. *A Textbook of Belief Dynamics. Theory Change and Database Updating.* Dordrecht: Kluwer, 1999.

[Hué et al., 2007] Julien Hué, Odile Papini, and Éric Würbel. Syntactic propositional belief bases fusion with removed sets. In Khaled Mellouli, editor, *ECSQARU*, volume 4724 of *Lecture Notes in Computer Science*, pages 66–77. Springer, 2007.

[Hué et al., 2008] Julien Hué, Odile Papini, and Éric Würbel. Removed sets fusion: Performing off the shelf. In Malik Ghallab, Constantine D. Spyropoulos, Nikos Fakotakis, and Nikolaos M. Avouris, editors, *ECAI*, volume 178 of *Frontiers in Artificial Intelligence and Applications*, pages 94–98. IOS Press, 2008.

[Hué et al., 2009] Julien Hué, Odile Papini, and Éric Würbel. Merging belief bases represented by logic programs. In Claudio Sossai and Gaetano Chemello, editors, *ECSQARU*, volume 5590 of *Lecture Notes in Computer Science*, pages 371–382. Springer, 2009.

[Inoue and Sakama, 1998] Katsumi Inoue and Chiaki Sakama. Negation as failure in the head. *Journal of Logic Programming*, 35(1):39–78, 1998.

[Katsuno and Mendelzon, 1991] Hirofumi Katsuno and Alberto Mendelzon. Propositional Knowledge Base Revision and Minimal Change. *Artificial Intelligence*, 52:263–294, 1991.

[Konieczny and Pino Pérez, 1998] Sébastien Konieczny and Ramón Pino Pérez. On the logic of merging. In *KR*, pages 488–498, 1998.

[Konieczny and Pino Pérez, 1999] Sébastien Konieczny and Ramón Pino Pérez. Merging with integrity constraints. In Anthony Hunter and Simon Parsons, editors, *ECSQARU*, volume 1638 of *Lecture Notes in Computer Science*, pages 233–244. Springer, 1999.

[Konieczny and Pino Pérez, 2002] Sébastien Konieczny and Ramón Pino Pérez. Merging information under constraints: A logical framework. *Journal of Logic and Computation*, 12(5):773–808, 2002.

[Konieczny and Pino Pérez, 2011] Sébastien Konieczny and Ramón Pino Pérez. Logic based merging. *J. Philosophical Logic*, 40(2):239–270, 2011.

[Konieczny et al., 2004] Sébastien Konieczny, Jérôme Lang, and Pierre Marquis. Da2 merging operators. *Artificial Intelligence*, 157:49–79, 2004.

[Konieczny, 2000] Sébastien Konieczny. On the difference between merging knowledge bases and combining them. In Cohn et al, pages 135–144.

[Krümpelmann and Kern-Isberner, 2012] Patrick Krümpelmann and Gabriele Kern-Isberner. Belief base change operations for answer set programming. In *JELIA*, pages 294–306, 2012.

[Kudo and Murai, 2004] Yasuo Kudo and Tetsuya Murai. A method of belief base revision for extended logic programs based on state transition diagrams. In *KES*, pages 1079–1084, 2004.

[Lafage and Lang, 2000] Céline Lafage and Jérôme Lang. Logical representation of preferences for group decision making. In Cohn et al, pages 457–468.

[Lehmann, 1995] Daniel Lehmann. Belief revision revised. In *IJCAI*, pages 1534–1540. Morgan Kaufmann, 1995.

[Leite, 2003] Joan Leite. *Evolving Knowledge Bases*. IOS Press, 2003.

[Leone et al., 2006] Nicola Leone, Gerald Pfeifer, Wolfgang Faber, Thomas Eiter, Georg Gottlob, Simona Perri, and Francesco Scarcello. The DLV system for knowledge representation and reasoning. *ACM Trans. Comput. Log.*, 7:499–562, 2006.

[Lifschitz et al., 2001] Vladimir Lifschitz, David Pearce, and Agustín Valverde. Strongly equivalent logic programs. *ACM Trans. Comput. Log.*, 2(4):526–541, 2001.

[Lifschitz, 2008a] Vladimir Lifschitz. Twelve definitions of a stable model. In *Proceedings of ICLP*, number 5366 in Lecture Notes in Computer Sciences, pages 37–51, 2008.

[Lifschitz, 2008b] Vladimir Lifschitz. What is answer set programming? In Dieter Fox and Carla P. Gomes, editors, *AAAI*, pages 1594–1597. AAAI Press, 2008.

[Lifschitz, 2010] Vladimir Lifschitz. Thirteen definitions of a stable model. In Andreas Blass, Nachum Dershowitz, and Wolfgang Reisig, editors, *Fields of Logic and Computation*, volume 6300 of *Lecture Notes in Computer Science*, pages 488–503. Springer, 2010.

[Lin and Mendelzon, 1998] Jinxin Lin and Alberto Mendelzon. Merging databases under constraints. *International Journal of Cooperative Information Systems*, 7:55–76, 1998.

[Lin and Zhao, 2004] Fangzhen Lin and Yuting Zhao. Assat: computing answer sets of a logic program by sat solvers. *Artificial Intelligence*, 157:115–137, 2004.

[Lin, 1996] Jinxin Lin. Integration of weighted knowledge bases. *Artificial Intelligence*, 83:363–378, 1996.

[Makinson, 1997] David Makinson. Screened revision. *Theoria*, 63(1-2):14–21, 1997.

[Nebel, 1991] Bernhard Nebel. Belief Revision and Default Reasoning : Syntax-based Approach. In *KR*, pages 417–428, 1991.

[Niemelä and Simons, 1997] Ilkka Niemelä and Patrik Simons. Smodels - an implementation of stable model and well-founded semantics for normal logic programs. In Jürgen Dix, Ulrich Furbach, and Anil Nerode, editors, *LPNMR*, volume 1265 of *Lecture Notes in Computer Science*, pages 420–429. Springer, 1997.

[Papini, 1992] Odile Papini. A complete revision function in propositionnal calculus. In B. Neumann, editor, *ECAI*, pages 339–343. John Wiley and Sons. Ltd, 1992.

[Pearce, 1996] David Pearce. A new logical characterisation of stable models and answer sets. In Jürgen Dix, Luís Moniz Pereira, and Teodor C. Przymusinski, editors, *NMELP*, volume 1216 of *Lecture Notes in Computer Science*, pages 57–70. Springer, 1996.

[Pearce, 2006] David Pearce. Equilibrium logic. *Annals of Mathematics and Artificial Intelligence*, 47(1–2):3–41, 2006.

[Przymusinski and Turner, 1997] Teodor C. Przymusinski and Hudson Turner. Update by means of inference rules. *Journal of Logic Programming*, 30(2):125–143, 1997.

[Revesz, 1993] Peter Z. Revesz. On the semantics of theory change: Arbitration between old and new information. In *Proceedings of PODS*, pages 71–82. ACM Press, 1993.

[Revesz, 1995] Peter Z. Revesz. On the semantics of arbitration. *Inernational Journal of Algebra and Computation*, 7:133–160, 1995.

[Sakama and Inoue, 1999] Chiaki Sakama and Katsumi Inoue. Updating extended logic programs through abduction. In Michael Gelfond, Nicola Leone, and Gerald Pfeifer, editors, *LPNMR*, volume 1730 of *Lecture Notes in Computer Science*, pages 147–161. Springer, 1999.

[Schaub, 2008] Torsten Schaub. Here's the beef: Answer set programming ! In Maria Garcia de la Banda and Enrico Pontelli, editors, *ICLP*, volume 5366 of *Lecture Notes in Computer Science*, pages 93–98. Springer, 2008.

[Slota and Leite, 2010] Martin Slota and João Leite. On semantic update operators for answer-set programs. In Helder Coelho, Rudi Studer, and Michael Wooldridge, editors, *ECAI*, volume 215 of *Frontiers in Artificial Intelligence and Applications*, pages 957–962. IOS Press, 2010.

[Syrjanen, 2002] Tommi Syrjanen. Lparse 1.0 user's manual, 2002.

[Turner, 2001] Hudson Turner. Strong equivalence for logic programs and default theories (made easy). In Thomas Eiter, Wolfgang Faber, and Miroslaw Truszczynski, editors, *LPNMR*, volume 2173 of *Lecture Notes in Computer Science*, pages 81–92. Springer, 2001.

[Turner, 2003] Hudson Turner. Strong equivalence made easy: nested expressions and weight constraints. *Theory and Practice of Logic Programming*, 3:609–622, 2003.

[Würbel et al., 2000] Eric Würbel, Robert Jeansoulin, and Odile Papini. Revision : An application in the framework of gis. In Cohn et al, pages 505–518.

[Zhang and Foo, 1998] Yan Zhang and Norman Y. Foo. Updating logic programs. In *ECAI*, pages 403–407, 1998.

7 Proofs

Proof (of proposition 1).

Inclusion For any strategy P, by definition $\Delta_{P,IC}^{\Pi RSF,1}(E) \subseteq \Pi_1 \cup \ldots \cup \Pi_n$.

Symmetry is rephrased as follows. From σ a permutation on indexes, we denote by $\bar{\sigma}$ a permutation on profiles such that let $E = \{\Pi_1, \ldots, \Pi_n\}$ be a profile $\bar{\sigma}(E) = \{\Pi_{\sigma(1)}, \ldots, \Pi_{\sigma(n)}\}$, the symmetry property leads to $\Delta_{P,IC}^{\Pi RSF,1}(E) = \Delta_{P,IC}^{\Pi RSF,1}(\bar{\sigma}(E))$. For any strategy P, the order with which the logic programs are given does not change the result of merging.

Strong Consistency follows from the definition of the ΠRSF merging operator.

Vacuity holds because if $(\Pi_1 \cup \ldots \cup \Pi_n) \cup IC$ is AS-consistent then $\mathscr{R}_P^1(E) = \emptyset$.

Global core-retainment is rephrased as follows. If $r \in R$ with $R \in \mathscr{R}_P^1(E)$ then there is a logic program Π such that $\Pi \subseteq (\Pi_1 \cup \ldots \cup \Pi_n) \cup IC$, Π is AS-consistent but $\Pi \cup \{r\}$ is AS-inconsistent.

If $\mathscr{R}_P^1(E) = \emptyset$ the postulates holds. If $\mathscr{R}_P^1(E) \neq \emptyset$, if $r \in R$ with $R \in \mathscr{R}_P^1(E)$ then $r \in \Pi_1 \cup \ldots \cup \Pi_n$. Let $\Pi = (\Pi_1 \cup \ldots \cup \Pi_n) \backslash R$, Π is AS-consistent by definition of strong removed sets and $\Pi \cup \{r\}$ is AS-inconsistent, because if $\Pi \cup \{r\}$ is AS-consistent, $R \backslash \{r\}$ is a strong potential removed set minimal w. r. t. set inclusion which contadicts (2) in definition 2.

For Σ and *Card*, $\Delta_{P,IC}^{\Pi RSF,1}(E)$ satifies *Congruence* postulate.

Congruence is rephrased as follows. Let $E = \{\Pi_1,\ldots,\Pi_n\}$ and $E' = \{\Pi'1,\ldots,\Pi'n\}$ be two profiles such that $\Pi_1 \cup \ldots \cup \Pi_n = \Pi'_1 \cup \ldots \cup \Pi'_n$ then $\Delta_{P,IC}^{\Pi RSF,1}(E) = \Delta_{P,IC}^{\Pi RSF,1}(E')$.

If a profile is partitionned according to two different ways, the result of merging remains the same. Indeed, the result holds for Σ and *Card* strategies because, let X be a removed set, $\sum_{1 \leq i \leq n} |X \cap \Pi_i| = \sum_{1 \leq i \leq n} |X \cap \Pi'i|$. However, for *Max* and *GMax* strategies the result does not hold because, let X be a removed set, it is not granted that $\max_{1 \leq i \leq n} |X \cap \Pi_i| = \max_{1 \leq i \leq n} |X \cap \Pi'i|$.

Proof (of proposition 2). The proof is similar to the proof of proposition 1.

Proof (of proposition 3). First we define the *counterpart* of an answer set S of $PL^1(E,IC)$ in $E \cup IC$ as $S_{E \cup IC} = \{a \in E \cup IC \mid a \in PL^1(E,IC)\}$

Now we proove two lemmas.

Lemma 1. *let S be an answer set of $PL^1(E,IC)$. A rule $r \in E$ is defeated by $S_{E \cup IC}$ if and only if $r \in Rules(S)$.*

Proof. $\boxed{\Rightarrow}$ Suppose that there exist a rule r which is defeated by $S_{E \cup IC}$ and such that $r \notin Rules(S)$. This means thar $rm_r \notin S$. According to the construction of the rules having rm_r as a consequence, either $head(r)$ is false, or $body(r)$ is false. In both cases, the rule r is not defeated, which contradicts the hypothesis.

$\boxed{\Leftarrow}$ Let us suppose that there exist a rule r which is not defeated by $S_{E \cup IC}$ and such that $r \in Rules(S)$. The latter means that $rm_r \in S$. According to the construction of the rules having rm_r as a consequence, $head(r)$ is false and $body(r)$ is true, but in this case the rule r is defeated, which contradicts the hypothesis.

Lemma 2. *Let S be an answer set of $PL^1(E,IC)$. There is no rule r in $(\bigcup E \setminus Rules(S))^{S_{E \cup IC}}$ such that r is defeated by $S_{E \cup IC}$.*

Proof. Suppose that there exist a rule $r \in (\bigcup E \setminus Rules(S))^{S_{E \cup IC}}$ such that r is defeated $S_{E \cup IC}$. If r is defeated $S_{E \cup IC}$, then by lemma 1 $r \in Rules(S)$. But then $r \notin (\bigcup E \setminus Rules(S))$ and finally $r \notin (\bigcup E \setminus Rules(S))^{S_{E \cup IC}}$, which contradicts the hypothesis.

Proof of the proposition: Added to the translation mechanism described in section 4, rules are added inorder to mimic each rule of E:

$$auth(c) \leftarrow not\ rm_r, auth(a_1),\ldots, auth(a_m),$$
$$not\ auth(b_1),\ldots, not\ auth(b_n).$$

The literals *not rm_r* deactivate the rules removed y the current interpretation, thanks to lemmas 1 and 2. The remaining program is equivalent to $(\bigcup E \setminus Rules(S))$ and for all atoms a, $auth(a)$ represents a. Moreover, rules of the form $\leftarrow a, not\ auth(a)$ guaranty that $S_{E \cup IC}$ is an answer set of $PL^1(E,IC)$. So, $Rules(S)$ is a strong potential removed set of E constrained by IC.

Proof (of proposition 4). $\boxed{\Rightarrow}$ For any set of literals S, we denote by $Rmr(S)$ the set of atoms rm_r which are in S, that is, $Rmr(S) = S \cap R^+$. For all subset $Y \subseteq Lits(E)$ and for all $x \in \{S, E\}$, we denote by Y_x the set $\{y_x \mid y \in Y\}$. For all $I \subseteq Lits(PL^2(E, IC))$, let $\sigma(I) = \{(X, Y) \mid X, Y \subseteq Atoms(E), X_S = I \cap Atoms(E)_S, Y_E = I \cap Atoms(E)_E\}$, and for any collection $\mathscr{I}$ of sets of literals of $PL^2(E, IC)$, let $\Sigma(\mathscr{I}) = \bigcup_{I \in \mathscr{I}} \sigma(I)$.

if S is an answer set of $PL^2(E, IC)$, and if E is SE-inconsistent ($SE(E) = \emptyset$), then $Rmr(S) \neq \emptyset$. Consider the program $\Pi' = PL^2(E, IC) \setminus \{r \mid head(r) \in Rmr(S)\}$. Obviously, we see that $S \setminus Rmr(S)$ is an answer set of Π', and that $\sigma(S)$ is a SE-model of $E \setminus Rules(S)$. By definition of a weak potential removed set, S is a weak potential removed set of E.

$\boxed{\Leftarrow}$ If S is such that $Rules(S)$ is a potential removed set of E, then, by definition of a weak potential removed set, $SE(E \setminus Rules(S)) \neq \emptyset$). But then, $\Sigma(AS(PL^2(E, IC))) = SE(E \setminus Rules(S))$.

Proof (of proposition 5).

$\boxed{\Rightarrow}$ We suppose that S is an answer set of $PL^1(E, IC) \cup \Pi^\Sigma(E, IC)$ (resp. $PL^2(E, IC) \cup \Pi^\Sigma(E, IC)$), but that $Rules(S)$ is not a strong (resp. weak) removed set of E according to Σ.

If S is an answer set of $PL^1(E, IC) \cup \Pi^\Sigma(E, IC)$ (resp. $PL^2(E, IC) \cup \Pi^\Sigma(E, IC)$), then S is an answer set of $PL^1(E, IC)$ (resp. $PL^2(E, IC)$), and there is no answer set Y of $PL^1(E, IC)$ (resp. $PL^2(E, IC)$) such that $|Y \cap R^+| < |S \cap R^+|$, thanks to the #*minimize* statement. But $Rules(S)$ is necessarily a strong (resp. weak) potential removed set of E, because of proposition 3 (resp. proposition 4). But, if $Rules(S)$ is not a strong (resp. weak) removed set of E according to Σ, then there exist Y such that Y is an answer set of $PL^1(E, IC)$ (resp. $PL^2(E, IC)$) and $|Y| < |S|$, which contradicts the hypothesis.

$\boxed{\Leftarrow}$ Let us suppose that $Rules(S)$ is a strong (resp. weak) removed set of E according to Σ, but that S is not an answer set of $PL^1(E, IC) \cup \Pi^\Sigma(E, IC)$ (resp. $PL^2(E, IC) \cup \Pi^\Sigma(E, IC)$). If $Rules(S)$ is a strong (resp. weak) removed set of E according to Σ, then there is no set of literals Y such that $|Rules(Y)| < |Rules(S)|$ (and thus $|Y \cap R^+| < |S \cap R^+|$) and $AS((E \cup IC) \setminus Y) \neq \emptyset$ (resp. $SE((E \cup IC) \setminus Y) \neq \emptyset$). But if Ss is not an answer set of $PL^1(E, IC) \cup \Pi^\Sigma(E, IC)$ (resp. $PL^2(E, IC) \cup \Pi^\Sigma(E, IC)$), on the other hand S is an answer set of $PL^1(E, IC)$ (resp. $PL^2(E, IC)$), because of proposition 3 (resp. proposition 4), but then there exist a set of literals Y such that Y is an answer set of $\Pi^1(E, IC)$ (resp. $\Pi^2(E, IC)$) and $|Y \cap R^+| < |S \cap R^+|$, because of the #*minimize* statement, which contradicts the hypothesis.

Proof (of proposition 6). The *Card* strategy being identical to the Σ strategy, except that we replace multiple occurrences of a rule by a single occurrence, the proof is strictly identical to the proof established above for the Σ strategy.

Proof (of proposition 7).

1. We know by proposition 3 (resp. proposition 4) that S is an answer set of $PL^1(E, IC)$ (resp. $PL^2(E, IC)$) if and only if $Rules(S)$ is a strong (resp. weak) potential removed set of E.

2. Let S be an answer set of $PL^1(E,IC)$ (resp. $PL^2(E,IC)$). The atom $size(U)$ is true in $PL^1(E,IC) \cup \Pi^{Max}(E,IC)$ (resp. $PL^2(E,IC) \cup \Pi^{Max}(E,IC)$) if and only if ther exists i such that $|Rules(S) \cap \Pi_i| = U$, by using an α_i rule. So, $negmax(W)$ is true if and only if there exists W and i such that $|Rules(S) \cap \Pi_i| = W$ and there exists V and j such that $|Rules(S) \cap \Pi_j| = V$ with $V > W$. In other words, $negmax(W)$ is true for all W such that $size(W)$ is true but non maximal according to the α_i and β_1 rules. Lastly, $max(U)$ is true if and only if $U = \max_{i=1}^n |Rules(S) \cap \Pi_i|$, according to the α_i, β_1 and β_2 rules.
3. Now let us show that if S is an answer set of $PL^1(E,IC) \cup \Pi^{Max}(E,IC)$ (resp. $PL^2(E,IC) \cup \Pi^{Max}(E,IC)$), then $Rules(S)$ is a strong (resp. weak) removed set of E according to Max.

   $\Rightarrow$ If S is an answer set of $PL^1(E,IC) \cup \Pi^{Max}(E,IC)$ (resp. $PL^2(E,IC) \cup \Pi^{Max}(E,IC)$), we denote by $val(S)$ its value for the #minimize statement, and we recall that $val(S) = \max_{i=1}^n |Rules(S) \cap \Pi_i|$. Suppose that S is such that $Rules(S) \notin \mathcal{R}^1_{Max}(E)$ (resp. $\mathcal{R}^2_{Max}(E)$). This means that there exist S' such that $Rules(S') \in \mathcal{R}_{Max,IC}(E)$ with $\max_{i=1}^n |Rules(S') \cap \Pi_i| < \max_{i=1}^n |Rules(S) \cap \Pi_i|$, and so $val(S') < val(S)$, which contradicts the #minimize instruction and the fact that S is an answer set of $PL^1(E,IC) \cup \Pi^{Max}(E,IC)$ (resp. $PL^2(E,IC) \cup \Pi^{Max}(E,IC)$).

   $\Leftarrow$ Suppose that $Rules(S)$ is a strong (resp. weak) removed set of (E according to Max, but that S is not an answer set of $PL^1(E,IC) \cup \Pi^{Max}(E,IC)$ (resp. $PL^2(E,IC) \cup \Pi^{Max}(E,IC)$). We know that S is an answer set of $PL^1(E,IC)$ (resp. $PL^1(E,IC)$) (see item 1. of the proof). So, there exists S' such that $val(S') < val(S)$, and consequently $\max_{i=1}^n |Rules(S') \cap \Pi_i| < \max_{i=1}^n |Rules(S) \cap \Pi_i|$. This contradicts the hypothesis.

Proof (of proposition 8).

1. We know by proposition 3 (resp. proposition 4) that S is an answer set of $PL^1(E,IC)$ (resp. $PL^2(E,IC)$) if and only if $Rules(S)$ is a strong (resp. weak) potential removed set of E.
2. Let S be an answer set of $PL^1(E,IC)$ (resp. $PL^2(E,IC)$). The atom $size(V,U)$ is true in $PL^1(E,IC) \cup \Pi^{GMax}(E,IC)$ (resp. $PL^2(E,IC) \cup \Pi^{GMax}(E,IC)$) if and only if $|Rules(S) \cap \Pi_V| = U|$. We define $p_S^V =_{def} |Rules(S) \cap \Pi_V|$. Thus, by using the rule α_0, wee know that $\max(X_1,\ldots,X_n)$ is true if and only if $(X_1,\ldots,X_n)$ represent a sequence sorted by decreasing order of the p_S^i values computed by the rule γ_i. Finally, the rules α_i isolate the X_1 values in order to use them in the #minimize instruction.
3. Let us show that if S is an answer set of $PL^1(E,IC) \cup \Pi^{GMax}(E,IC)$ (resp. $PL^2(E,IC) \cup \Pi^{GMax}(E,IC)$), then $Rules(S)$ is a strong (resp. weak) removed set of E according to $GMax$.

   $\Rightarrow$ if S is an answer set of $PL^1(E,IC) \cup \Pi^{GMax}(E,IC)$ (resp. $PL^2(E,IC) \cup \Pi^{GMax}(E,IC)$), then we denote by $val(S)$ its value for the #minimize statement, and we recall that $val(S) = \sum_{i=1}^n (p_S^i) \times (m+1)^{n-i}$. Suppose that there exists S such that $Rules(S) \in \mathcal{R}^1_{GMax}(E)$ (resp. $Rules(S) \in \mathcal{R}^2_{GMax}(E)$) buth

that S is not an answer set of $PL^1(E,IC) \cup \Pi^{GMax}(E,IC)$ (resp. $PL^2(E,IC) \cup \Pi^{GMax}(E,IC)$). This means that there exists S' such that $val(S') < val(S)$. This implies by construction that:
- $\exists i$ such that $p^i_{S'} < p^i_S$;
- $\forall j < i$, we have $p^j_{S'} = p^j_S$;

This implies in turn that $(p^1_{S'}, \ldots, p^n_{S'}) <_{lex} (p^1_S, \ldots, p^n_S)$, and thus $Rules(S) \notin \mathscr{R}^1_{GMax}(E)$ (resp. $Rules(S) \notin \mathscr{R}^2_{GMax}(E)$), which contradicts the hypothesis.

$\boxed{\Leftarrow}$ Suppose that there exist S such that S is an answer set of $PL^1(E,IC) \cup \Pi^{GMax}(E,IC)$ (resp. $PL^2(E,IC) \cup \Pi^{GMax}(E,IC)$), and that $Rules(S) \notin \mathscr{R}^1_{GMax}(E)$ (resp. $Rules(S) \notin \mathscr{R}^2_{GMax}(E)$). This means that there exists S' such that $(p^1_{S'}, \ldots, p^n_{S'}) <_{lex} (p^1_S, \ldots, p^n_S)$, and, as a consequence:
- $\exists i$ such that $p^i_{S'} < p^i_S$;
- $\forall j < i$, we have $p^j_{S'} = p^j_S$;

And thus $val(S') < val(S)$, which contradicts the hypothesis that S is an answer set of $PL^1(E,IC) \cup \Pi^{GMax}(E,IC)$ (resp. $PL^2(E,IC) \cup \Pi^{GMax}(E,IC)$).

www.ingramcontent.com/pod-product-compliance
Lightning Source LLC
Chambersburg PA
CBHW051038160426
43193CB00010B/986